DOLIA

Dolia

THE CONTAINERS THAT MADE ROME AN EMPIRE OF WINE

CAROLINE CHEUNG

PRINCETON UNIVERSITY PRESS
PRINCETON & OXFORD

Published by Princeton University Press
41 William Street, Princeton, New Jersey 08540
99 Banbury Road, Oxford OX2 6JX

press.princeton.edu

Library of Congress Cataloging-in-Publication Data

Names: Cheung, Caroline, author.
Title: Dolia : the containers that made Rome an empire of wine / by
Caroline Cheung.
Description: Princeton : Princeton University Press, [2024] | Includes
bibliographical references and index.
Identifiers: LCCN 2023034342 (print) | LCCN 2023034343 (ebook) | ISBN
9780691243009 (hardback ; acid-free paper) | ISBN 9780691242996 (ebook)
Subjects: LCSH: Dolia. | Wine industry—Rome. | BISAC: HISTORY / Ancient /
Rome | SOCIAL SCIENCE / Anthropology / Cultural & Social
Classification: LCC TP792 .C43 2024 (print) | LCC TP792 (ebook) | DDC
658.7/85—dc23/eng/20230915
LC record available at https://lccn.loc.gov/2023034342
LC ebook record available at https://lccn.loc.gov/2023034343

British Library Cataloging-in-Publication Data is available

Editorial: Rob Tempio, Chloe Coy
Production Editorial: Elizabeth Byrd
Jacket: Chris Ferrante
Production: Erin Suydam
Publicity: William Pagdatoon (US), Charlotte Coyne (UK)
Copyeditor: Ashley Moore

Jacket Credit: Image by Manuel García Ávila.

This publication is made possible in part by the Barr Ferree Foundation Fund for Publications, Department of Art and Archaeology, Princeton University.

This book has been composed in Arno Pro

Printed in the United States of America

10 9 8 7 6 5 4 3 2 1

CONTENTS

List of Illustrations vii
List of Tables xi
Preface xiii
Acknowledgments xvii
Abbreviations xxi

1 Food Storage, Containers, Empire 1
Studying Dolia and Tracing Their Development 1
Urban Growth and the Food Supply 9
An Empire Full of Containers 12
Organization of Chapters 16
The Data 19

2 Building Big: A New Craft Industry 24
Dolium Production: A Specialist Craft 25
Cosa: Early Developments 33
Pompeii: Improvements in the Craft 38
Ostia and Rome: New Standards 43
Opportunity and Profit in the Ceramic Valley 49

3 Dolia on the Farm: Conspicuous Production and Storage 52
Investing in Storage 53
Dolia for Wine and Olive Oil 56
Dolia and the Development of Villas in Central Italy 65
Celebrating Surplus 69

4 Dolia Abroad: Innovations in Transport 74
Breaking into Markets 75
Dressel 1 Amphorae and Trade: The Sestius Family's Amphora Enterprise 83
Bulk Transport: The Piranus Family's Dolium Tanker Ships 87
Profitable Packaging 97

5 Dolia in Iberia and Gaul 101
Agriculture and Storage in Iberia and Gaul 103
Dolium Development in the Northwest 106
Villas and Agricultural Production: A New Scale of Production 111

6 Dolia in *Urbe*: Expanding Urban Storage and Consumption 117
Trials and Tribulations with Technology: The Case of Cosa 119
Pompeii: Dolia in Urban Retail and Service 122
Rome and Ostia: The Warehouse of the World 130
Concentrating Wine and Wealth 142

7 Mending Costly Investments 144
Damaged Dolia 146
Dolium Repairs and the Repairers 147
Cosa: A Motley of Dolium Repairs 155
Pompeii: Experimentation in the Field and within the Workshop 160
Ostia and Rome: A Trend toward Production Repairs 164
Reinforcing and Repairing Dolia 168

8 From Valued to Trash: The Disappearance of Dolia 173
Dolium Reuse and Discard 174
Moving Away from a Specialized Container System 179
Barrels and a New Container System 193

9 Dolia: The Storage Container of the Roman Empire 197
Investors, Workshops, and Personnel 198
Choosing Container Technologies 201
The Legacy of Dolia 207

Guide to the Appendixes 213
Appendix 1. Tables 215
Appendix 2. Descriptions of Select Dolia 245
References 255
Index 293

Color plates follow page 154

ILLUSTRATIONS

1.1. Dolium (I.22 no. 5), Pompeii; and dolia compared with amphorae. 2

1.2. Dolia defossa in Caseggiato dei Doli, Ostia; and in Villa Regina, Boscoreale. 3

1.3. Iron Age Cretan pithos from Knossos; and seventh-century BCE Etruscan pithos, Brooklyn Museum. 4

1.4. Dolium production sites in Tuscany, Latium, and Campania. 6

1.5. Evidence for production and repair of dolia. 8

1.6. Supply chain, and containers, of wine. 15

1.7. Map of Italy with case study sites marked. 20

2.1. Funerary altar for a *doliarius*. 25

2.2. Cracks between coils in forming dolia, Pompeii. 28

2.3. Evidence of joining dolium coils, Pompeii. 29

2.4. Evidence of joining dolium rim to vessel, Pompeii. 31

2.5. Profile drawings of dolia from Cosa, Pompeii, and Ostia and Rome. 34

2.6. Dolium lids from Cosa. 34

2.7. Microphotographs of dolium ceramic fabrics from Cosa, Pompeii, and Ostia. 35

2.8. Stamped dolium rims, Cosa. 37

2.9. Large dolium fragment from Cosa with stamp and incision. 38

2.10. Interior dolium lid (*operculum*), Villa Regina, Boscoreale; and system of closing dolia. 39

2.11. Exterior dolium lid (*tectorium*), Pompeii. 40

2.12. Dolium rim with stamps, Pompeii. 41

2.13. A dolium and a dolium with capacity incision, Ostia. 44

2.14. Dolium rims with stamps, Ostia. 46

2.15. Other products made alongside dolia in workshops in Tuscany, Latium, and Campania. 47

3.1. Plans of various wine and oil cellars in central Italy. 58

3.2. Egg-shaped fermenters and wine circulation. 60

3.3. Moving and applying pitch to a dolium from Rustic Calendar mosaic, Saint-Romain-en-Gal. 64
3.4. Villas and farms with dolia in central Italy. 66
3.5. Scene of cupids sampling wine, House of the Vettii, Pompeii. 72
4.1. Painting depicting workers filling amphorae with wine brought in a culleus. 78
4.2. Dressel 1B and 2/4 amphorae. 80
4.3. Dolium tanker ship model and drawings. 89
4.4. Map of dolium shipwrecks in western Mediterranean. 90
4.5. Plans of dolium tanker shipwrecks and profile drawing of ship dolium. 93
4.6. Shipwrecked dolium found off the coast of Italy. 95
5.1. Dolia from Iberia. 107
5.2. Dolia from Gallia Narbonensis. 109
5.3. Plans of the villa estate Saint-Bézard à Aspiran and various cellae vinariae in Iberia. 110
5.4. Cellae vinariae of the Roman villa of Arellano (Navarra) and 28 Place Vivaux, Marseille. 114
6.1. Depiction of a Roman winery, World Museum Liverpool. 118
6.2. Plans of Cosa and Cosan tavern. 120
6.3. Plan of Pompeii with dolia and storage containers. 123
6.4. Pompeian dolia and cylindrical storage jars. 125
6.5. Drawing of a lararium, Hospitium of Hermes, Pompeii, and sign of wines and prices, Herculaneum. 127
6.6. Thermopolium of Vetutius Placidus, Pompeii. 128
6.7. Plans of Ostia and cellae with dolia. 131
6.8. Magazzino Annonario and Magazzino dei Doli, Ostia. 133
6.9. Funerary relief of Lucifer and Roman mosaic, *A Banquet in the Open Air*, Detroit Institute of Arts. 141
7.1. Relief depicting the encounter between Alexander the Great and Diogenes, Villa Albani. 145
7.2. Examples of double dovetail made during use and production. 148
7.3. Various dolium repairs. 150
7.4. Double dovetails, Ostia, and crack that formed at clamp pin hole, Pompeii. 151
7.5. Production phase drill hole for clamp pins and double dovetails, Pompeii. 153
7.6. Tool marks on surface of double dovetail tenon, Pompeii. 153
7.7. Clamps on a dolium fragment and *mortarium*, Cosa. 156

7.8. Dolium rim repaired with screws, Cosa. 157
7.9. Lead clamp on dolium shoulder, Cosa. 158
7.10. Dolium base and lower wall mended with lead clamp and double dovetail, Cosa. 158
7.11. Dolium rim with drill hole and half double dovetail, Cosa. 159
7.12. Lead fills on dolia, Pompeii. 161
7.13. Lead triangular clamp and hybrid mortise-and-tenon staple, Pompeii. 162
7.14. Dolium mended by double dovetail tenons and additional clay, Pompeii. 163
7.15. Dolia mended with lead double dovetail tenons, Ostia. 165
7.16. Lead double dovetail tenons, Museo Nazionale Romano. 165
7.17. Interior walls repaired with lead (alloy) fills, Ostia. 166
7.18. Lead double dovetail tenon on dolium interior, Ostia. 167
7.19. Embellished lead repair, Museo Nazionale Romano. 168
8.1. Bar with five jars installed, three of which were reused dolia, Pompeii. 177
8.2. Gauloise 4 amphora, Spello amphora, and wooden barrel. 183
8.3. Reliefs showing porters handling barrels. 186
8.4. Funerary monuments depicting barrels. 188
8.5. Plans of the villa at La Maladrerie in Saillans. 190
8.6. Barrel terminology and the process of cooperage. 192
9.1. Stone epitaph of merchants moving wine. 206
9.2. Mosaic showing wine production, Mausoleum of Santa Costanza; sarcophagus representing a Dionysiac vintage scene, J. Paul Getty Museum; and Miracle of Cana, Basilica of Santa Sabina, Rome. 208
9.3. Mauro Gandolfi, *Alexander and Diogenes*, Harvard Art Museums/Fogg Museum; and frontispiece of *Cornelianum dolium*, Princeton University Library. 209
9.4. *Diogenes*, by Jean-Léon Gérôme, Walters Art Museum; and *The Vintage Festival*, by Lawrence Alma-Tadema, National Gallery of Victoria, Melbourne. 210

TABLES

2.1. Logistics of building large fermentation jars. 32
2.2. Hypothesized opus doliare workshops producing multiple products. 49
3.1. Legal maximum vehicle load weights based on the *Codex Theodosianus* 8.5.8. 55
4.1. Estimated volume of wine transported on various ancient ships. 97
6.1. Selection of properties (primarily production) in Pompeii with dolia. 127
6.2. Properties with dolia in Ostia. 137
7.1. Proportion of repaired dolia at Cosa, Pompeii, Ostia, and Rome. 155
8.1. Comparison of barrels, ceramic containers, and cullei. 185
8.2. Comparison of the logistics of barrels, dolia, amphorae, and cullei. 195
A1.1. Cosa dolium and other storage jar dimensions. 215
A1.2. Dolium stamps from Cosa. 216
A1.3. Pompeii dolium dimensions. 217
A1.4. Dolium stamps from Pompeii. 221
A1.5. Ostia dolium dimensions. 224
A1.6. Rome dolium dimensions. 228
A1.7. Dolium stamps from Ostia. 229
A1.8. Dolium production sites in west-central Italy. 232
A1.9. Villas and farms with dolia in west-central Italy. 234
A1.10. Volume incisions on dolia at Ostia. 237
A1.11. Possible craftspeople of dolium repairs, according to stage of execution. 238
A1.12. Dolium repairs at Cosa. 239
A1.13. Types of dolium repairs at Pompeii. 239
A1.14. Dolium repairs at Pompeii. 240
A1.15. Types of dolium repairs at Ostia. 241
A1.16. Dolium repairs at Ostia. 242
A1.17. Dolium repairs in Rome. 243
A1.18. Comparison of dolium repair types. 243

PREFACE

HAILED AS ONE of the largest food supply systems in the premodern world, the Roman food trade has never failed to captivate. The scale of operations for the grain supply alone was extraordinary: the city of Rome imported four hundred thousand tons of grain per year from Egypt, North Africa, Sardinia, and other breadbaskets around the Mediterranean basin. The Romans also consumed a variety of foods, a substantial portion of which was exotic. A contender for the most extravagant example was the emperor Vitellius's favorite dish, a monstrous hodgepodge known as the shield of Minerva, which consisted of "the livers of parrotfish, the brains of pheasants and peacocks, with the tongues of crimson-winged flamingos, and the innards of lampreys, which had been brought in ships of war from as far as Parthia to the Spanish Straits" (Suetonius *Vit.* 13). The empire's foods reached ordinary people too. Down in Campania, Pompeians regularly ate and drank not only massive quantities of locally produced wine, olive oil, fruits, and vegetables but also a variety of imported foods, such as dates from the Near East and sesame seeds from Egypt. The empire's food supply apparatus was truly colossal, extending well beyond Italy. From the first century CE onward, even far-flung Britain witnessed a revolution in food unseen since the introduction of agriculture. New and exotic foods from the Mediterranean such as figs and olives reached urban markets, military forts, and wealthy country villas throughout the island. Even more impressive is the journey of black pepper all the way from India. The story of these traveling foods is compelling, but only part of the story. Without a way to store these comestibles reliably, such large-scale movements across entire continents would have been simply impossible, and perhaps not even attempted.

Fifty years ago, Geoffrey Rickman's *Roman Granaries and Store Buildings* illustrated one tangible but underdiscussed aspect of Roman imperial power: storage facilities. Across the empire, from cities in Italy to military settlements in the north, massive store buildings dotted the landscape. Within these almost monumental structures, people placed grain, wine, olive oil, pepper, papyrus, parchment, candles, and other goods for safekeeping. Rickman's study highlighted how effective these buildings were at protecting their contents from not only theft but also environmental conditions, an especially daunting challenge in the fickle Mediterranean climate. Astrid Van Oyen's recent *The Socio-economics of Roman Storage: Agriculture, Trade, and Family* drives forward our understanding of storage and its wider significance in the Roman world. Her survey of a wide range of materials in storage demonstrates how the practice of storage shapes social relations, economic possibilities, and history. While most studies on storage since Rickman's have focused on the buildings, the grain supply, or the institutional apparatus, the present book joins the chorus with a different perspective that centers on storage technology for wine (and, to a lesser extent, olive oil) through the lens of the storage vessels themselves.

Wine (and olive oil) was just as crucial to the Mediterranean diet as cereals. In fact, wine and olive oil are two of the three members of what is known as the Mediterranean trinity or triad—wheat, wine, and olive oil—those quintessential elements of the ancient Mediterranean diet. When we envision ancient Rome, we think about all the different kinds of wine that banqueters drank at lavish parties in abundance, bordering on excess. Painters from antiquity through the twentieth century frequently portrayed aspects of the ancient world, from humble taverns to posh parties, with wine front and center. The average ancient Roman drank 250 liters of wine per year, almost an entire bottle each day. Wine was always in demand, but grapes were harvested and processed during the autumn season, and often aged for at least some time. Yet people were able to produce and store so much wine without refrigeration, making it available year-round thanks to their storage technology.

Romans developed and used a massive ceramic storage jar known as a dolium. Capable of holding over a thousand liters, dolia claim the title for the largest type of pottery in the ancient world. They were also extremely challenging and labor intensive to produce, move, repair, and maintain. Today only a handful of potters have the skills and expertise to construct a jar comparable to these ancient giants. In many settlements, storerooms filled with dolia were used to hold over a hundred thousand liters of wine. But the dolia were more than just receptacles; they were designed for viticulture. Not only did they protect and insulate their contents, dolia actually promoted and contributed to wine's transformations during fermentation. Because of their shape and material, dolia influenced the flavors and textures of wines. The dolium constituted a type of storage container technology unique to the Roman Empire, becoming the backbone of the Roman wine trade. As dolia became more widespread, people invested heavily in the production and use of the supersized pots, which opened doors to new opportunities. Dolia changed not only people's fortunes but also how they made them.

Although many dolia have been found at various archaeological sites, they have rarely received the attention they deserve. Systematic studies of dolia are few, partly because of the challenges in studying them and their preservation, but also because their scholarly value has been overlooked. Archaeologists commonly misidentify fragmentary dolia they recover during the excavation process, and the pieces are either thrown away or condemned to the purgatory of noninventoried artifacts. This starts as early as in the excavation or survey phase of an archaeological project: because of the similarity in their ceramic fabric, it can be difficult to distinguish a dolium sherd from a piece of brick or tile, with the result that dolia are often not even identified. But even when they are properly identified, there are other challenges. Those that are recognizable are usually buried in the ground or embedded in an architectural feature, making it impossible to get a full view of the vessel. Furthermore, unlike decorated table pottery, which potters changed rapidly over time for developing tastes and preferences, dolia were utilitarian vessels that remained mostly unchanged, so they are impossible to date precisely based on form alone. Described disparagingly as boring, ugly, and (ironically) a storage problem for museums and archaeological sites, dolia are both the elephant in the room and the underdog of Roman material culture. They are found in abundance, sometimes well preserved and often still in situ, but only cursorily studied, usually to identify an ancient site and its activities. They are rarely studied as objects in their own right.

This book looks at the dolium not only as a unique, supersized storage container that was both a product and driver of Roman economic expansion but also as an industry and technology that cast a wide net: craftspeople developed and refined a craft industry, various investors poured resources into the dolia's production and use, and different entrepreneurs and communities adopted them to expand their wine supplies and retail power. This book serves as an exhortation to understand the enormous efforts that went into quenching an empire's thirst, and to consider how a container technology could make one rich.

ACKNOWLEDGMENTS

DURING THE SUMMER of 2014, Ted Peña introduced me to a set of curious dolia at Regio I Insula 22 at Pompeii, entrusting me to design my own research project as part of the Pompeii Artifact Life History Project; little did I know that those afternoons meticulously documenting dolia would lead to years of fascination with these vessels. The project began to expand and take shape in 2015 as a doctoral dissertation, and it has grown and benefited enormously over the years from the generosity and spirit of many individuals and communities. It is a pleasure to acknowledge them here. The Graduate Group in Ancient History and Mediterranean Archaeology, as well as the Center for the Tebtunis Papyri and the Department of Classics at the University of California, Berkeley, formed my home base during the early years and encouraged me from day one to expand my horizons and pursue multi- and interdisciplinary approaches. Ted's vast knowledge of Roman pottery and his critical eye and attention to detail helped me better clarify and support my observations. Carlos Noreña provided invaluable guidance and helped frame and shape the project, pushing me to expand on the broader significance of the data.

I was most fortunate to have spent 2016–2017 as the recipient of the Andrew W. Mellon Pre-doctoral Rome Prize at the American Academy in Rome. They say being at the American Academy in Rome changes your life and perspective for a reason. Fellows and friends at the academy provided good company, cheer, and a warm and intellectually stimulating community. Seeing and exploring the city through their eyes helped me notice new things and present my findings to a broader audience. Three generous scholars I had the good fortune to meet through the American Academy in Rome greatly influenced this project: Kim Bowes, Lisa Fentress, and Lynne Lancaster gave generous feedback, and their insights prompted me to confront why this project matters and how to make that come across most effectively. Lindsay Harris organized helpful writing groups to workshop our ideas and drafts. Giulia Barra secured permissions to access different sites and materials during the early, critical stages of the project. The American Academy library, especially the support of Sebastian Hierl, was instrumental in accessing the Cosa Archive. The academy remained my home base in my final year as a graduate student, supported by the University of California Dissertation Year Fellowship, and has continued to be an invaluable resource for me and my project over the years.

Since I joined the faculty in 2018, Princeton University has provided an intellectually stimulating community. I would like to thank my colleagues in the Department of Classics for their support, encouragement, advice, and enthusiasm for my project over the years, especially Yelena Baraz, Josh Billings, Marc Domingo Gygax, Denis Feeney, Andrew Feldherr, Harriet Flower, Michael Flower, Melissa Haynes, Daniela Mairhofer, Dan-el Padilla Peralta, Brent Shaw, and Katerina Stergiopoulou. Special thanks go to my Roman historian colleagues: Harriet has been most helpful in her sage guidance and mentorship and Dan-el has shared useful resources over the years. Colleagues across campus—Nathan Arrington, Peter Brown, Dimitri Gondikas, Sam Holzman, Janet Kay, Michael Koortbojian, AnneMarie Luijendijk, and Debbie Vischak—have also shared helpful feedback, advice, and resources. William Dingee carefully checked the ancient texts and my translations, and Jamie Wheeler helped proofread the manuscript. Our wonderful library, and Dave Jenkins especially, has procured publications at record speed to bolster my research. The Department of Classics, Department of Art & Archaeology and its Barr Ferree Fund, the Stanley J. Seeger Center for Hellenic Studies, the University Committee on Research in the Humanities and Social Sciences, and the Center for Digital Humanities provided financial support for this project and publication.

I am grateful to several individuals for their contributions to this project: Gina Tibbott for her beautiful illustrations and insights for this project (and all things pottery related)—readers will have a better understanding of the material thanks to her thoughtful and informative illustrations; David Stone for reading and copyediting the manuscript so thoroughly—it is much stronger thanks to his critical eye and expertise; Allison Emmerson and Jared Benton, who generously read early drafts of the manuscript and provided feedback; Kevin Moch for his philological and agricultural expertise in reading portions of the manuscript; Massimo Betello, who helped with the logistics of living and researching in Italy; Stanley Chang for patiently working with imperfect data to estimate vessel capacities; and Andrea Carpentieri for analyzing samples for residues and to identify metals used in dolium repairs, supported by the Archaeological Institute of America's Kathleen and David Boochever Endowment Fund for Fieldwork and Scientific Analyses.

My project benefited from many conversations over the years, and I would like to thank for their insights the following people: Gregory Bailey, Hilary Becker, Jared Benton, Seth Bernard, Dorian Borbonus, Valeria Brunori, Evelyne Bukowiecki, Maureen Carroll, Angelos Chaniotis, Kristi Cheramie, Sophie Crawford-Brown, Janet DeLaine, Kevin Dicus, Emlyn Dodd, Steven Ellis, Christine Hastorf, Todd Hickey, Carl Knappett, Ann Olga Koloski-Ostrow, Jenny Kreiger, Sarah Levin-Richardson, Archer Martin, Victor Martínez, Michael McCormick, Guy Métraux, Alessia Monticone, Sarah Murray, Salvatore Ciro Nappo, Esin Pektas, Lisa Pieraccini, Eric Poehler, Nicholas Purcell, Amy Richlin, Andrew Riggsby, Jordan Rogers, Michele Salzman, Philip Sapirstein, Christy Schirmer, Jane Shepherd, Irene Soto Marín, Marcello Spanu, Deborah Steiner, David Stone, Rabun Taylor, Steven Tuck, Astrid Van Oyen, Eeva-Maria Viitanen, Alex Walthall, Morgan Williams, and Michael Zellmann-Rohrer. I am especially appreciative to Maureen Carroll, Emlyn Dodd, and Astrid Van Oyen for discussing dolia, wine, and storage with me and sharing new (and sometimes unpublished) work.

I am indebted to staff and local friends at several archaeological sites for facilitating my research: the Soprintendenza Archeologica di Pompei, Massimo Osanna, Gabriel Zuchtriegel, Grete Stefani, Annamaria Sodo, Giuseppe Di Martino, Domenico Busiello, Ulderico Franco, Vincenzo Sabini, Patrizia Tabone, Laura D'Esposito, and Luana Toniolo at Pompeii, as well as members of the Pompeii Artifact Life History Project team and friends in Pompeii: Allison Emmerson, Leigh Lieberman, Laure Marest, Aimée Scorziello, Ambra Spinelli, and Gina Tibbott; the Parco Archeologico di Ostia Antica, Alessandro D'Alessio, Mariarosaria Barbera, Marina Lo Blundo, Paola Germoni, Tiziana Sorgoni, Angela

Amoresano; the Polo Museale Regionale della Toscana, Susanna Sarti, Stefano Casciu, and Graziano Bannino at Cosa and Andrea De Giorgi, Russell T. Scott, and the team of the Cosa Excavations Project; Daniela Porro, Stéphane Verger, and Agnese Pergola at the Museo Nazionale Rome; and Alberto Danti and Angela Carbonaro at the Musei Capitolini.

The project has benefited from the feedback of organizers and audiences of different conferences and venues: "Technologies, Humans, and Environment in the Greek and Roman World" at the 2021 Society for the History of Technology (organized by Rabun Taylor and Ann Olga Koloski-Ostrow); "Roman Anticipations: Material, Cognitive and Affective Histories of the Roman Future" at the 2021 Archaeological Institute of America and Society for Classical Studies (AIA/SCS) Annual Meeting (organized by Anna Bonnell Freidin and Duncan MacRae); "Making Value and the Value of Making: Theory and Practice in Craft Production" at the Nineteenth International Congress of Classical Archaeology in Cologne/Bonn, Germany (organized by Helle Hochscheid and Ben Russell); "Recent Work in Vesuvian Lands: New Projects, Practices, and Approaches" at the Villa Vergiliana (organized by Steven Tuck); "On Outgroups and Muted Groups: A Conference in Honor of Amy Richlin" at the University of California, Los Angeles, Department of Classics; "Work/Life: Institutions, Subjectivities, and Human Resources in the Roman World" at the New York University Center for Ancient Studies (organized by Jordan Rogers and Del Maticic); "Climate and the Roman Conquest of Italy" at the University of Toronto, Department of Classics (organized by Seth Bernard); and various talks at the annual meetings of the AIA/SCS, the University of Toronto, the University of Pennsylvania, and the American Academy in Rome.

Many thanks go to the team at Princeton University Press—especially Rob Tempio and Chloe Coy—for their support and interest in the project, the readers who provided instrumental feedback in shaping the book, and to Melody Negron for carefully and patiently overseeing the book's production.

I am grateful to my family for their love and support, and especially my father, who taught me the values of hard work, dedication, and perseverance. Lastly, my husband Sven's tireless patience, support, and sharp eye made this project better. I am looking forward to future adventures together.

ABBREVIATIONS

ABBREVIATIONS OF ANCIENT authors are based on the *Oxford Classical Dictionary,* 4th edition. Abbreviations of publications are based on the guidelines of the *American Journal of Archaeology*. The following abbreviations are also used:

AE	*L'année épigraphique: Revue des publications épigraphiques relatives à l'antiquité romaine,* 1888–
CIL	*Corpus Inscriptionum Latinarum,* ed. T. Mommsen et al., 1863–
FUR	*Forma Urbis Romae*
HAE	*Hispania Antiqua Epigraphica,* Suplemento anual de *Archivo Español de Arqueología,* Instituto de Arqueología y Prehistoria "Rodrigo Caro," Madrid: Consejo Superior de Investigaciones Científicas, 1950–
IG	*Inscriptiones Graecae,* 1873–
PPM	*Pompei: Pitture e Mosaici,* 1990–
TLG	*Thesaurus Linguae Graecae: A Digital Library of Greek Literature* (www.tlg.uci.edu)
TLL	*Thesaurus Linguae Latinae,* 1900–

1

Food Storage, Containers, Empire

TWO THOUSAND years ago, the residents of the city of Rome drank so much wine that a year's worth would have overfilled the Pantheon.[1] But almost none of that wine came from Rome, a metropolis too large and densely occupied to produce its own food and drink.[2] Instead, Rome's residents depended on a large-scale food supply system that not only served the city but also sustained the empire's expanding territory and population. This was an enormous feat. Without climate-controlled airplanes or trucks, moving and protecting so much food was a challenge. Adding to this, consumers expected a constant supply of wine, yet the grape harvest and wine production occurred once a year starting in the autumn season and could be a lengthy activity if the vintner wished to age the wine. Having enough wine available for a large population year-round required not only high production but also effective and plentiful storage.

Under the Roman Empire, food storage technology and infrastructure reached new heights. Massive warehouses (*horrea*) lined rivers and coasts and punctuated the ancient capital and numerous other urban and military settlements. The productive power of vineyards, farms, and villas was expanded by large wine cellars filled with *dolia* (singular *dolium*), enormous ceramic pots the scale of which was almost never found again after the Roman period (Figure 1.1; Plate 1).

Studying Dolia and Tracing Their Development

Dolia claim the title for being by far the largest vessels in antiquity. Capable of holding anywhere from hundreds to as much as three thousand liters, dolia were massive jars often taller than a person. They primarily held wine, though they occasionally stored other foods, such as oil, grain, and fish sauce; olive oil storage is discussed intermittently in this book because dolia were only occasionally used to hold olive oil in central Italy, due to the low yields and consumption of olive oil compared with wine, other storage jar options for oil, and production trends of the area. Thousands of *dolia defossa*, dolia buried to their shoulders to keep their contents cool, can still be found in ancient houses, farms, warehouses, and port facilities in special storerooms known as *cellae* (Figure 1.2; Plate 2);

1. Frier 1983, 257n3.
2. Morley 1996.

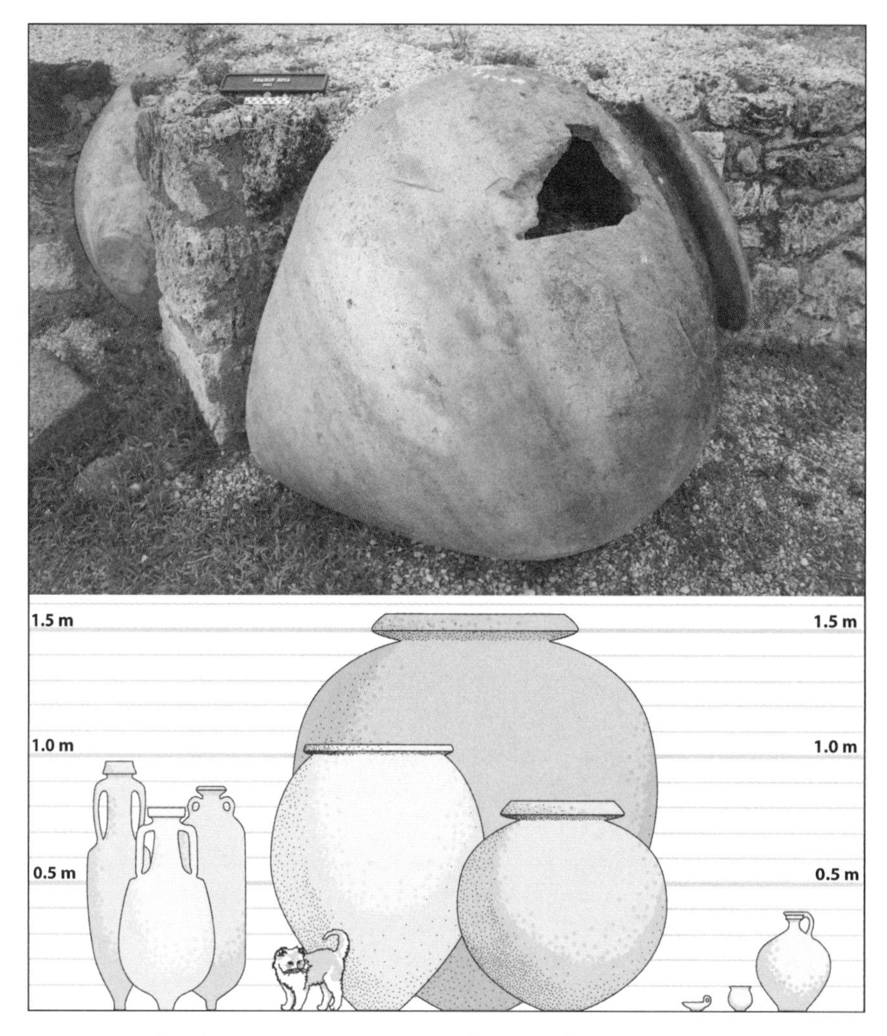

FIGURE 1.1. (*Top*) Dolium lying on its side (I.22 no. 5), Pompeii. Courtesy of the Ministry of Culture—Archaeological Park of Pompeii. Reproduction or duplication by any means is forbidden. (*Bottom*) Dolia compared with amphorae (large bulk transport jars, *left-hand side*) and other pottery (*right-hand side*). Illustration by Gina Tibbott.

some were even cemented into the hulls of ships.[3] Because they were such widely used storage vessels, dolia can be found throughout the Roman world, and multiple complete sets can still be found in situ, thus providing direct evidence for how people in antiquity stored wine and olive oil.

Despite their large numbers and preservation in the archaeological record, dolia remain understudied. One of the greatest challenges for a comprehensive study of dolia is that there has not even been a general scholarly consensus on what a dolium is. The term *dolium*

3. Marlier and Sibella 2008; Gianfrotta and Hesnard 1987; Carrato and Cibecchini 2020; Heslin 2011.

FIGURE 1.2. Dolia defossa (*top*) in Caseggiato dei Doli (I.4.5), Ostia, by Jamie Heath, 2009; and (*bottom*) in Villa Regina, Boscoreale. Courtesy of the Ministry of Culture—Archaeological Park of Pompeii. Reproduction or duplication by any means is forbidden.

is casually employed to describe any giant jar that is not easily identified, and dolia are often conflated with other big vessels, but it is crucial to begin with distinguishing dolia as vessels distinctive from generic large storage jars of an earlier tradition. Storage jars, known as *pithoi* (singular *pithos*), were instrumental for conserving food in the Mediterranean for centuries (Figure 1.3).[4] Capable of holding several hundred liters, pithoi were large terracotta jars used to store a variety of foods such as cereals, legumes, wine, and olive

4. For Etruscan *pithoi*, see Perkins 2021; Ridgway 2010. For Greek *pithoi*, see Christakis 2005, 2008; Giannopoulou 2010.

FIGURE 1.3. (*L*) Iron Age Cretan pithos from Knossos, by Jastrow, 2005. (*R*) Seventh-century BCE Etruscan pithos, Brooklyn Museum, Gift of Robin F. Beningson, 88.202.6.

oil. Their thick terracotta walls kept their contents cool and dry and protected them from extreme and fluctuating temperatures and humidity, as well as moisture and pests. Typically cylindrical or piriform, pithoi featured wide rims for easy access to their contents. They were usually placed on the ground or only partly submerged to stabilize the jars. Often decorated with geometric or figural moldings and incisions (especially during the early Iron Age), occasionally even endowed with ornamental but nonfunctional handles (pithoi were much too heavy and bulky to lift by these tiny handles), pithoi also had high symbolic value. Associated with surplus, food, and even life, pithoi were status objects for households, communities, and even the deceased who were buried or deposited in the vessels;[5] their decoration and placement further enhanced their visibility. The pithos' general utility was crucial for a wide range of consumers from individual households to communities to palatial and aristocratic complexes. In fact, scholars of the Aegean Bronze Age have identified pithoi as instrumental in the palace's ability to collect, store, and redistribute agricultural surplus, the foundation of the palace's power.[6] Pithoi also directly contributed to the increased wealth in central Italy. Phil Perkins' recent synthesis of Etruscan pithoi, for example, persuasively shows that pithoi enabled large-scale economic growth

5. E.g., Knappett 2020; Knappett et al. 2010; Zeitlin 1997; Steiner 2013; Sissa 1990; Dubois 1988.

6. E.g., Christakis 2005, 2008; Pullen 2010, 2011; Halstead 2011; Nakassis et al. 2011; Privitera 2014.

and urbanism across Etruria during the seventh to fifth centuries BCE by increasing agricultural productivity and towns' and cities' abilities to feed their residents.[7]

Although Mediterranean settlements had a long history of using large ceramic jars to store food, dolia were different. Identifying the moment when potters developed dolia as jars distinctive from pithoi and other storage jars, however, is impossible given how sparse the evidence is.[8] There was probably not one particular moment or place this happened, but a gradual process as potters learned new skills and demand increased. Discerning the difference between dolia and pithoi can also be difficult since Latin authors often used the term *dolia* and Greek authors *pithoi* interchangeably because these terms overlapped conceptually as the largest type of pottery, though they had their differences. In antiquity, the term *dolium* also designated a specific ceramic storage vessel for wine that diverged from other jars;[9] although dolia varied in size, ancient writers and archaeological evidence show that they were large, capable of holding on average approximately 550–750 liters.[10] *Pithos* was used early on in the Greek speaking world to describe a similar type of large ceramic storage container; a pithos, however, was a general food storage container, and not associated with any particular content. On the other hand, *dolium* came much later, appearing for the first time in a text in Plautus' *Pseudolus* and Cato's *De Agri Cultura*.[11] Cato's discussion of managing a farm included extensive lists of equipment, chief among which was the dolium for wine and oil cellars.[12] In Plautus' *Pseudolus*, on the other hand, the enslaved protagonist Pseudolus said, "we are loading words into a perforated dolium" (l. 369: *in pertusum ingerimus dicta dolium*), using the saying "to load something into a perforated dolium" (*ingerere aliquid in pertusum dolium*) to mean to waste one's effort or to labor in vain. By the early second century BCE, then, the dolia were well known enough in the cultural imagination and day-to-day vocabulary that one could speak about them proverbially. Varro also tells us that, before *dolium*, there existed an ancient word, *calpar*, a vessel associated with wine; *calpar* came from the Greek word *kalpis*, which was a term for a specific wine vessel and also meant "new wine," because the vessel's primary function was to hold sacrificial wine.[13] About a century later, Pliny the Elder echoes this idea, claiming that "dolia [were] invented for wine" (*NH* 35.56: *doliis ad vina excogitatis*). A dolium was therefore designed for a primary purpose: to hold wine.

Archaeological evidence also suggests that the dolium had a particular design and became established by the late third or early second century BCE. While pithoi and similar vessels had a more cylindrical shape, dolia were strawberry-shaped or spherical, without any decoration whatsoever, to facilitate wine fermentation. Thanks to their thick ceramic walls and their placement, partly buried in the ground (dolia defossa), they kept wine cool.

7. Perkins 2021.

8. Perkins 2021, Carrato 2017, and Salido Domínguez 2017 take the same stance.

9. K.D. White 1975, 145ff. Iul. *Dig.* 50.16.206 classifies dolia as wine containers.

10. Diocletian's *Price Edict* 15.97: a dolium holding 1,000 *sextarii* (550 liters); Vitr. *Arch.* 6.6.3: one *culleus* (ox-hide container, ca. 518 liters); Columella *Rust.* 12.18.7: *sesquiculleari*s (one and a half ox-hide containers, ca. 750 liters); Palladius 10.11: two hundred *congii* (ca. 650 liters).

11. Some scholars have hypothesized that a fragment of Ennius (Fest. 278) includes *pertusum dolium*.

12. See Terrenato 2012 for pitfalls of using Cato's *De Agri Cultura* as a historical source for villaculture.

13. *TLL* entries on *dolium* and its synonym *calpar*; Varro fr. Non. p. 547.

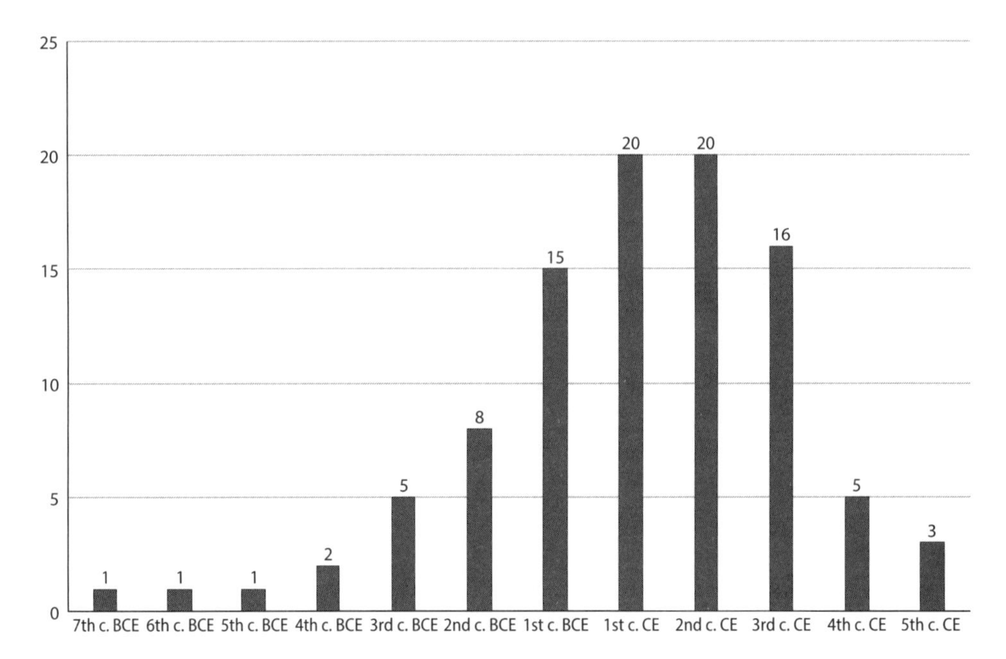

FIGURE 1.4. Dolium production sites in Tuscany, Latium, and Campania.

The earliest securely identified dolia come from contexts dated to the middle Republican period.[14] The production of dolia, at least in Latium, Tuscany, and Campania, is not well attested until the third century BCE or later, before which only a handful of dolium production sites were tenuously identified (Figure 1.4).[15] Elite villas with dolia, which Chapter 3 will discuss in greater detail, do not appear in significant numbers until the second century BCE.[16] According to both textual and archaeological evidence, then, dolia were a distinct kind of vessel with a specific design and purpose, and they were first developed by specialized potters around the third century BCE for the expanding Italian wine industry.

Dolia were unlike other types of pottery. With their cumbersome size and shape, they were not considered *portable* material culture and were moved only when necessary: from production facility to place of use, and then again if their owners sold the property where they were installed.[17] (Moving a large dolium called for the help of several people.)[18] Although they were considered a class of pottery, they were often produced alongside brick and tile products, as well as other heavy terracotta objects, in the same workshops that

14. Nicoletta 2007; Bergamini 2007. Possible rim fragments from the late fourth/early third century BCE from the Auditorium Villa in Rome (Carandini et al. 2006) and from the mid-third century BCE found at Ostia (Olcese and Coletti 2016, 455–456).

15. Olcese 2012; Bergamini 2007; Tol and Borgers 2016.

16. The single dolium from the Iron Age has not been confirmed; the original publication (Piccarreta 1977, #8–14) included no photographs or drawings; see also Attema and van Leusen 2004, 88.

17. E.g., dolia of the villa of N. Popidi Narcissi Maioris (De Spagnolis 2002, esp. 273–274); Pompeii I.22 (Cheung and Tibbott 2020); Villa Magna (Fentress et al. 2017); Apul. *Met.* 9.5–7.

18. Eight men are required to move a single *qvevri* (Georgian wine vessel similar to dolia) (Slatcher 2017).

supplied the building industry of Rome, known from stamped bricks and tiles as *opus doliare* workshops.[19] The law classified them as fixed architectural elements of a property and the defining feature of a wine cellar.[20] Considered both ceramic containers and architectural elements, dolia bring together various aspects of society normally studied separately: pottery, agriculture, wine, trade, construction and architecture, and craft production.

Dolia were highly valued, useful, and, when well made, incredibly robust. As a result, they were often used for at least several decades, sometimes bearing evidence of not only their use but also repairs to extend their lives.[21] Although dolia are rarely considered participants in the ancient Mediterranean economy, this book shows that dolia can be informative in multiple ways and at multiple scales and argues that they expanded the Roman wine trade. While double-handled ceramic transport jars known as amphorae have been studied to trace and quantify the scale and expansion of Roman trade, dolia and the potential insights they offer on the wine trade have been overlooked.[22] When dolia are studied, they are usually used only to identify the function of a room and perhaps underpin estimates of the scale of wine and olive oil production.[23] The emergence and growing numbers of dolia throughout central Italy and the Mediterranean reflected and supported a changing scale of the economy and of the wine industry in particular, becoming features in both moral and economic arguments.[24] Dolia can also feed into a different narrative, however, each vessel with a story of its own to tell. To the trained eye, the physical condition of the vessel reads like a history of its interactions, shedding light also on the people who came into contact with it (Figure 1.5). Following the "life" of a dolium can uncover the different skills, resources, and labor invested in it, as well as how the vessel's value or purpose changed over time.[25]

Let us briefly look at one hypothetical example. Sometime in the first century CE, a specialist potter manufactured a dolium; a stamp found on the dolium tells us that the potter worked in a workshop in Campania that also supplied bricks, tiles, and amphorae to other sites in the region. The potter formed the dolium through coil building over the course of several weeks. After a lengthy period of air drying to remove moisture from its thick walls, the workshop's kiln operators fired the vessel with a batch of other heavy terracotta objects; temperature changes had to be gradual, and the process took several days (and overnight shifts) from loading, firing, and cooling to unloading the kiln. The pot's production was challenging, time consuming, and prone to failure, but it fetched a high profit. A customer from Pompeii ordered the vessel and arranged its transport with two contract drivers who packed the big jar carefully with straw onto a large mule-drawn cart.

19. For ceramic fabric, see Orton and Hughes 1993. On the similarity between the fabrics of bricks, tiles, and dolia, see Lazzeretti and Pallecchi 2005.

20. Ulp. *Dig.* 33.6.3.

21. Dolium-like vessels today, such as Portuguese *tinajas* and Georgian *qvevri,* are used for at least several decades, some even for over a century.

22. See Komar 2021 for recent discussion of amphorae and economy. On dolia, see Brenni 1985; Carrato 2017; Salido Domínguez 2017; Carrato and Cibecchini 2020.

23. E.g., De Caro 1994, 63–69, on the Villa Regina at Boscoreale; De Simone 2017.

24. Van Oyen 2015a; 2020b, esp. 50–53.

25. For the life history of dolia, see Peña 2007b, 324–325; also 213–227, 35, 46–47, 194–196. See also Skibo 1992; Schiffer 1972, 1996; Dobres 1999; Dobres and Hoffman 1999.

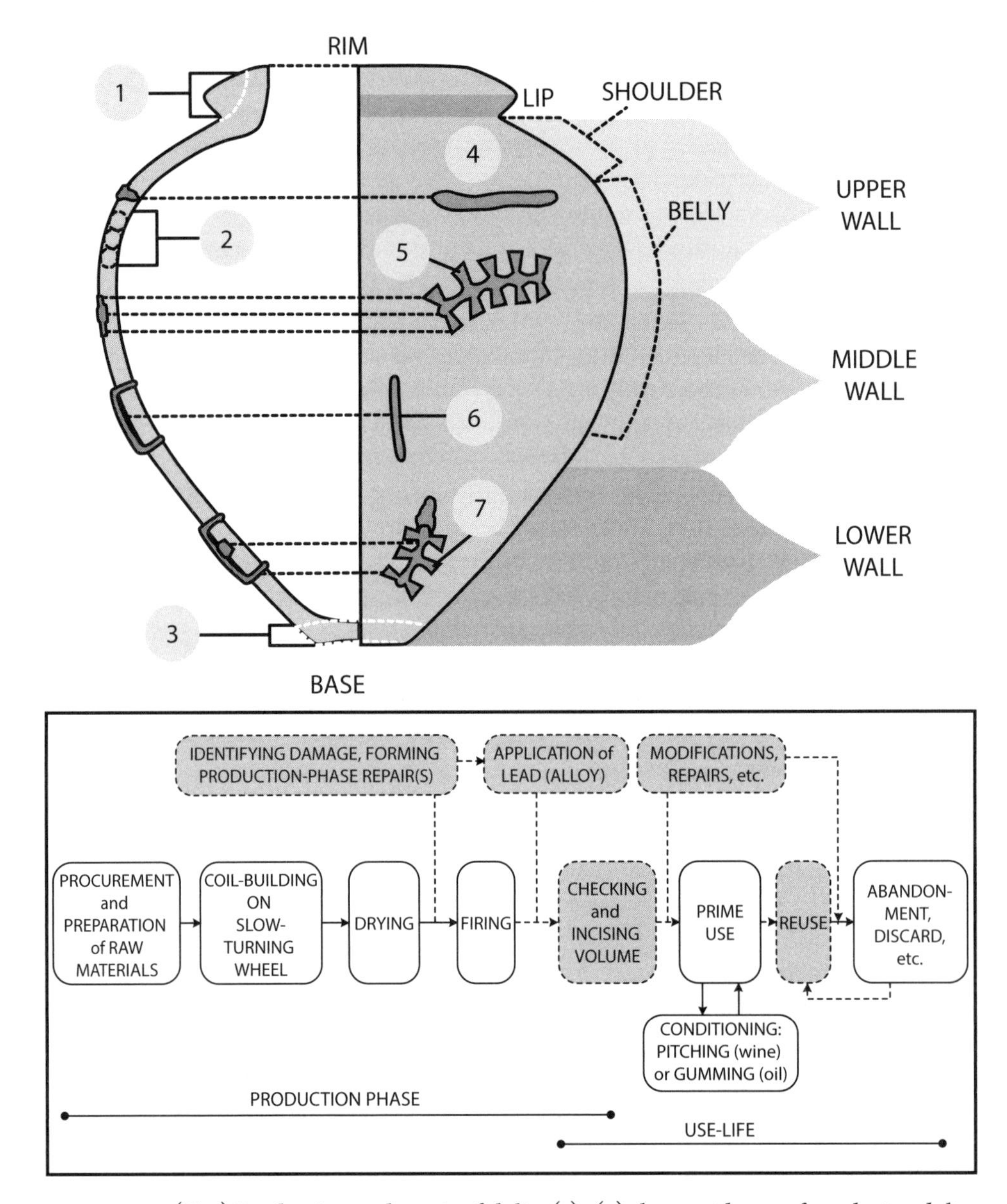

FIGURE 1.5. (*Top*) Production and repair of dolia; (1)–(3) show evidence of producing dolia (coil building); (4)–(7) show evidence of dolium repair: (4) lead fills, (5) double dovetail tenons, (6) staples, and (7) hybrid repairs. (*Bottom*) Dolium "life cycle"; shaded boxes represent optional stages. Illustrated by Gina Tibbott.

They navigated the cart from the workshop to its destination, a bar in Pompeii, and unloaded the dolium from the wagon. Later, they dug a hole, placing the jar halfway in up to its shoulder to stabilize it. Every few days, workers from the countryside dumped wine they brought in a large ox-hide container into the dolium; because it was partly buried in the ground, the dolium kept the wine cool, and workers regularly ladled out wine for customers. The workers also kept nuts, lentils, beans, vegetables, and eggs warm in other storage jars cemented in the counter, serving the local clientele. One day, cracks formed on our dolium around the rim and shoulder. Luckily, the workers found a pottery mender, who was able to scale up her technique and materials to mend the large jar. After drilling holes on either side of the crack, she formed lead clamps to bridge the crack. The repair was mostly successful, though the workers had to be careful not to fill the dolium too full, just up to where the crack started, and lined it with pine resin regularly to keep it from leaking. A couple of years later, a powerful earthquake shook the town, damaging many houses and other buildings. Our jar's crack grew, and the jar lost most of its wine. The workers were able to find the pottery mender again, but this time she not only gave a higher quote because it would be a more challenging job but also could not guarantee the repairs would work. The owner decided to cut his losses and found a neighbor willing to pay a few denarii to take the dolium off his hands. The neighbor ran a perfume workshop out of a flower garden just down the street. He had already salvaged a few severely cracked wine dolia from other properties for dirt cheap. He buried our dolium in the garden to store rainwater next to another broken jar that he had sawn in half and mounted on four low pillars as a house for the garden's canine guardian.[26] Not long afterward, Mount Vesuvius erupted and covered the entire town of Pompeii. Our dolium would not be discovered until almost two thousand years later.

The cracks, discoloration, repairs, and stamp on the dolium tell us about the jar's trajectory: from how a local potter skillfully manufactured it to the care in its transportation and installation, how useful it was for retail, and how people found ways to mend and reuse the dolium when it broke. This is the story of just one dolium out of many. Studying multiple dolia together, we can learn more about this type of storage technology and systematic patterns of how people invested in and profited from this type of vessel. With a large, diverse dataset, we can trace regional differences and chronological developments in how dolia were made and used. Dolia in west-central Italy and beyond over the course of approximately four hundred years were products of a developing ceramic craft, which specialist potters established and continuously refined. Using and maintaining dolia also required particular procedures, knowledge, and resources. Dolia were an expensive type of hardware. For those who could afford them, they became a tool integral to systems of trade that enabled economic strategies previously unattainable. This book therefore considers dolia a type of storage container *technology*—that is, a type of tool or instrument and the skills by which people in antiquity produced and used it.[27] As a type of supersized storage container and technology, dolia not only emerged during a period of intensifying large-scale and long-distance trade across expanding imperial territory but also propelled it.

Urban Growth and the Food Supply

Technology does not develop or exist in a vacuum. People develop technological innovations to solve problems or streamline a process, and the success of these technologies and innovations depends on how well integrated they are in their cultural context.[28] Technologies reflect

26. For identification of the jar as a doghouse, see Jashemski 1979a.

27. Bain's (1937, 860) classic definition of technology: "includes all tools, machines, utensils, weapons, instruments, housing, clothing, communication and transporting devices and the skills by which we produce and use them. Social institutions and their so-called non-material concomitants such as values, morals, manners, wishes, hopes, fears and attitudes are directly and indirectly dependent upon technology and are mediated by it."

28. Schatzberg 2018; Rogers 2003.

their times, and dolia were no different.[29] Dolia were designed, refined, and implemented to expand production, storage, and trade across the Mediterranean basin.

The emergence of Rome's sweeping territorial empire during the final two centuries BCE fundamentally transformed the Mediterranean region.[30] The population within Italy itself experienced major demographic shifts, with many from the countryside and abroad streaming into the city of Rome.[31] By the first century BCE, Rome became the largest metropolis in the Mediterranean world, with a population of one million.[32] The dramatic growth of the city and the attendant demands for resources reverberated throughout Italy and stimulated developments in both agricultural production and urbanization. Changing patterns of land use and ownership, the increasingly uneven distribution of wealth, large influxes of enslaved people, and the growing use of enslaved labor in agriculture disrupted and even displaced many free peasants, who poured into the city of Rome for work.[33] As the population of the city of Rome grew, so did demands for a reliable food supply.[34] The hinterland and adjacent territories of Rome were areas where these demands had the greatest impact on agricultural and horticultural transformations in the landscape. This new settlement pattern required a sophisticated regime for the production, storage, and distribution of agricultural products to feed the city, often requiring the state to facilitate and maintain this new system.

Ensuring a supply of food was always paramount for the Roman state, and numerous attempts were made to provide and guarantee grain for the urban populace.[35] The state offered several incentives to those willing to transport grain to Rome, such as tax exemptions, social privileges, and even citizenship.[36] Moreover, institutional developments and technological advances during this period enabled large-scale and long-distance merchant shipments at major ports in the Mediterranean, especially along the west-central coast of Italy.[37] In fact, from the outset of Rome's expansion, the coast of Italy was of prime strategic importance for Rome, and the establishment of Roman colonies at Cosa to the north and Paestum to the south in 273 BCE, as well as the general oversight of the Tiber River, safeguarded this vital region.[38] Between the two colonies, a series of ports dotting the coast

29. E.g., Finley 1965, 1973; Greene 2000; A.I. Wilson 2002; Taylor 2010.

30. For climate history during Rome's expansion, see Bernard et al. 2023.

31. Hopkins 1980; Hin 2013; Morley 1996; Witcher 2005; Scheidel 2004, 2005.

32. Scheidel 2007, 2001; Morley 2013, 1996; Parkin 1992; Hopkins 1978; Hermansen 1978; Storey 1997. For summary, see Morley 2013, 1996, which estimates a population between 850,000 and 1,000,000.

33. Studies include Morley 1996 on developments in agriculture in Rome and its hinterland; Purcell 1994 on the *plebs urbana*; and Witcher 2005 for demographic changes in Rome's immediate hinterland.

34. Morley 1996; Hopkins 1980, 1978.

35. Rickman 1980; Erdkamp 2005; Sirks 1991; Garnsey and Morris 1989; Garnsey 1998, 1999; Mattingly and Aldrete 2000; Vitelli 1980; Geraci 2018; Holleran 2019.

36. On the role of the state in lowering transaction costs, increasing agricultural productivity, and protecting farmers and landholders, see Kehoe 2013.

37. See D. Robinson et al. 2020. A.I. Wilson 2011 attributes the high frequency of long-distance commercial shipping to institutional developments, the eradication of piracy, the use of a single currency, reduced transaction costs, a greater integration of markets, and the consolidation of the Mediterranean Sea under one political entity.

38. Vell. Pat. 1.14; Livy *Epi. Per.* 14.

served Rome's military and economic interests, permitting vast quantities of goods to enter the city. Roman conquests enabled the extraction of grain from acquired territories such as Egypt, Sicily, and North Africa, allowing some farmers and landowners in Italy to turn to large-scale wine production.

Viticulture was potentially profitable, but also highly risky.[39] Wine featured in all aspects of daily life in ancient Rome, from casual dining to lavish banquets to religious festivals, and many considered it a staple food. With many different types of vintages and varying grades of quality, the price of wine could fluctuate wildly, but unlike for cereals, the state made no attempt to regulate it. Yet viticulture could also lead to financial ruin. Ancient writers discussed examples of vineyards increasing in value exorbitantly, or coming to a crash and emptying an investor's coffers. Pliny the Elder recounts a story about a man who acquired a neglected vineyard for a low price to revive it in ten years and sell it at four times more than the original price.[40] Stories of flipping vineyards or losing an investment weave across discussions of wine production, cautioning the reader that financial ruin could await a negligent or unlucky farmer. Cultivating grapes and producing wine was an expensive activity fraught with potential failures from as early in the process as establishing a vineyard to the storage and sale of wine. Setting up a vineyard required time, patience, and money: newly planted vines do not begin to produce fruit until at least three years after their planting, during which farmers would sink labor and equipment into the vineyard without making any profit. Their cultivation also drew on specialized skills and knowledge as they required proper support and pruning, topics popular among agronomists who codified best practices for viticulture in their agricultural manuals for maximum yields.[41] The labor regime for wine production was also uneven, necessitating vast amounts of labor focalized during the harvest season, which required the landowner to hire seasonal farmhands to harvest and process grapes quickly. But the most disappointing failure was when wine, after processing and aging, spoiled in storage, dashing a vintner's hopes of reaping a fortune.

The key to Rome's wide-reaching, massive food supply was storage. The storage of agricultural surplus was the building block of every society.[42] Communities and households in antiquity processed and stored their foods to access these items of sustenance throughout the year. Scholars working across different areas and time periods have identified a range of storage infrastructure from silos and pits to storage bins and jars to specially built storerooms. Communities practiced large-scale storage of agricultural surplus to buffer against periodic

39. On the development, profitability, and risks of viticulture in Italy, see Purcell 1985.

40. Plin. *NH* 14.48ff. Shaw 2019, 535.

41. On types of supports for vines, see Varro *Rust.* 1.8; Columella *Rust.* 1.4–6; Cato *Agr.* 1.7. In the early history of viticulture in Egypt, Greek landowners who leased their agricultural properties to local Egyptians still took care of viticulture themselves. *P.Ross.Georg.* 2.19; *P.Oxy.* 47.3354; Rowlandson 1996, 231–236; Kloppenborg 2006, 516–521, 528–534; Langellotti 2020, 188–193; Vandorpe and Clarysse 1997; Rowlandson 1999, 139ff.

42. For studies in the Roman world, see Erdkamp 2005; Morley 1996; Garnsey 1988; Bowman and Wilson 2013; Rickman 1980; Virlouvet 1995; Van Oyen 2020b. In the Greek world, see Foxhall 1993; Foxhall and Forbes 1982; Garnsey and Morris 1989; Halstead and Frederick 2000; Riley 1999; R. Palmer 2001; van Andel and Runnels 1987; Alcock et al. 1994; Halstead 1987, 1989; Halstead and O'Shea 1982, 1989; Wells 1992; Barret and Halstead 2004; Howe 2008. See J.C. Scott 2017; Chankowski et al. 2018; Forbes and Foxhall 1995. For the Andes and Greece, see Hastorf and Foxhall 2017.

variations in food availability, an issue particularly problematic in the Mediterranean.[43] Astrid Van Oyen has recently shown that storage as a practice is both universal and historical: storage is a practice all societies engage in, but it varies across time and space and shapes history.[44] As a result, modern scholars have been fascinated with the large storehouses featured in many ancient Mediterranean settlements. Storehouses, especially granaries, are some of the most conspicuous and distinctive buildings in Roman settlements. Often large with two or more floors, thick walls, and multiple single rooms, sometimes around a large central courtyard, these structures dominated the landscape and imposed a sense of controlled surplus.[45] Security, organization, and supervised access were shared concerns across different sites, regions, and stored foods. For grain, the raised floors of the horrea promoted ventilation and the thick walls helped keep temperatures cool and stable. Storing wine and olive oil, however, entailed additional requirements. Critical to their storage was proper packaging (and containers) in order to hold, protect, identify, and transport them.[46] Storage, in other words, lies between production and consumption—where most scholarly attention has been focused. Yet it was nonetheless essential, and its infrastructure often required specialized expertise, particular modes of organizing labor, and, most importantly, proper containers.[47] Wine and oil cellars (*cellae vinariae* and *cellae oleariae*) were architecturally and functionally distinct from other types of storerooms thanks to the containers within: the dolia.

An Empire Full of Containers

In a world without refrigeration, the containers in which food was stored and transported were the superstars of a large food supply system. Containers are not only capable of holding something for transport and storage; they can also package and "brand" goods.[48] People in the ancient Mediterranean used various types of containers to protect, store, and distribute different foods. The importance of these containers in antiquity can hardly be overestimated. They transported olive oil and wine over long distances. More importantly, though, each container protected its contents, ensuring that the product's quality would be preserved throughout its journey.[49] But two important points must be clarified immediately. First, not all containers are created equal.[50] Different types of containers had their own properties that

43. On variation in food supply in the Mediterranean, see Garnsey 1988; Bintliff 1997; Halstead and O'Shea 1989; Horden and Purcell 2000, ch. 6. Responses to food scarcity included diversification of agricultural products and production, storage, and redistribution of surplus. For agriculture and political economy, see Earle 2002; M.E. Smith 2004; Foxhall 1995; D'Altroy and Earle 1985; LeVine 1992.

44. Van Oyen 2020b, 1–18. See also Bevan 2020.

45. Rickman 1971; Van Oyen 2020b, 2015a. On the efficiency of horrea, see Pagliaro et al. 2014, 2016.

46. Cheung 2020; Curtis 2015, 2016.

47. E.g., Dietler 2010b; Kehoe 2007; Jongman 2007; Morley 2007.

48. On containers and packaging, see Klose 2015, 323–341. On container types, see Knappett 2020, 130–166; Hunter-Anderson 1977; papers in Shryock and Smail 2018a, esp. Shryock and Smail 2018b; Bevan 2018; Robb 2018.

49. See Bevan 2014 for a *longue durée* study of containers in the Mediterranean basin; see McCormick 2012 for amphorae and barrels. Shipping containers today transformed production and consumerism; see George 2013; Levinson 2006; Klose 2015; Shryock and Smail 2018a.

50. Cf. Twede's work on packaging, the history of packaging, and packaging performance: Twede 2002a, 2002b, 2005a, 2009; Twede and Harte 2011; Twede et al. 2000a, 2000b.

made them advantageous or ineffective for certain products, modes of transportation, or steps in the supply chain. It would not make sense, for example, to package a commodity normally sold in bulk in fragile containers—for example, shipping grain in glass bottles. Some containers were more effective for storage—that is, maintaining ideal conditions to stop or delay the deterioration of their contents. Second, containers only functioned as people expected if they were made well and were handled properly before, during, and after each usage. If a jar had a production defect, it had to be either repaired or replaced to ensure its contents would be protected. Moreover, the choice of container was influenced not only by accessibility and costs of materials but also by cultural preferences and workforces and industries in the area or the ability to import them. People might even expect to receive certain types of containers with their products.

Containers were some of the most essential objects and actors in an intricate system of storage and packaging that made food available year-round and in far-flung destinations, one of the most remarkable traits of the Roman Empire. They are the products of the traditions and behaviors of storage and packaging, and reflect some of the deepest cultural mentalities and preferences. In the United States, for example, milk is packaged in paper cartons, plastic jugs (often with a single handle), or glass bottles, whereas in Canada milk is sold in bulk in large plastic bags. The availability of certain containers in antiquity too depended not only on natural resources but also on cultural preferences, social expectations, labor, and economic conditions. Agricultural workers expected specific types of equipment and containers to process and package their goods. Wine was supposed to have a particular taste and texture, and it was expected to be stored, packaged, presented, and labeled a certain way. Lastly, containers were so widely used for all types of distribution and consumption—just think of how many bags, cartons, bottles, and cans are picked up when buying groceries—that their production and distribution could make people rich. (Case in point: the Uihlein family, owners of ULINE, a company that produces packaging and shipping materials, is worth over $4 billion.)[51]

Studying containers and their biographies, itineraries, or trajectories helps us recognize the vast array of craftspeople, skills, manpower, and organization of labor required for making and using these containers; the social and cultural meanings ascribed to them; and their role in shaping labor, the economy, and agricultural practices.[52] At the peak of the Roman Empire, the dolium reigned supreme. The heavyweight ceramic storage vessel was uniquely placed among different containers, fulfilling special roles that no other container did in antiquity and becoming a concept that symbolized abundance and wealth. In order to understand what exactly that role was, and what the potential payoff is from studying it, it will be useful to review briefly the process of making wine and olive oil, and the other types of containers that operated in the same system.

Sometime in early autumn, droves of farmhands, usually contract laborers, freed ripe grapes from the vine, placing them into baskets (*fisci*).[53] It was an urgent time. As soon as

51. Saul and Hakim 2018. In 2020, ULINE generated almost $6 billion in revenue.

52. For the (changing) value of objects, see Appadurai 1986; Kopytoff 1986. For benefits of the term "trajectory," see Van Oyen and Pitts 2017b, 13ff.: object biographies are useful for studying single objects; trajectories give objects a role to play. See also Joyce and Gillespie 2015a, 2015b; Hodder 2012; Bennett 2009.

53. On contract vineyard work, see Kloppenborg 2006. On tenancy in Byzantine Egypt, see Hickey 2012.

grapes reached the peak of ripeness, they were at their sweetest but also most vulnerable state. Workers had to move quickly and harvest the grapes, while taking care not to bruise them or puncture their skins; after gathering the grapes, workers would then tread them in vats or press them, often in a bag or sack (*saccus*), and the freshly pressed juice (*mustum*, "must") would be collected in open vats (Figure 1.6).[54] After the initial fermentation period, which lasted a few days, vintners could move wine from vats into dolia and seal the dolia. After at least thirty days of fermentation, the wine was separated from its sediments (lees).[55] From fermentation until the wine was packaged for sale, winemakers could employ different sorts of treatments to protect the wine's quality and alter its taste, and they could even reserve some to age into a more expensive vintage. To sell and distribute wine, workers transferred the wine from the dolium into other containers, such as amphorae, which were often used to export products, especially overseas, or an ox-hide container known as a *culleus*, which could then be carted to its final destination, where the wine could be decanted into other vessels, or to a bottling facility where wine could be poured into amphorae and shipped overseas.[56] Variations to this schema were possible, but these were the typical stages and containers in the fermentation, storage, and packaging of wine.

Olives too were harvested and pressed in the fall, though a significant crop generally developed only every two years due to the olive's biennial cycle.[57] Olives also required processing. Anyone who has ever tasted an olive off the tree knows it has to be treated before it is edible. The olive is a bitter fruit that requires brining or other preparations before it can be consumed. In antiquity, the process for pressing olives for olive oil was similar to that for making wine and, in fact, they required much of the same equipment, though the processing of wine and olives usually did not share equipment at elite production facilities to avoid contamination.[58] Olives were generally harvested in late autumn, though there was a wider window of time than that for grapes: harvesting less ripe olives at the earlier end of the spectrum would yield smaller amounts of high-quality oil, whereas harvesting ripe olives later would increase quantities of lower-quality oil.[59] After collecting and cleaning olives, workers used a stone olive mill (*trapetum*) to crush the olives, after which they placed the pulp, flesh, and fragmented seeds into baskets for pressing, usually with a lever or screw press.[60] After pressing, workers separated the oil from the bitter, aqueous part of the olive

54. This is a simplified account of wine production. Three batches of wine could be produced: (i) from treading, (ii) from the first pressing, and (iii) from pressing the skins, known as *lor(e)a*, which would be given to enslaved workers for rations. For discussion, see Curtis 2001, 375ff.; Thurmond 2006. For overview of equipment, see K.D. White 1975; Hilgers 1969.

55. Cato *Agr.* 25: wine should ferment for at least thirty days. For variations and transformative qualities of wine, see Thurmond 2017; Dodd 2022; Van Oyen 2020b, 50–53.

56. Villa B of Oplontis is a unique example of a bottling facility; M.L. Thomas 2015, 2016. For wine and amphorae production in regional networks, cf. Peña and McCallum 2009a, 2009b.

57. See Waliszewski 2014 for how Romans might have tinkered with olive groves to get annual yields.

58. See Rossiter 1981; Curtis 2001; Marzano 2013a. On distinguishing between facilities, see Brun 1993. Peasant farms probably did not have separate equipment; see Vaccaro et al. 2013, 140–142.

59. Plin. *NH* 18.320: olives are harvested and oil produced after the vintage. Thurmond 2006, ch. 2. Rowan 2019a.

60. For different types of presses used in the Roman world, cf. Curtis 2001, 381ff.; Lewit 2020.

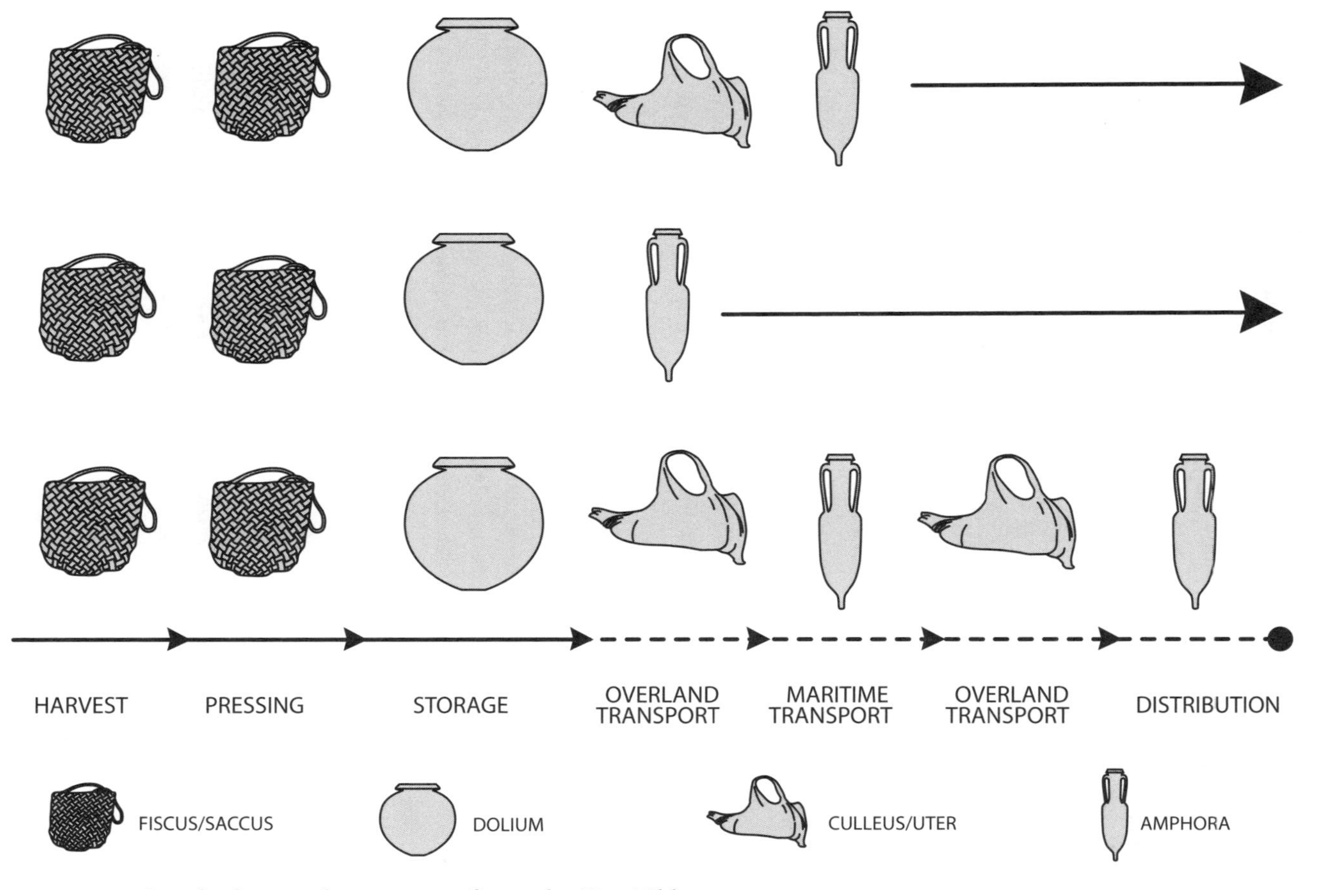

FIGURE 1.6. Supply chain, and containers, of wine, by Gina Tibbott.

(*amurca*).[61] Depending on the scale of production, they placed the oil in a dolium or smaller jars to settle for a few months, and later transferred the oil to another container for distribution.[62]

Although wine and olive oil production had their own procedures, timelines, and concerns, both required the storage conditions that a dolium offered. Foods in general deteriorate from exposure to light, oxygen, and high temperatures, so they need protection from heat, air, and light, as well as pests. In general, the various stages of the supply chain for wine or oil relied on different containers, but dolia often held wine and olive oil for long (sometimes the longest) periods of time and during the most formative stage of the process.[63] It is thus no surprise that farmers and merchants invested in and installed thousands of these supersized jars across the ancient Mediterranean.

Organization of Chapters

The dolium-based storage technology was in full force in central Italy during a period of imperial expansion, circa 200 BCE–200 CE, when Rome was becoming the largest and most populated city and when the Mediterranean was increasingly unified economically and politically.[64] In seeking to better understand this unification, a fuller picture of the investments, skills, labor, and people involved can emerge by tracing the development of dolia not only as objects but as an industry and a type of technology in their own right. This book evaluates the economic and social realities of Roman imperialism for the individuals living in the shadow of the epicenter of a Mediterranean-wide empire through the lens of dolia. Situated at the intersection of pottery, craft production, agriculture, and the construction industry, dolia bring to light interactions and relationships between elites and subelites alike among seemingly disparate activities. By studying the nuts and bolts of this commerce, the book opens a new window on a whole series of uncharted interactions in the ancient world. The following chapters show how the expansive, highly profitable wine trade so distinctively characteristic of the Roman period was only able to emerge and grow thanks to the dolium storage technology. Dolia enabled the large-scale production, storage, and distribution of wine to supply the ancient capital, and their increasing use offered new opportunities for wealth. In other words, the dolium storage technology was simultaneously a manifestation, product, and instrument of Roman economic expansion.

Because of the uneven nature of the evidence, and in order to explore the importance of dolia from several distinct angles, this book is organized thematically, rather than chronologically or regionally, and primarily follows the life cycle of (1) the dolium, dolium technology, and dolium industries—from development, through use and maintenance, to demise—and, in discussing the use of dolia, (2) the supply chain of wine, from its production to its storage, transport, and retail. Chapter 2 starts from the beginning of the dolium's story and traces the dolium industry. The chapter explores how potters developed and refined a challenging craft for a new, specialized storage technology. Although dolium

61. See Curtis 2001, 394, for techniques of oil separation.

62. Smaller jars—e.g., *labra* and *seriae*—stored olive oil; cf. K.D. White 1975. Varro *Rust.* 3.2.8: *serias olearias.*

63. Many olive oil production sites did not have dolia (see Chapters 3 and 5).

64. Bevan 2014, 392. Morley 1996 discusses broad agricultural developments to supply Rome.

producers used the same technique to make the vessels, the scale of production differed drastically between regions and among workshops. Dolium production was a challenging, long-term process that required at least several weeks and substantial upfront costs, which most could not afford. Over time, investors in multiple large-scale opus doliare (heavy terracotta and ceramic products; the term *opus doliare* is based on attestations on brick stamps) workshops in the Tiber River Valley successfully and profitably included dolium production in their repertoire of mass-produced bricks and tiles, becoming attractive "one-stop shops." Dolium production became a lucrative industry, garnering not only financial gain for workshop owners but also status, power, and control over resources essential to viticulture and being a good farmer.

The next part of the book (Chapters 3–6) reviews the various uses of dolia for wine and olive oil, from their production to their transport and sale. Chapter 3 looks at how farmers in central Italy used dolia for viticulture, sometimes amassing capacious wine cellars to produce for a market. Although dolia were designed for and used in viticulture, they were practical as multiuse vessels too, commonly implemented in large-scale olive oil production. All farms needed storage equipment and facilities, but dolia appeared in great numbers on estates, especially large ones, supporting and enabling profitable viticulture and olive oil production for sale on the market, and often for export too. Dolia became so instrumental in expanding an estate's ability to produce and store wine that they came to represent good farming and abundance, and some villas even celebrated viticulture and storage by embellishing their wine cellars and dolia.

The next two chapters expand the geographical lens to look at how communities adopted dolia beyond Italy in a highly connected Mediterranean world. The massive capacities of dolia became attractive for an unexpected purpose: bulk transport. Markets farther afield offered potentially huge profits, luring merchants to undertake more long-distance trade. Bringing goods to more-distant destinations, however, called for containers that could not only hold large amounts but also protect their contents on potentially long journeys. Chapter 4 takes stock of the different containers for the local and long-distance trade in wine (and olive oil) during the Roman period. Although traders had traditionally packaged wine, olive oil, and other foods in amphorae for centuries, some traders forged a new and highly specialized form of bulk shipping. Synthesizing recent discoveries in underwater archaeology, this chapter sheds light on innovations in the dolium industry as a group of entrepreneurs merged dolium construction with shipbuilding to create a new vessel to deliver tons of wine more efficiently to markets in the western Mediterranean. Thanks to the ships' unique design, traders could quickly dump massive quantities of wine into dolia built directly into tanker ships and sail to both maritime and fluvial ports where lucrative markets were based within a narrow "open" sailing season.

Chapter 5 shows that, in an increasingly connected economy, lucrative opportunities in the wine and olive oil markets took off and fostered different knock-on effects. Landowners in the northwestern Mediterranean, where dolium ships visited, also invested in dolia as local villas developed and diversified their agricultural portfolio to include wine and olive oil. Over time, producers in Gaul and Iberia supersized their own production and storage capabilities, tapping into the dolium-based infrastructure, as they expanded their operations and delivered huge amounts of wine and olive oil to the capital's doorstep. As a result, expansive villas specializing in wine and olive oil production spread across Gaul and Iberia, often surpassing Italian villas in scale. In addition to markets in the northwest Mediterranean, Rome was a destination and consumer of those wines as demand and infrastructure for a more robust food supply system grew.

Although dolia were originally used for food production and storage on farms and villas, their effectiveness as storage containers opened new possibilities for urban retailers, traders, and consumers, especially around the imperial capital. The next part of the book shifts its gaze back to urban infrastructure in west-central Italy. Chapter 6 examines how different urban communities in central Italy used these vessels to support urban populations near the capital. Thanks to their design, dolia could be installed to maximize food storage in densely populated areas. Their labor-intensive use and maintenance, however, also meant that their adoption was not always successful, compatible, or long-lived. In places where dolia became worthwhile investments, these vessels were built into shops, bars, and warehouses and occasionally featured in dining establishments where additional services could be provided. Dolia helped bridge town and country, and their role in urban storage fostered the rise of urbanism and increasing specialization in the urban economy. In Ostia and Rome, wine warehouses were especially massive and specialized structures, and probably ventures only the very wealthy and powerful could afford to bankroll. Those who had the financial and social capital to set up cellae vinariae not only controlled wine supplies; they also wielded considerable negotiating power, influence, and prestige.

The next part of the book casts light on the maintenance, repair, and longevity of these valuable vessels. Chapter 7 explores how dolium owners protected their investments and how the dolium industry became increasingly specialized as different parties developed dolium repair techniques. Dolia were listed among essential farm equipment and were expensive, yet they were susceptible to damage and prone to break. Many dolium users chose to have their costly vessels repaired to prolong their use rather than just throw them away. Mending dolia, however, posed new challenges. Although they are ceramic, they were much bulkier and heavier than other types of pottery, and traditional pottery repair materials and methods often failed to stabilize and hold together these hefty pots. Craftspeople experimented with and devised new techniques, finding ways to make stronger repairs. Traditionally, nonspecialist craftspeople fixed dolia when they became damaged in use. As the dolium industry developed, however, dolium repair in some areas became more specialized within the workshop itself, drawing on techniques from the architectural industry. Regional discrepancies in dolium repairs reveal different resources and organizations of labor available for urban food storage. In the area around the capital, where demand and specialization were higher, opus doliare workshops directed significant skill and labor into preemptively reinforcing their products. Owners of profitable urban warehouses not only procured well-made and reinforced dolia from opus doliare workshops; they also hired designated workforces to maintain and routinely provide upkeep for their costly investments. The various, and often specialized, repair techniques and workforces simultaneously extended the reliability and longevity of dolia, as well as the success and stability of the dolium industry, but involved substantial resources and new skills.

Despite diligent maintenance and repairs, dolia inevitably broke or fell out of use. Chapter 8 surveys different ways dolium owners might have tried to recuperate their investments, as well as why and how some eventually discarded them. Because dolia were so large and unwieldy, people found creative ways to reuse and jettison the vessels. The chapter then notes widespread abandonment of the technology too, as farmers and merchants shifted their priorities when they faced new opportunities. For some urban communities, using dolia and their specialized system no longer seemed worthwhile, and they stopped moving and storing wine in large dolia. The dolium-based system enjoyed success for several centuries, but merchants and vintners began to abandon the technology across the board. The chapter considers broader changes to the industry and to storage and packaging as some farmers and merchants switched to a radically new container technology that would be in place for almost two millennia, a more generalized system that revolved around the barrel. Their growing use to deliver wine to the capital was not a simple or straightforward replacement of dolia but sheds light on the pitfalls of the specialized storage system that had sprung up around the dolium as well as new economic strategies wine traders were pursuing to supply Rome and other communities across the empire. Chapter 9 zooms out to consider the broader implications of a dolium-based container system for investors, workshops, and personnel before discussing container systems and the enduring legacy of dolia.

The Data

Much of the book focuses on west-central Italy, the area around the capital, and is informed by published material and further augmented by unpublished material from four sites that provide a detailed view of dolium industries in that area from the second century BCE to the second century CE (Figure 1.7): Cosa (second century BCE to second century CE, though most of the material is from the final two centuries BCE), Pompeii (third quarter of the first century CE), Ostia, and Rome, the capital itself (second century CE). The towns and cities are well-known archaeological sites of central Italy that have been relatively well published. The dolia, however, have not been studied in depth, and this study, the first to document the dolia in great detail, thus integrates their analysis within a richer, contextually informed discussion. The data on these dolia—their dimensions, stamps, markings, and repairs, among other types of evidence—have been compiled in Appendix 1 (A1), and more detailed descriptions of individual dolia are included in Appendix 2 (A2). (The Guide to the Appendixes will provide more information on the assembly and organization of the data.) Overall, the data show that the trajectories in both the development of dolia and their industries diverged between the sites, highlighting the multiple ways dolia could be adopted in a range of urban areas: the towns of Cosa and Pompeii occupied a place in agriculturally rich hinterlands known as wine exporters, whereas the densely occupied cities of Ostia and Rome relied on foods produced elsewhere.[65]

The first case study settlement is the town of Cosa, a port colony founded in 273 BCE in southern Tuscany with a thriving wine industry that dominated the western Mediterranean from the mid- to late Republic.[66] Perched on a hill about 110 meters above sea level and just approximately 150 km northwest of Rome, Cosa overlooked both the Tyrrhenian Sea and a hinterland speckled with multiple villas engaged in lucrative activities. Hundreds

65. Erdkamp 2001.

66. F.E. Brown 1951, 1980.

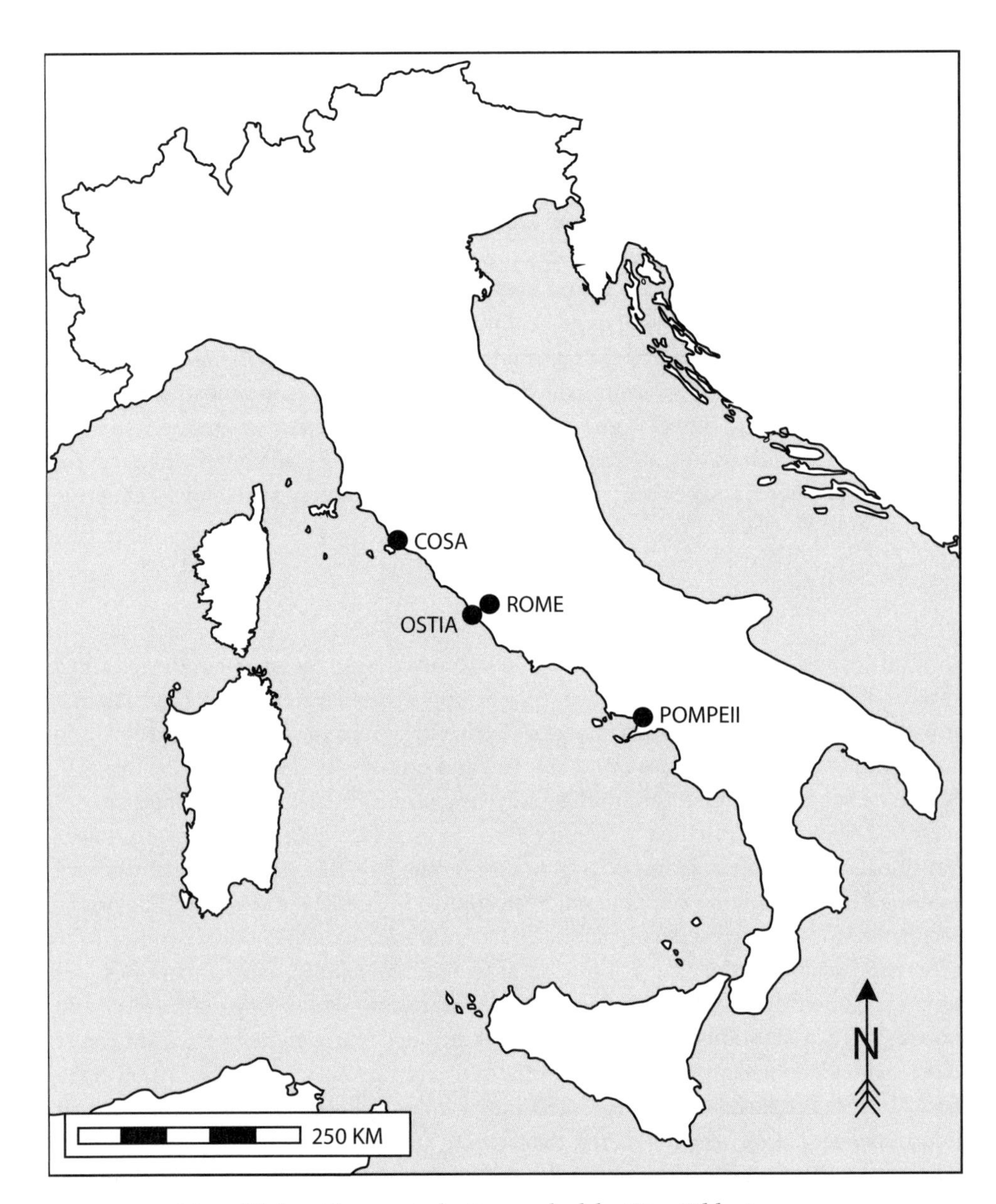

FIGURE 1.7. Map of Italy with case study sites marked, by Gina Tibbott.

of wine amphorae originating from Cosa have been found in large concentrations in areas as distant as southern Gaul, testifying to its large-scale wine enterprise. Many of these amphorae bore stamps linked to the prominent Sestius family, offering an opportunity to explore the developing relationships between industries for wine, agricultural production, and pottery in this region. From the end of the Republic through the imperial period, Cosa increasingly imported, rather than exported, wine.[67] Cosa and its

67. Will 1987.

hinterland, the *ager Cosanus*, have been the focus of many archaeological projects and studies, and excavations of the town over the years have explored the forum, several temples, some houses, and now the bath complex. Among the many artifacts recovered from excavations of the town were nearly fifty dolium fragments;[68] though they are low in quantity, often from reuse or discard contexts (only a few are from primary use contexts), and not well preserved, these dolium fragments are among the earliest datable dolia from an urban site. The majority of dolia and dolium-related objects were discovered in the early excavations sponsored by the American Academy in Rome and formed part of a publication on the utilitarian pottery finds, but only diagnostic fragments (sherds with features such as rim and base) were published.[69] The current Cosa excavations of the bath complex discovered other dolium fragments, which were reused as fill, that inform this study.[70] In general, most of the diagnostic dolium fragments date from the early second to late first century BCE, and some fragments, primarily body sherds, were reused in second-century CE contexts. With materials spanning from the early to later history of the town, the Cosan evidence highlights the wax and wane of dolium use in an agriculturally productive colony.

The second case study site, Pompeii, offers a detailed view of storage during the first century CE, a period notable for global trade and a "consumer revolution."[71] Pompeii, an urban settlement in Campania, was founded sometime in the seventh or sixth century BCE and was granted colonial status in 80 BCE after Sulla's conquest. Ancient authors noted that Pompeii was particularly fertile thanks to its rich volcanic soil and was hence known as a region that produced plentiful fruits and well-known wine.[72] Archaeological evidence, including architectural and archaeobotanical remains, confirms that Pompeii itself was a productive agricultural town, with farmhouses and villas clustered densely not only outside the town but even within the town walls.[73] Due to the eruption of Vesuvius in 79 CE, Pompeii offers a unique opportunity to study Roman agricultural production and its integration within an urban fabric. The southeastern sector (Regio I and Regio II), notably the town's verdant sector, along with several "villas" outside the town walls such as the Villa of the Mysteries and Villa Regina of Boscoreale, illuminates the storage and packaging behaviors of Pompeii and its *ager*.[74] Most of the dolia are well preserved and have been left in situ since the eruption of Mt. Vesuvius in 79 CE, but some were moved from the town and surrounding areas to the storerooms on-site. Pompeian dolia were typically found in shops and in planted areas, such as gardens, vineyards, and groves for

68. Some dolia were published (Dyson 1976), but many remain unstudied.

69. Nondiagnostic fragments such as body sherds were often discarded; catalog cards in the Cosa Archive at the American Academy in Rome indicate dolia were discarded over the years.

70. R.T. Scott et al. 2015; De Giorgi 2018.

71. Wallace-Hadrill 2008 places Italy's "consumer revolution" to a time when the population began to recover from civil war and demands for consumer goods spurred production in Italy.

72. Plin. *NH* 14.35, 14.38; Flor. *Epitome* 1.16. Vesuvian wine was shipped overseas. M.L. Thomas (2015) and Peña and McCallum (2009a, 2009b) have posited that a wine packaging facility was located on the coast near wine production centers.

73. For discussion and evidence of cultivation in Pompeii, see Jashemski 1979b, 1993.

74. See Nappo 1997 for growth and reorganization of Pompeii's southeastern sector: housing developments in the late third/early second century BCE, then agricultural production in the first century CE.

the storage, fermentation, or processing of agricultural products. Approximately one hundred dolia and dolium fragments, and another hundred of a cylindrical storage jar, mostly found where they were used in antiquity, reveal how integral and specialized dolium storage was for an agricultural town's food production and retail in the first century CE.

The urban populace of Rome constitutes, of course, the major beneficiary of these long-distance movements, and the city's infrastructure constantly evolved to store masses of commodities flowing into the city.[75] The third case study focuses on the area of the imperial capital and examines both Ostia and the city of Rome. As one of the capital's most crucial ports, Ostia shows how the "local" territory was affected by Rome's own growth from the late first to the beginning of the third century.[76] Ancient authors credited the foundation of Ostia, situated twenty-five kilometers southwest of Rome, to Ancus Marcius, the fourth king of Rome, in the late seventh century BCE.[77] For most, if not all, of its occupation, Ostia was a naturally strategic harbor for Rome, and as the city and empire of Rome grew, so did Ostia.[78] Ostia underwent several major renovations during the early second century CE, when various parts of the settlement were expanded or rebuilt. Among the enhancements in the harbor district were several warehouses containing dolia defossa to hold wine. Ostia had over two hundred dolia installed, but they have only been briefly mentioned in publications.[79] The roughly 125 dolia across three storerooms still accessible today provide valuable insights into the development of the dolium craft and the large-scale enterprise for storing wine to supply the capital. The city of Rome, a densely populated metropolis, was also equipped with a plethora of warehouses, including wine and oil cellars with dolia, though we mostly know of them through inscriptions rather than physical remains. The continuous occupation of the city has meant that only a small portion of the ancient city has been excavated, but several well-preserved dolia have been recovered, approximately twenty of which were documented for this study. In antiquity, the numerous cellae vinariae of the capital were surely filled with many dolia, three dozen of which are now scattered in various museums or set up as décor in gardens, parks, and even the US embassy to the Italian Republic. Most of the dolia were moved from their ancient contexts to museums early on and now lack provenance, though they likely came from several wine warehouses built in the second century CE. Because Rome and Ostia were serviced by the same workshops, their dolia were mostly found in the same types of contexts, and the uneven sets of evidence from the two cities inform each other, they are discussed as a single case study in the following chapters.

A close examination of dolia brings to light the ingenuity, cross-craft fertilizations, collaborations, and social and economic constraints of largely invisible craftspeople whose remarkable products stored and transported wine across a Mediterranean-wide empire. But this is not just a book on pottery, and not a traditional ceramics study. This book does not present a comprehensive typology of dolia (and it would probably be impossible to

75. Rickman 1971, 1980, 2002; Virlouvet 1995, 2011; Van Oyen 2020b.

76. Ostia was not continuously inhabited and built over and helps understand Rome's urban layout.

77. Livy 1.33.9; Enn. *Ann.* 2, fr. 22; Cic. *Rep.* 2.5, 2.33; Dion. Hal. *Ant. Rom.* 3.44.4; Isid. *Orig.* 15.1.56; Plin. *NH* 3.56, 31.89; Meiggs 1973, 16–17.

78. Meiggs 1973, 1–10, 16–50, 479–482.

79. Peña 2007b is a notable exception.

produce one given how much these hand-built vessels varied from one another), nor does it catalog or delve into thick descriptions of all individual unpublished dolia (though there is more information on select examples in Appendix 2).[80] Almost none of the dolia themselves are newly or recently excavated specimens. With the exception of a dozen or so fragments at Cosa, the other dolia and dolium fragments have long been excavated. New discoveries, and more scientific analyses, will augment our understanding and interpretations of the dolia, but small sets of data, no matter how high resolution, limit comparisons across sites and time. Instead, this study interrogates large sets of previously excavated material by asking new questions from old artifacts. By looking at seemingly mundane materials closely, we can tease out how a new food storage technology promoted investments, labor, trade, and even urban infrastructure.

Although wine was only part of the ancient economy, the wine trade was a characteristic feature of ancient Rome; with its colossal scale, immense variety, and extensive reach, the Roman wine trade was unique in the history of the Mediterranean. This study builds on previous work on the Roman food supply but provides a fresh perspective from the bottom up, one that focuses on what I argue to be the "keystone" container of the Roman wine trade: the dolium.[81] Various containers made possible the range of economic activities and storage and movement of goods, at the center of which was the dolium. The following chapters highlight the role that the dolium storage technology played in the wine trade, especially in supplying the capital. But this was not a direct route. Dolia shuffled resources, and hence possibilities for profit and wealth. As the common thread running through wine production, storage, and distribution (both wholesale and retail), dolia could expand a farm's ability to produce wine, reduce bottlenecks in the shipping process, and hold wine until merchants could sell it for higher prices. In becoming the backbone of the Roman Empire's complex, large-scale, long-distance, and highly profitable wine trade, dolia also opened new channels for economic opportunities and strategies. Some people became rich(er) investing in dolium production, producing more wines, operating shops or restaurants, shipping wines, or all the above. There were those who also found new, ancillary work that supported the dolium-based infrastructure. This was not just about money, though. Financial gain was certainly an incentive that drove people to have a hand in the food supply. Stories of Romans making a fortune on hoarding and price gouging wheat (and surely other foods) run rampant in ancient sources. But having a direct hand in the food supply—whether through agricultural production or distribution—also guaranteed access to valuable, and often coveted, resources, clinching one's power and influence over supplies, and even other people.

80. Carrato (2017, 2020) produces a typology based on dolia in Gaul, Iberia, and Italy.

81. On the "material turn" and need to incorporate material culture into (economic) history, see Bowes 2021c; Van Oyen and Pitts 2017a.

2

Building Big

A NEW CRAFT INDUSTRY

SOMETIME IN THE MID-FIRST to late second century CE, a Lucius Aurelius Sabinus commissioned a funerary altar along the Via Appia near the ancient town of Calatia in Campania (Figure 2.1).[1] In the Latin portion of the bilingual Greek and Latin text, L. Aurelius Sabinus called himself a *doliarius*, an unusual and rare Latin word meaning "dolium maker."[2] Also engraved into the stone funerary altar, below the bilingual inscription, was a depiction of three jars: a large, double-handled jar on its side, likely an amphora, and, in the background, two handleless pots standing upright, perhaps the dolia that the *doliarius* fabricated. Although we do not know much about L. Aurelius Sabinus from this short inscription—how old was he? where and when did he live? was he married?—his commemoration was a testament that he was at least wealthy enough to set up a stone altar on the Via Appia for himself and his family. L. Aurelius Sabinus' self-identification as a *doliarius* also suggests that his source of wealth sprang from his occupation. Romans regularly indicated their occupations on their funerary epitaphs, and L. Aurelius Sabinus likely included his as the only detail of himself because he was proud of his abilities as an expert dolium maker.[3]

The prestige and pride associated with an occupation, and its requisite skills, were generally rooted in the high value of the artisan's craft product. For L. Aurelius Sabinus, pride in his profession was tied to the value placed on dolia. But a dolium was also considered pottery, which was generally cheap and widely available.[4] How was a dolium considered a high-value object? To understand why L. Aurelius Sabinus chose to call himself a *doliarius* (rather than

1. *CIL* 10.403 = 483; *IG* 14 71; Mandowsky and Mitchell 1963, 118 no. 127, plate 71a; Zimmer 1982, 201n146. I thank Michael Zellmann-Rohrer for discussing the inscription and names with me.

Λ(ΟΥΚΙΩ) ΑΥΡΕΛΙΩ ΛΑ	For Lucius Aurelius
ΠΥΝΩ ΟΝΑΓΡΩ	Lapynos Onagros
ΚΑΙ ΑΥΡΕΛΙΩ̣	and Aurelius (or Aurelia?)
L(UCIUS) AVRELIVS SABI	L(ucius) Aurelius
NVS DOLIARI	Sabinus, dolium-maker,
VS FECIT SIBI	made (this) for himself
ET SVIS	and for his own.

2. *Gloss.* 3.309, 13.

3. Joshel 1992, esp. 62–91. See also Flohr 2016; Tran 2016; Treggiari 1975, 1976, 1980.

4. Plin. *NH* 35.52.

FIGURE 2.1. Drawing of a funerary altar for a *doliarius*, by Gina Tibbott after Zimmer (1982).

a *figlinus*, "potter"), then, we need to investigate a series of questions: How was dolium production different from making other types of pottery? What was the role of the *doliarius*, and how did it fit into the larger ceramic workshop? As we review the dolium craft, we will begin to understand not only what a challenging and risky venture it was but also how lucrative and rewarding it could be for those who could afford to invest in it. Dolium production was a specialized and time-consuming craft, requiring immense upfront investments. Those who became successful not only offset the production risks; they also cut costs by merging dolium production with the architectural industry.

Dolium Production: A Specialist Craft

Dolium production was the most challenging, time-consuming, and specialized type of ceramic production. Although dolia were considered the largest type of pottery, making dolia was not a process that simply scaled up ordinary pottery production. In order to achieve the dolia's massive size and distinctive shape, dolium production required mastery over a different set of materials and skills. This section reviews a range of evidence for dolium production to consider not only how they were manufactured but also the requisite skills and challenges in the process. Some ancient texts mention the status of the dolium craft industry, how expensive a vessel might be, and what a potential buyer should do to select a storage jar, but there is no account from the ancient world that can tell us exactly how risky or demanding this process was from the perspective of the dolium maker. To reverse engineer the steps, we turn to archaeological evidence as well as ethnographic studies on dolium-like vessels still made and used today for fermentation, such as Greek pithoi, Georgian *qvevri*, Spanish and Portuguese *tinajas* and *talhas*, and Korean *onggi*. The few places that still practice this scale of traditional pottery production shed light on how people in antiquity constructed these vessels, as well as the logistical strategies they employed.[5] Dolium production required particular clay and clay mixtures, masterful forming, and proper firing to make the vessel a suitable wine container, concerns and material constraints that spanned across time.[6]

Dolia were by far the most expensive pottery one could buy in antiquity. According to Diocletian's *Price Edict*, issued in 301 CE, these vessels were much pricier than their smaller ceramic cousins, costing 1,000 denarii, about 2,500 times more than a ceramic lamp (4 denarii for ten lamps).[7] A new dolium was probably out of reach for most of the population, because it would cost at least forty days' worth of wages for farmhands (25 denarii per day) and over eighty days' worth for weavers (12–16 denarii per day).[8] Dolia fetched such high prices not only because of the great quantity of material needed to build these massive vessels (hundreds of kilograms of clay) but also because of the high levels of skill, the amount of time required for this enterprise, and the risk of production flaws or even failure.[9] Unlike other types of ceramic containers, dolia were only produced by experienced specialists. Even as early as circa 380 BCE, Plato's Socrates referenced a saying:[10]

> In truth would it not be ridiculous that men should come to such folly that, without first practicing privately, sometimes with indifferent results, sometimes with success, and so getting adequate training in the art, they should, as the saying is, attempt to learn pottery by starting on a pithos, and attempt public service themselves and summon others of their like to do so?

> οὐ καταγέλαστον ἂν ἦν τῇ ἀληθείᾳ, εἰς τοσοῦτον ἀνοίας ἐλθεῖν ἀνθρώπους, ὥστε, πρὶν ἰδιωτεύοντας πολλὰ μὲν ὅπως ἐτύχομεν ποιῆσαι, πολλὰ δὲ κατορθῶσαι καὶ γυμνάσασθαι ἱκανῶς τὴν τέχνην, τὸ λεγόμενον δὴ τοῦτο ἐν τῷ πίθῳ τὴν κεραμείαν ἐπιχειρεῖν μανθάνειν, καὶ αὐτούς τε δημοσιεύειν ἐπιχειρεῖν καὶ ἄλλους τοιούτους παρακαλεῖν; (*Grg.* 514e)

According to Plato's Socrates, mastering a craft took practice, and pithos production was the most difficult kind of pottery production, the type only the most experienced potters should

5. E.g., pithoi production in Greece (Blitzer 1990), *tinajas* in Spain and Portugal (Romero and Cabasa 1999), and *qvevri* in the Republic of Georgia (Barisashvili 2011).

6. Workshops are near clay sources that have ideal properties for the type of pottery production the workshops specialize in. For dolia clay mixtures in southern Gaul, see Carrato et al. 2019.

7. Diocletian's *Price Edict* 15.97–101.

8. Diocletian's *Price Edict* 7.1a, 20.12–13.

9. Modern-day potter Andrew Beckham uses nine hundred pounds of clay to build a dolium with a capacity of ca. 750 liters, comparable to the largest dolia at Pompeii and smaller than the average dolium at Ostia. Cheung et al. 2022.

10. See also Curtis 2016, 589.

attempt. His reference insinuates that a potter learned the craft by ascending levels of difficulty, reaching pithoi at the height of one's training. Dolia were larger than pithoi, and the dolium's strawberry shape and top-heaviness posed even greater challenges in manufacture, with more risks of production-based defects such as a collapse of the vessel wall, cracks, or other problems.[11] Experiments have confirmed that a subpar vessel could be too porous and result in aeration and too much oxygen contact, unhygienic conditions, and even losses of wine.[12]

Dolium production required knowledge and skills, even from the initial steps of clay collection. Getting and preparing the raw materials drew on some practical knowledge and differed little from clay preparation for brick and tile production, which was known as a seasonal task:[13] clay digging generally took place between the late summer and early autumn, and workers left clay to weather until the spring to become more workable and less prone to flaws and defects down the line.[14] The clay minerals had to break down to allow the material to be further tempered and processed more easily, with both the removal of large impurities and addition of stable ballast. Although the groggy clay body (coarse clays high in alumina and silica, such as clays consisting of ground-up fired clay particles—e.g., discarded pottery, bricks, and tiles) used for the building of dolia added strength and warp resistance and was a stable material, an impurity could lead to radiating cracks either during the air-drying process or, if it was small enough to make it through drying without affecting the pot, during the firing process. Workshops developed their own clay mixes with different clay additives, such as grog, sand, or plant fibers, to make the clay more workable and structurally strong.[15]

Forming dolia was also time-consuming and onerous skilled work. Ancient evidence and contemporary pottery production indicate that these vessels were generally too large to throw on a standard potter's wheel and were coil-built over the course of at least several days or even weeks.[16] The fourth-century CE author Anatolios, whose agricultural advice is among those compiled in the tenth century *Geoponika*, noted that "potters do not raise all pithoi on the wheel, only the small ones; they build up the bigger ones placed on the ground in a warm room day by day, and make them large" (*Geoponika* 6.3.4: οὐ πάντας δὲ τοὺς πίθους ἐπὶ τὸν τροχὸν ἀναβιβάζουσιν οἱ κεραμεῖς, ἀλλὰ τοὺς μικρούς· τοὺς μέντοι μείζους χαμαὶ καιμένους ὁσημέραι ἐν θερμῷ οἰκήματι ἐποικοδομοῦσι, καὶ μεγάλους ποιοῦσιν). Several dolia at the House of Stabianus (I.22) in Pompeii that were serendipitously freestanding at the time of the eruption offer the most revealing evidence for dolium production in central Italy and beyond.[17] Cracks and seams indicate that dolium makers built the dolia gradually with coils on a slow-turning wheel or turntable. The potter started with a disc of clay to form a small base, usually between six and ten centimeters thick, and progressively added

11. Kang (2015) describes the trials and failures in making an *onggi* with a belly.

12. Caillaud (2020) notes difficulty in using ceramic jars for wine fermentation and storage.

13. DeLaine 1997. This was the same season as shipping and agricultural activities, so labor shortages were possible. Erdkamp 1999, 2015; Horden and Purcell 2000; Shaw 2013; Hawkins 2017.

14. Clay for *onggi* is gathered and left to weather for one year before use.

15. For one workshop's superior clay mixture for dolia in southern Gaul, see Carrato et al. 2019.

16. The rates of shrinkage in slab building are problematic for very large vessels; slab building is also more time consuming. For the possibility of slab building, see Rando 1996; Peña 2007b, 35, 218. For explanations of pottery terminology and techniques, see Hamer and Hamer 2004.

17. See Cheung and Tibbott 2020 for discussion of dolia from the House of Stabianus (I.22). See A2: P I.22 nos. 5, 7, 9; VII.6.15 no. 1.

FIGURE 2.2. (*Top*) Crack between dolium base and first body coil (I.22 no. 7); and (*bottom*) horizontal cracks between coils (I. 22 no. 5), Pompeii. Courtesy of the Ministry of Culture—Archaeological Park of Pompeii. Reproduction or duplication by any means is forbidden.

hand-squeezed or rolled coils, likely just a single coil or two per day to allow it to dry sufficiently to support the weight of the next one (Figures 1.5 top, 2.2; Plate 3; A2 P I.22 nos. 7, 5); a good clay body enabled faster coil building and lessened the likelihood of production flaws. A few areas difficult for the potter to reach show that the potter generally smoothed the seams between the coils as the vessel was built up and sometimes paddled or scored surfaces of the coils that would come in contact with one another to encourage joining (Figure 2.3;

FIGURE 2.3. (*Top*) Seam between coils on dolium interior wall (I.22 no. 5); and (*bottom*) scored or paddled coil edge, Pompeii. Courtesy of the Ministry of Culture—Archaeological Park of Pompeii. Reproduction or duplication by any means is forbidden.

Plate 4).[18] If the previous coil did not bond properly with the next, horizontal cracks could form between the seams or even lead to the vessel breaking. After forming the vessel body, the potter would shape the rim, which was usually thick and sturdy to support large dolium lids. The rim offered the greatest opportunity for variation: the potter could build a smaller coil, onto which he or she molded the lip of the rim after scoring the two pieces, or make a coil for the rim and then pull and fold part of it onto itself to shape the lip (Figure 2.4; Plate 5; A2 P I.22 no. 9, VII.6.15 no. 1).

Dolium production required long-term and advance investments for the materials, equipment, and labor. Traditional pottery production of vessels of similar size and shape, such as pithoi in Greece and *qvevri* in the Republic of Georgia, indicate that the production of dolia likely required several days to even weeks for the forming process alone (Table 2.1). The vessel's massive size, and the large amount of ceramic material used to construct it, necessitated long drying times and controlled firing. Air-drying dolia (and other pottery) is a crucial step to minimize cracking or breaking during the kiln-firing process, since a wet or damp piece would explode. Given their large size, dolia probably required one to three months to air-dry before they reached a bone-dry state and could be gradually fired in the kiln over the course of several days. Anatolios also had some insight on the firing process for pithoi: "Firing is also an important part of the manufacture: the fire must be neither too low nor too high, but at the proper heat" (*Geoponika* 6.3.5: οὐ μικρὸν δὲ τῆς κεραμείας ἐστὶ μέρος ἡ ὄπτησις. δεῖ δὲ μήτε ἔλαττον, μήτε πλέον, ἀλλὰ μεμετρημένως τὸ πῦρ ὑποβάλλειν); some estimates identify a firing temperature around 850°C.[19] Indeed, Augustine's reference to the biblical saying "The kiln tests the vessels of the potters, and the trial of tribulation the just men" (*De civ. D.* 21.26: *vasa figuli probat fornax et homines iustos temptatio tribulationis*) builds on an understanding that the firing process often revealed the limitations of not only the kiln operators' skills but also the potters' when fired vessels exhibited production defects.[20] Although potters attempted to minimize flaws and defects, plenty could go wrong. Uneven drying could lead to cracks, especially between coils. A dolium could distort or suffer dunting (vertical cracking from the firing process). A dolium could even break during drying or firing if temperatures fluctuated too quickly, and wasters (pieces damaged or deformed during firing) were common by-products of dolium production.

Most, if not all, dolia were coil-built, but the investment, scale, and organization of labor for their manufacture could vary widely among workshops. Workshops, in general, afforded some control over the conditions of the production, because the fixed space and facility enabled the installation of permanent structures and equipment, such as vats and kilns, and the accumulation and transmission of knowledge. Workshops, especially larger ones, were often in areas with ample space and access to materials such as high-quality clay, water, and wood. They could be remote, but their success hinged on relationships, networks, and access to stable markets, consumer demand, and reliable transportation. Dolium production centers could serve a minor area, manufacturing dolia for local destinations, or entire regions and perhaps even export vessels to more far-flung destinations. Reputation mattered. Already in the second century BCE, when early signs of dolium use surface in central Italy, Cato

18. See Rando 1996; Peña 2007b. Carrato (2017, 135ff.) notes scoring/paddling marks on dolia in Gaul. Peña 2007b, 218: one dolium at Ostia was perhaps slab-built.

19. Carrato et al. 2019, 75. *Qvevri* potters fire at 900–1,300°C. Temperatures depend on the clay body.

20. *Ecclesiasticus* 27.5.

FIGURE 2.4. (*Top*) Seam between rim coil and lip (I.22 no. 9); and (*bottom*) paddling marks on rim coil (VII.6.15 no. 1), Pompeii. Courtesy of the Ministry of Culture—Archaeological Park of Pompeii. Reproduction or duplication by any means is forbidden.

recommends Trebla Alba and Rome as major production centers for dolia.[21] The concern for a quality dolium endured as long as the craft. Half a millennium later, Anatolios advised a pithos customer to visit the workshops, assess their clay, and even perform a test on the

21. Cato *Agr.* 135. Cato's readers were probably mostly elite landowners; small, rural sites removed from these commercial networks were likely served by small, local workshops or itinerant potters. Taglietti and Zaccaria 1994; Uboldi 2005; Lazzeretti and Pallecchi 2005.

TABLE 2.1. Logistics of building large fermentation jars.

Production	Logistics
Pithoi, Greece (Blitzer 1990)	One large pithos required twenty days to form, ten days to dry indoors, and another ten days to dry outdoors before it could be fired in the kiln. With nineteenth-century kiln technology, a load of six pithoi could be fired, requiring ca. 12–14 hours of firing and several days to cool.
Dolia, Italy (Artenova n.d.)	A five-hundred-liter dolium required fifteen to twenty days to form, one month to air-dry, and three days of firing in a kiln at 1,000°C.
Tinajas (Romero and Cabasa 1999)	*Tinajas* require two months to dry.
Qvevri, Republic of Georgia (Barisashvili 2011; Caillaud 2020)	To coil-build one *qvevri* takes one month; a large batch of *qvevri* takes three months. A *qvevri* takes one month to dry, seventy-two hours or more to fire, and seventy-two hours or more to cool before removal from the kiln.
Onggi, Korea	To fire *onggi* requires three to five days to preheat the kiln and gradual cooling over at least two days.

vessel before purchase: "There are those who are content with the examination of whether a pithos is well made, if it gives out a sharp and piercing note when struck. But this is not sufficient. The client must be there at the pottery workshop, must see to it that the clay has been well worked" (*Geoponika* 6.3.2–3: τινὲς μὲν οὖν ἀρκοῦνται ⟨ἐν⟩ τῇ δοκιμασίᾳ τοῦ καλῶς κεκεραμευμένου πίθου, τῷ κρουσθέντα αὐτὸν ἀποδοῦναι ἦχόν τινα ὀξὺν καὶ τορόν. οὐκ ἔστι δὲ τοῦτο αὔταρκες, ἀλλὰ χρὴ τὸν κατασκευάζοντα παρεῖναι τῇ κεραμείᾳ, καὶ ὅπως ὁ πηλὸς καλῶς εἰργασμένος εἴη προνοῆσαι.). Anatolios' test was one way a layperson could tell if a dolium was made well and worth the investment, but he urged potential customers to make the trip to the workshop and ensure the clay was good and potters' skills sound. For the price of the dolium, a customer would want to make sure the dolium would last.

The following sections survey the material evidence for dolium production at the case study sites to track developments in the craft in west-central Italy—namely, by observing the dolia's size, shape, and quality, as well as whether there were attempts to produce standardized products (Tables A1.1, A1.3, and A1.5–6 include the recorded dimensions). Standardization testified to not only the quality and functionality of an object but also the industry standards that workshops, craftspeople, and consumers widely accepted.[22] Dolium stamps from the region are also revealing, showing how and why dolium production became intertwined with the manufacture of other terracotta objects (Tables A1.2, A1.4, and A1.7 compile dolium stamp information, noting provenance and, when available, names and workshops that appear across different sites). Many workshops stamped their ceramic and terracotta products, including fineware pottery, amphorae, bricks, tiles, *mortaria* (a type of open bowl with thick, heavy flanges and a gritty surface used for pounding

22. Osborne 2017, 123.

or grinding food), and even dolia. Dolium stamps, found on the rim of 5–20 percent of dolia across the different sites, were generally short, with the name of one person, usually in the genitive, who was most likely the owner of the workshop; some of the dolia featured a second stamp with another name or pictorial symbols. They were often stamped as part of the production phase, as a form of internal control over a workshop's production, which was especially important for large-scale workshops run by employees responsible for various tasks and operating on different hierarchies. Stamps often also affirmed and guaranteed the quality of a product, and included information for the customer, tracing the dolium back to the workshop and giving a sense of the workshop's organization.[23]

Cosa: Early Developments

The evidence at Cosa suggests that local workshops supplied dolia to the town, with dolium makers standardizing certain parts of their products. All fifty Cosan dolia broke in antiquity, and many featured fractures on the rim or shoulder, suggesting the Cosan dolia had production weaknesses. The dolia were generally heavy vessels with thick walls and bulky rims (Figure 2.5A). Some rim fragments broke off from large dolia that are typically found in wine production or storage facilities, while houses kept smaller storage jars for household provisions (Table A1.1). Several diagnostic dolium fragments are similar in size, suggesting that potters generally made dolia of such a scale with a certain rim size in mind. Potters formed several massive dolia (inner rim diameter ca. 55–65 cm) and some that were slightly smaller (inner rim diameter ca. 40–50 cm), which would have fit well with the large dolium lids that were found on-site (Figure 2.6), as well as a few vessels that were much smaller (inner rim diameter 15–25 cm). Although it is impossible to standardize fully the production and dimensions of such massive hand-built vessels, parts of the vessel, such as the base and rim, were easier to control.

The variation between the vessels and their general (low) quality point to several workshops with different techniques and clay recipes supplying Cosa from the second through the first centuries BCE.[24] These large dolia were mostly similar in scale but exhibited a wide range in the types of ceramic fabric as well as the form of the rims (Figure 2.7A; Plate 6A).[25] Two fabric types, reddish yellow in color, contain grog with large and medium inclusions, the first with inclusions that appear to be feldspar, calcite, and quartzite, and the second of sandstone, quartzite, and feldspar (without petrographic analysis, which requires removing a sample from the object, we cannot be certain of their identification; here we will focus on general traits of the fabric and how they contribute to the production of the dolia). These two common fabric types at Cosa feature a noticeably greater percentage of grog overall, especially of larger grain size, a characteristic of dolium clay bodies. Adding more grog built substantial strength into the clay body and reduced the risk of

23. Manacorda 1993, 44–45.

24. Only one dolium securely identified to the first century CE is preserved and was higher in quality than earlier dolia (Cosa n. 19); the sample size (ca. fifty) is too small to form any conclusions.

25. If dolium production of the region was part of brick and tile production, it is possible that dolia were produced in the same workshops; Gliozzo (2013) found that bricks and tiles of Cosa were made with clay from three sources.

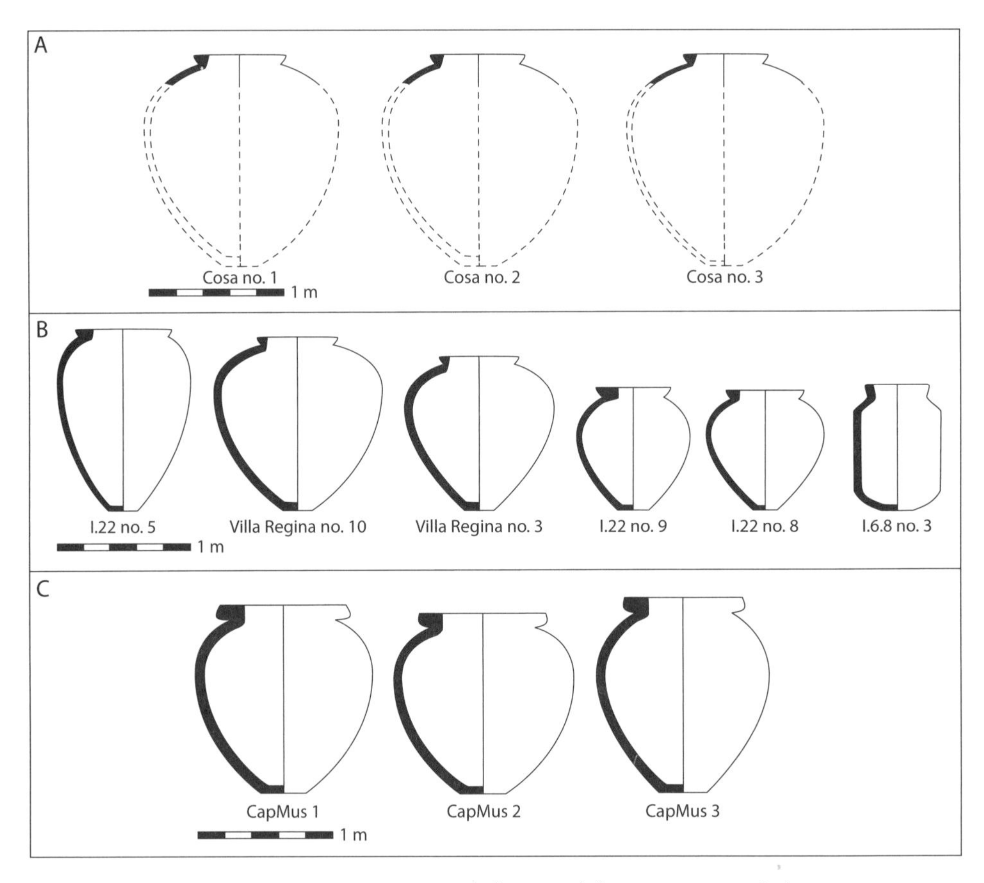

FIGURE 2.5. Profile drawings of dolia from (A) Cosa, (B) Pompeii, and (C) Ostia and Rome, by Gina Tibbott.

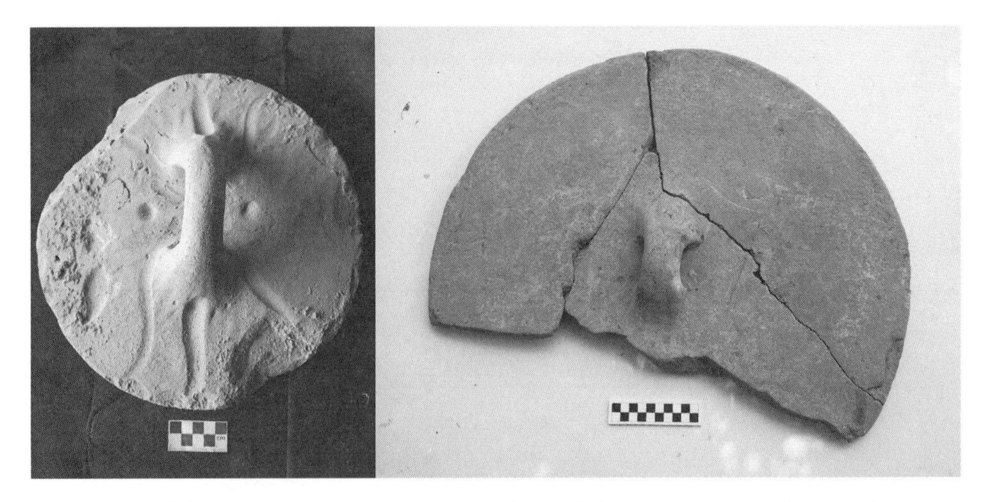

FIGURE 2.6. (*L*) A small dolium lid (CE1633/23) and (*R*) large dolium lid (C70.485), Cosa. Courtesy of the National Archaeological Museum and Ancient City of Cosa (Regional Directorate of Museums of Tuscany). Reproduction or duplication by any means is forbidden.

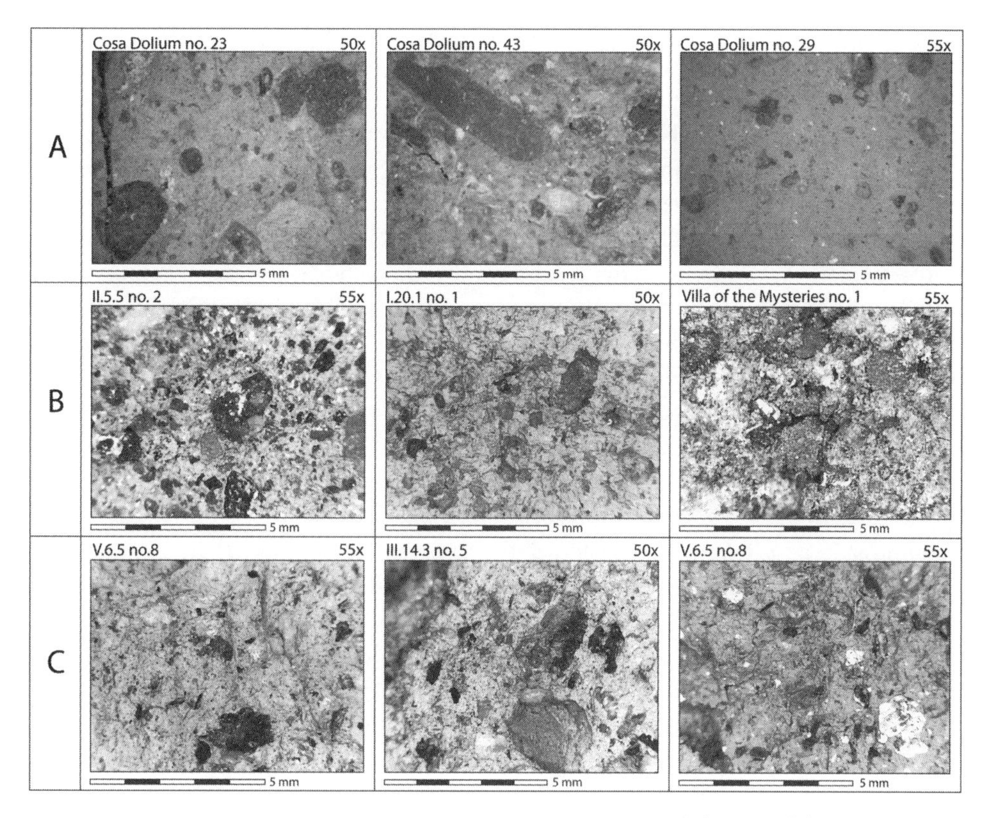

FIGURE 2.7. Microphotographs of dolium ceramic fabrics from (A) Cosa, (B) Pompeii, and (C) Ostia.

collapse or cracking during firing, on the one hand, and decreased the amount of clay needed for each piece to help conserve valuable clay, on the other. The third most common fabric, a light red ceramic makeup, featured finer inclusions such as mica, which would have increased the plasticity of the clay. Size and shape among the smaller vessels varied even more, suggesting less standardization with smaller storage jars; the smaller version was likely a side product that dolium producers could quickly manufacture and that could be used by dolium potters in training to practice and develop their skills. It is possible that at least some dolia were made to order and perhaps customized (rather than standardized and stockpiled) at this early stage in the dolium craft. A workshop would not want to spend time and resources making jars for no customers to buy them. There was even variation among the dolium lids: although they featured loop handles, some were hand-built (made from a slab or coil) and others were pulled handles. Moreover, most of the dolia were not fired well, resulting in either friable material or surfaces that flaked or rubbed off easily—not the type of vessel that would have given a sharp, piercing ring when struck as Anatolios advised; evidence from other contemporary dolium production sites in central Italy south of Rome, such as a couple of remote dolium workshops in the Pontine region, also reveal failures in dolium production.[26]

26. Tol and Borgers 2016: some rural workshops in the Pontine plain marketed second-rate products.

Three of the largest dolia preserve stamps, but they are abbreviated and how to interpret them is unclear, pointing to a lack of convention in stamping among workshops in the area (Table A1.2). A rim fragment, found in the horreum and tentatively dated to the second century BCE, bears a small letter *H* surrounded by a triangular border (Figure 2.8 top; Plate 7 top).[27] No other example with this type of stamp is known, but studies of cursory brick stamps in Hispania have suggested that shorthand stamps were probably used as a rapid way to document batches of products in a workshop.[28] The two other dolium stamps both feature names. One dolium rim fragment from the late second or early first century BCE, found at the Temple of Jupiter, bears a stamp with the text "L·REMIO·C·F" (Figure 2.8 bottom; Plate 7 bottom).[29] The name could be related to a Remmius of the Republican period who has been documented in Etruria.[30] The name of this stamp is in the dative or ablative, whereas most dolium stamps from later periods feature a name in the genitive or nominative. It might have been in the dative case to indicate possession, "for Lucius Remius, son of Gaius," perhaps because the customer ordered the vessel in advance and had it marked, or the name was in the ablative to indicate origin; brick stamps from later periods featured the preposition *ex* or *de* followed by a name in the ablative to signify "from [the estates or clay beds of] so-and-so." Indicating the workshop seems to have been a common concern, as confirmed by the third stamped dolium, which was the town's largest, highest-quality, and latest datable dolium (first century CE), with wide shoulders, a less bulky rim, and a well-fired ceramic body: on the rim was a small rectangular stamp with "C·TVRI," which most likely referred to the estate or workshop owner, Gaius Turus or Turius (Figure 2.9A-B; Plate 8A-B). The stamp might have attested to the quality of the vessel by identifying its workshop, been used in internal bookkeeping, or marked ownership. The name is otherwise unattested, as many individuals on ceramic stamps are. Just below the stamp was an incision of a pictorial symbol, depicting an anchor or stylized phallus (Figure 2.9C; Plate 8C). Overall, the stamps on the Cosan dolia lacked consistency and differed from dolium stamps of the Vesuvian region and the urban area of Rome and Ostia, suggesting chronological or regional differences in the scale and organization of workshop operations.

The fragmentary dolia of Cosa, plus those of contemporary workshops and other sites, shed light on local workshops' early and varied attempts to standardize vessels. The dolia and other various storage vessels of Cosa were not robust or particularly well made, especially compared with dolia from later periods found in Pompeii, Ostia, and Rome; workshops supplying Cosa shared some common product designs, but the craft had not developed highly enough to guarantee a standard product or quality that would ensure long-term use. A significant number of workshops from this period were based at rural sites and villa estates, suggesting that many early dolium workshops directly supplied estates practicing viticulture. With the overall paucity of dolium fragments throughout the history of the town, the urban settlement of Cosa was probably only a minor destination and market for the dolium industries, or received leftover, second-rate products from local workshops.

27. F.E. Brown 1984.

28. Roldán Gómez and Bustamante Álvarez 2015, 2017.

29. Bace 1984, 172.

30. *CIL* 1^2 2063; Bace 1984, 172.

FIGURE 2.8. (*Top*) Dolium rim, stamped with "H" in triangular border; and (*bottom*) dolium rim, stamped with L·REMIO·C·F, Cosa. Courtesy of the National Archaeological Museum and Ancient City of Cosa (Regional Directorate of Museums of Tuscany). Reproduction or duplication by any means is forbidden.

FIGURE 2.9. (A) Large dolium fragment; (B) dolium stamp on rim: "C·TVRI"; and (C) incision of an anchor or stylized phallus on shoulder, Cosa no. 19. Courtesy of the National Archaeological Museum and Ancient City of Cosa (Regional Directorate of Museums of Tuscany). Reproduction or duplication by any means is forbidden.

Pompeii: Improvements in the Craft

The regularity and common features of the Pompeian dolia suggest that local workshops specializing in the manufacture of large ceramic and terracotta objects supplied standardized dolia to the town in the first century CE. Pompeian dolia had small bases and wide shoulders. The larger dolia had an especially pronounced strawberry shape, demonstrating the incredible skill of the dolium potters (Table A1.3; Figure 2.5B); the rims were also not as thick as the Cosan dolia, requiring less clay for a more streamlined shape. Overall, the average volume of smaller vessels clocked in at approximately 175 liters and large wine fermentation dolia at roughly 500–700 liters.[31] Although it was impossible to build such large vessels with identical dimensions, dolium makers were able to control several aspects of the vessels. Dolium bases were always small (diameter ca. 20–25 cm), while the thickness of the dolium's walls (ca. 3–5 cm) provided the structural support that allowed the vessel to reach its large size.[32] Controlling a dolium's size and volume was challenging. Dolium makers formed rims in three distinct sizes to fit with standardized lids known as *opercula* (Figure 2.10): (i) small dolia, probably better identified as *doliola*, that could be covered with smaller lids (diameter ca. 30 cm); (ii) dolia that could be covered with larger lids (diameter ca. 50–52 cm); and (iii) only a few examples of extremely massive dolia that were closed with very large lids

31. Cheung et al. 2022.

32. A sampling of several dolia indicates that the widest part of the vessel was usually twenty-five to thirty times the vessel's wall thickness (I.8.15 no. 1; I.20.5 no. 1; I.21.2 no. 1; I.22 no. 8; IX.9.10 no. 1; Villa of the Mysteries no. 2).

FIGURE 2.10. (*L*) Interior dolium lid (*operculum*), Villa Regina, Boscoreale. Courtesy of the Ministry of Culture—Archaeological Park of Pompeii. Reproduction or duplication by any means is forbidden. (*R*) System of closing dolia with two types of lids, illustration by Jared T. Benton, after Pasqui (1897).

(diameter ca. 85 cm). The owner (or owners) of the Villa Regina of Boscoreale, about 2.5 km from Pompeii, was able to acquire eighteen standardized and well-made dolia for the wine cellar.[33] Although the dolia varied in how much they could hold—a difference in volume of almost 40 percent between the largest and smallest dolia—they all had the same size rims and their inner and outer lids, the *operculum* and *tectorium*, respectively, were also standardized, with a difference of at most just 10 percent (Figure 2.11).[34] Keeping the rims and lids standardized not only provided buyers set options and sizes but also allowed them to replace dolium lids easily. In fact, dolium lid production often operated as a separate industry from dolium production, because dolia and dolium lids broke often and easily during the fermentation process (see Chapters 3 and 7).[35]

The similar sizes and shapes yet different ceramic fabrics and workshop stamps of the Pompeian dolia suggest that dolium potters were standardizing dolia. The large dolia were mostly well fired with a red or light red ceramic fabric; some of the very massive dolia, such as one in the Villa of the Mysteries just outside the town, were not as well fired and had a crumbly surface. On the other hand, the small and medium-size spherical dolia all exemplified well-fired dolia and had a light red ceramic fabric or a red or yellowish-red ceramic

33. De Caro 1994, 66–69.

34. On sealing dolia, see Pasqui 1897; Annecchino 1982; Cheung 2020. De Caro 1994, 68: the largest dolium had a capacity of ca. 712.72 liters and smallest dolium 480.82 liters. All dolia (except the two *doliola*) had an interior rim diameter 41.5–42.5 cm and exterior rim diameter 60–61 cm. The *opercula* diameters varied 46–51 cm and *tectoria* 79–87 cm. Wicker or basketry lids were probably popular in southern Gaul, Hispania, and other parts of the Mediterranean; Brun 2003, 79; Carrato 2017, 205–206. On *tectoria,* see Cato *Agr.* 11.2; Taglietti 2015, 276ff.; Annecchino 1982. Lids have been found at Villa Regina at Boscoreale (De Caro 1994, 63–69) and Pompeii.

35. Taglietti 2015.

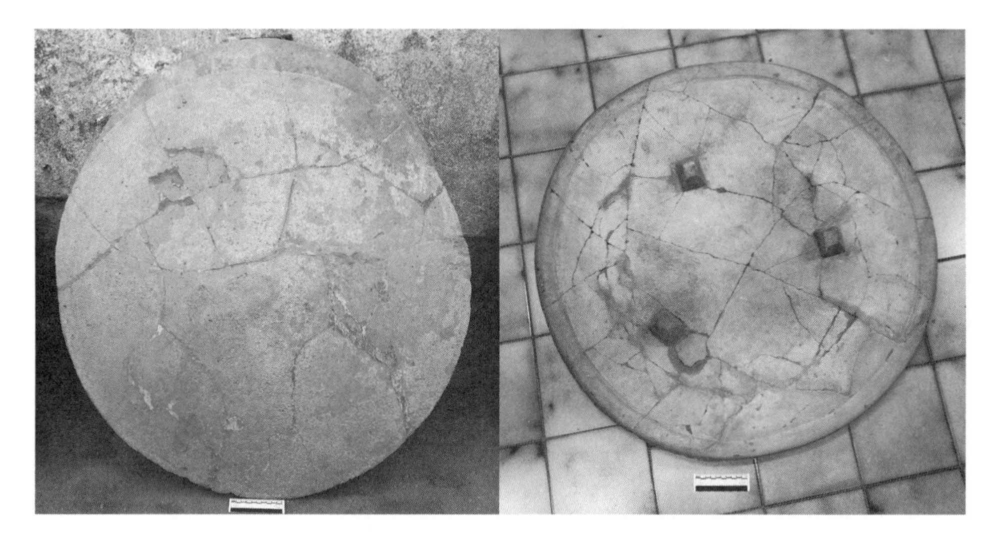

FIGURE 2.11. (*L*) *Tectorium*, Villa Regina, Boscoreale; and (*R*) underside of *tectorium*, Pompeii. Courtesy of the Ministry of Culture—Archaeological Park of Pompeii. Reproduction or duplication by any means is forbidden.

fabric with grog consisting of a dense concentration of small and medium-size inclusions of mica and feldspar to endow the clay body with both strength and plasticity, suggesting that local dolium makers exercised control and expertise when manufacturing and firing dolia (Figure 2.7B; Plate 6B). The vessels generally had the black sand characteristic of local clay sources, though some can be traced to workshops farther afield in the Bay of Naples and beyond by their stamps.

Fewer than 10 percent of the Pompeian vessels were stamped, but the stamps were uniform in their information and appeared on the large, well-made dolia (Table A1.4).[36] The stamps found on the Pompeian dolia consistently provided several pieces of information in a formulaic way regarding their manufacture and workshop. Most of the dolium stamps featured one name in the genitive case, which referred to the workshop owner (Figure 2.12; Plate 9). Several stamps included additional information, such as a name in the nominative of the person who "produced" (*fecit*) the vessel; this figure, known as an *officinator* based on later evidence, was sometimes an enslaved individual (indicated by *S* or *SER*) and could have been the person actually forming the dolium or, more likely, someone in a managerial role who was tasked with (and perhaps had enough experience for) overseeing batches of workshop products.[37] One of the larger dolia in town stamped with M LVCCEI QVARTIONIS, for example, also bore an incision with the Roman numeral XLIII to note that the vessel could hold forty-three *urnae* (563.3 liters). In fact, most

36. The table includes stamps found on cylindrical jars (see Chapter 6) as well, since *CIL* and Bloch (1948) do not distinguish between the two different vessels and their production was probably related. The figure given in the text (10 percent) for stamped dolia is based on the dolia I examined in person.

37. E.g., *CIL* 10.8047, 7: *D(e) F(iglinis) C(aius) CLVENTI/AMPLIATI/CORINTHVS S(ervus) F(ecit)*; *CIL* 10.8047, 15: *PHILEROS/M(arci) FULVI SER(vus)*.

FIGURE 2.12. Dolium rim with two stamps (VI.14.36 no. 4), C NAEVI/VITALIS, Pompeii. Courtesy of the Ministry of Culture—Archaeological Park of Pompeii. Reproduction or duplication by any means is forbidden.

of the dolia with stamps had capacities that ranged between 450 and 800 liters.[38] Some workshops were particularly successful, with products that reached various destinations in the Bay of Naples. The Appulei, for example, seem to have been prominent in the region. Several storage jars produced by at least two members of the family made their way to Pompeii, a villa in Gragnano, and Stabiae.

Numerous workshop owners named in dolium stamps have also been found on other heavy terracotta objects, testifying to the range of products their workshops supplied. The practice of stamping dolia fell in line with that of stamping bricks and tiles, as they were all produced in workshops known as opus doliare workshops. As Pliny the Elder noted, heavy terracotta objects such as dolia, bricks, tiles, and pipes were in high demand in the Roman world:

> Nor do pottery products satisfy with their constant supply, with dolia invented for wine, pipes for water, conduits for baths, tiles for roofs, fired brick for house-walls and foundations, or things made on a wheel, on account of which King Numa established a seventh Guild of the Potters.
>
> <n>eque adsiduitate satiant figlinarum opera, doliis ad vina excogitatis, ad aquas tubulis, ad balineas mammatis, ad tecta imbricibus, coctilibus laterculis fundamentisque aut quae rota fiunt, propter quae Numa rex septimum collegium figulorum instituit. (Plin. *NH* 35.56)

According to Pliny, a range of terracotta products served and fulfilled different needs, from viticulture to water management to construction. Although *figlinarum opera* constituted a range of material, Pliny's description specified terracotta objects—that is, utilitarian objects made of coarse clay that were undecorated and unglazed, and usually fired at lower temperatures than other types of ceramic pottery. And in fact, these objects were produced in opus doliare workshops, which specialized in heavy terracotta materials. Because these

38. An exception was IV.14.36: all four vessels were stamped, and from different workshops.

products shared the same groggy ceramic makeup and required the same equipment (kilns, tanks, et al.), some opus doliare workshops consolidated their production, perhaps to diversify their manufactory and cut costs. A larger workshop setup also opened the possibility of workers performing the different processes in manufacturing various opus doliare products through a division of labor or training. Because tiles, bricks, dolia, and *mortaria* were all made of the same ceramic body, but each product required varying degrees of specialized craft knowledge and skill, workshops with larger workforces could have designated workers responsible for certain tasks (collecting clay, kiln firing) or products (bricks, tiles, *mortaria*, and dolia), and perhaps even opportunities to train and advance.[39] For example, Corinthus, the enslaved potter working in the workshop of C. Cluentius Ampliatus, not only stamped two dolia found in Pompeii but had also signed a flanged tile.[40] Vittorio Spinazzola also identified him as the enslaved potter of Publius Cornelius Corinthus who made a large, decorated terracotta jug; if the last identification holds, the evidence suggests that Corinthus not only oversaw general opus doliare production but was also responsible for different ceramic goods—jugs, tiles, and dolia—and probably learned and ascended the ranks in the process.[41]

The dolium stamps at Pompeii show both the wide range and reach of opus doliare workshops. Workshops that provided dolia to Pompeii also produced dolium lids and tiles, and many of these workshops supplied farmhouses in the hinterland of Pompeii and other settlements in Campania. A few also furnished dolia to farms and warehouses as far afield as Neapolis, and even Rome:[42] for example, dolia of Phileros, slave of Marcus Fulvius, were found in Rome as well as in Pompeii.[43] A few workshops operated in or around Pompeii, such as a small opus doliare workshop in town during the third into the mid-second century BCE (around the time the town became more urbanized) and a suburban villa known as *in pompeiano in figlinis Arriani Poppeae Aug*, as well as workshops supplying other large agricultural equipment such as millstones.[44] Overall, the stamped dolia in Pompeii and environs were likely acquired from large-scale opus doliare workshops in northern Campania and southern Latium that sold their wares to a broader region of clients in both town and country.[45]

Unlike the dolia at Cosa, the dolia at Pompeii were mostly well made and maintained, and most were still in use at the time of the eruption. Although the dolia at Pompeii featured signs of a craft industry that had developed standards, some of the dolia still suffered from production weaknesses. Some of the large dolia exhibited vulnerabilities around the wide shoulder, where cracks often formed. Some also suffered from crumbly and friable surfaces, which stemmed from lime spalling (when lime absorbs moisture, it expands and

39. See Van Oyen 2020a for a recent study on innovation and investment in *terra sigillata* at Marzuolo.

40. For terracotta jug, see Maiuri 1927, 12; Spinazzola 1953, 687–689, 1011n453. For flanged tile, see Spinazzola 1953, 689. For summary of discussion, see Peña and McCallum 2009a, 62–63. On craft specialization, see Costin 1991.

41. On learning crafts and apprenticeships, see papers in Wendrich 2012a, especially Hasaki 2012; Wendrich 2012b; Miller 2012; Creese 2012.

42. L SAGINI appears on dolia in Pompeii and on tiles in Naples (*CIL* 10. 8047 17 and 8042 93).

43. *CIL* 15.2446: *Phileros // M.Fulvii [s(eruus) f(ecit)]*.

44. Braconi and Lanzi 2020; Arangio-Ruiz and Carratelli 1954, 56. Cf. Peacock 1980.

45. See Steinby 1993 for discussion of stamped opus doliare materials in Pompeii and their origins.

pushes off the ceramic material above it, the severity of which depends on the calcium carbonate content of the clay body and the firing temperature of the kiln). The production weaknesses suggest that the technology and perhaps also the resources for the manufacture of these vessels reached a certain limit, which local potters were not able to, or at least decided not to, exceed without great risk. Although dolium potters faced numerous challenges in producing dolia, the dolia in Ostia and Rome demonstrate that the challenges were not insurmountable.

Ostia and Rome: New Standards

The over two hundred dolia of Ostia and Rome reflect new heights in the craft. The dolia found in Ostia and the city of Rome are enormous, ranging from 1.25 to 1.75 meters tall *and* wide, representing the pinnacle of dolium production. Although the capacities of dolia varied to some degree, their shape, scale, rim sizes, wall thickness, and ceramic fabric exhibit remarkable standardization (Tables A1.5–6). Their wide rims (ca. 60–70 cm interior rim, ca. 85–95 cm exterior rim) were perched on top of incredibly pronounced shoulders, supported by thick walls (ca. 5 cm), that tapered down to tiny bases (ca. 25 cm), giving them an extreme strawberry shape (Figure 2.5C). Their rims, also closely standardized, fit specifically with common large lids, and opened so widely that a person could easily crawl inside the vessels to clean and maintain them.[46] Unlike dolium rims at Cosa and Pompeii, the rims of the Ostian and Roman dolia were more compact with a flat upper surface, the dolium makers' solution to reduce material use and to minimize the risk of dunting (Figure 2.13L; Plate 10L; A2 O V.11.5 no. 67, I.4.5 no. 16). Indeed, damage seems to stem mostly from the eventual abandonment of the dolia, not production defects, and their high-quality manufacture and robustness are still remarkable today.

The dolia of the capital not only were large but also featured markings identifying how much they could hold. Because Ostia and Rome were densely populated cities, their storehouses held large amounts of wine and olive oil, and the dolia often each had a capacity of over one thousand liters, far exceeding those of the dolia in Cosa and Pompeii:[47] many of the dolia could hold twice to three times as much as the largest dolia in Pompeii. While capacity incisions have not been found on dolia at Cosa and only rarely at Pompeii, almost all the dolia at Ostia and Rome featured incisions of Roman numerals on the shoulder or rim to indicate their capacities in units of amphorae (ca. 26.1 liters), with fractional units in *sextarii* (0.546 liters) (Figure 2.13R; Plate 10R; A2 O I.4.5 no. 12). Since many dolia around the capital were marked with these incisions, this systematic method of measuring and labeling dolia might have become part of the production process so workshops could verify their products' sizes and quality, which Chapter 6 will discuss.

The homogeneity of the many dolia of Ostia and Rome suggests a development in the craft where dolium potters closely aligned their techniques and followed certain standards.

46. Dolium lids are no longer in situ. One lid at Ostia is in a storehouse (Magazzino dei Doli, III 14 3) and has a diameter of ca. 70–75 cm.

47. Capacities ranged 774–1,231.4 liters (average 1,009 liters, including the few anomalous smaller dolia, or 1,026.3 liters excluding the smaller dolia).

FIGURE 2.13. (*L*) A dolium from Magazzino Annonario (V.11.5 no. 67) and (*R*) dolium with capacity incision XLIIↃII (42 amphorae + 2 *sextarii*, 1,101.5 liters) (I.4.5 dolium no. 12), Ostia. Courtesy of the Photographic Archive of the Archaeological Park of Ostia Antica.

The vessels had the same forms, dimensions, and types of capacity labels. The dolia, however, did *not* originate from the same workshops, even though they were all made with similar clay mixtures, resulting in reddish-yellow ceramic fabric with a range of grog that featured inclusions such as sparse black sand, quartzite, calcite, and mica, which reduced drying time and shrinkage (Figure 2.7C; Plate 6C). While almost all the dolia at Ostia had similar, if not the same, morphology and ceramic fabric, the stamp evidence points to different potters and workshops using similar clay mixtures to form dolia of comparable shape and size. Dolium potters had developed an advanced level of mastery in their craft and were performing the same procedures and processes, even across different workshops.

According to stamps found on approximately 20 percent of the dolia, Ostia and Rome procured some of the largest and most robust vessels from opus doliare workshops along the Tiber Valley.[48] "Urban" opus doliare workshops situated along the Tiber River within 50 km of Rome supplied not only the city of Rome and its markets but also more-distant communities by the first century CE.[49] The different dolium stamps point to multiple workshops with more complex organization, almost all of which also produced bricks and tiles (Table A1.7). Even some of the earliest workshops producing dolia were also making heavy terracotta objects, most commonly bricks, tiles, amphorae, and coarseware pottery.[50] Dolium production on its own was rare. Instead, the evidence points to the entanglement of dolium production with other ceramic and terracotta objects from its inception. Dolium stamps can be difficult to interpret and usually provide names of individuals unattested

48. Clay beds have been located based on toponyms, epigraphic evidence, and archaeometric analyses; see Graham 2006, 10–16, for summary. For a map, see Manca et al. 2016, image 1b.

49. Steinby 1981. E.g., Ebla (see Manca et al. 2016) and Puglia (see Montana et al. 2021). Bianchi 2016.

50. The only exception seems to be a Republican-period pottery production region in the Pontine plain consisting of several workshops: one specializing in just dolia, one in dolia with building materials, and another in building materials and pottery: Tol and Borgers 2016.

elsewhere, but show complex organization within and across opus doliare workshops.[51] To better understand the dolium stamps in Ostia and Rome, we should turn first to the brick stamps of second-century opus doliare workshops, which provide the most detailed information among stamped heavy terracotta objects. Second-century opus doliare workshops regularly utilized stamps with formulaic information about the *figlinae, dominus,* and *officinator.*[52] The relationships among them have been widely debated, but the prevailing view is that the *figlinae* should be interpreted as "clay lands" and referred to territorial districts, which were owned or managed by the *domini,* "landlords." *Officinatores* were entrepreneurs who could move from brickyard to brickyard and rent from different *domini,* but their status, role, and relationship to the *dominus* could vary: "The term *officinator* still may describe anything from an enslaved foreman in his master's service to a powerful industrialist of equestrian rank."[53] For bricks and pottery, stamps seem to have been used for both advertising and internal purposes—that is, to distinguish one *officinator*'s products from another's within large workshops with several *officinatores.*[54] In general, stamps on opus doliare products could trace defected or flawed products back to their workshops.[55] Although stamps of different opus doliare objects did not necessarily share the same format, they often overlapped in the type of information that was conveyed. Dolium stamps were similar to those found on *mortaria,* which featured cursory stamps of two registers on the rim.[56]

Dolium stamps at Ostia and Rome featured names of men and women in the genitive or nominative case and were usually decorated with pictorial images, such as a palm branch, staff, or ox head, the function of which is still unclear (Figure 2.14; Plate 11).[57] Names in the genitive case identified the owner of the workshop and estate, akin to the *dominus* or *domina* featured on brick and *mortarium* stamps. Names in the nominative case were followed by *f(ecit)* or *fec(it)* to denote the dolium makers or *officinatores;* according to the stamps, many of them (at least 30 percent) were enslaved, and their names were occasionally accompanied by a name in the genitive to identify the enslaver, but some of these figures had been manumitted and continued to work in the workshop.[58] The epigraphic evidence for the Tossius family gives us a glimpse of the internal dynamics, opportunities, and staff within a prominent opus doliare workshop.[59] The dolia from the Tossius workshop were distributed far and wide. Many of them were found in Ostia and

51. Bloch 1947, 1948; Steinby 1974/1975, 1981, 1982, 1993; volumes of the *CIL;* Carrato 2017; Carre and Cibecchini 2020.

52. The literature on this subject is vast; see Steinby 1982; Bodel 1983; Graham 2006, 12ff.

53. Bodel 1983, 4.

54. Manacorda 1993; Parca 2001, 68.

55. Steinby 1993; Aubert 1994, 234; Graham 2006, 12–16.

56. For an overview, see Frazzoni 2016. On *mortaria,* see Pallecchi 2005. See Manacorda 1993 for discussion of passages from the *Digest* (18.6.1.1–2, 19.1.6.4, 21.1.33).

57. Various images are included in Bloch 1948 and the *CIL* entries.

58. On manumission and the former enslaver's control over a freedman's labor, see Hawkins 2016, 130–191.

59. See Gregori 1994 for a summary of Cimber's career and discussion on a dolium rim bearing two stamps, one of Redemptus and one of Euphrastus, both of which associate the slaves with Cimber.

FIGURE 2.14. (*L*) Stamp on dolium rim (III.14.3 no. 10): C VIBI FORTVNATI/C VIBI CRESCENTIS; and (*R*) stamps on dolium rim (III.14.3 no. 1): PYRAMI ENCOLPI/AVG DISP·ARCARI (*left-hand side*) and AMPLIATVS·VIC·F (*right-hand side*), Ostia. Courtesy of the Photographic Archive of the Archaeological Park of Ostia Antica.

Rome, and in surrounding neighborhoods and towns; some of the dolia reached more distant destinations, such as Umbria, Tuscany, Marche, and even southern Gaul. The opus doliare stamps from the (Q.) Tossius workshop document at least six individuals in the business.[60] One figure, Q. Tossius Cimber, was first attested as a slave of Q. Tossius Ingenuus, then was freed and became an *officinator*, with two slaves of his own, Redemptus and Euphrastus. Opus doliare workshops likely offered different types of training and opportunities to rise through the ranks.

Both the epigraphic and archaeological evidence indicates that dolium production became increasingly intertwined with the production of construction materials. Archaeological projects in central Italy have identified various workshops and their products.[61] In earlier periods, potters often produced dolia in workshops that also manufactured other heavy terracotta objects such as amphorae, coarseware pottery, lamps, bricks, tiles, and architectural terracottas (Figure 2.15; Table A1.8). The earliest identified workshops manufactured dolia alongside other pottery objects made of groggy clays such as amphorae. This worked well for the producer since dolia and amphorae required the same materials and equipment and drew on similar skills and amounts of strength; in fact, dolium potters likely mastered the craft of building amphorae, which entailed throwing the large vessel as separate pieces and then joining them, before advancing to dolium production (this is also how potters from different areas and periods train to manufacture large dolium-like vessels). The combination of dolium and amphora production was convenient for the customer too, since vintners buying dolia likely produced wine on a large scale for the market and would package the wine in amphorae (see Chapter 4). Workshops continued to produce dolia alongside amphorae and other coarseware pottery, but also construction materials such as bricks and tiles. Bricks and tiles were essential components of Roman architecture, often forming the core of buildings when coupled with concrete.

60. Q. Tossius Clarus, Q. Tossius Ingenuus, Q. Tossius Iustus, Q. Tossius Priscus, Q. Tossius Proculus, and Q. Tossius Cimber.

61. Olcese (2012) compiles evidence from Tuscany, Latium, Campania, and Sicily.

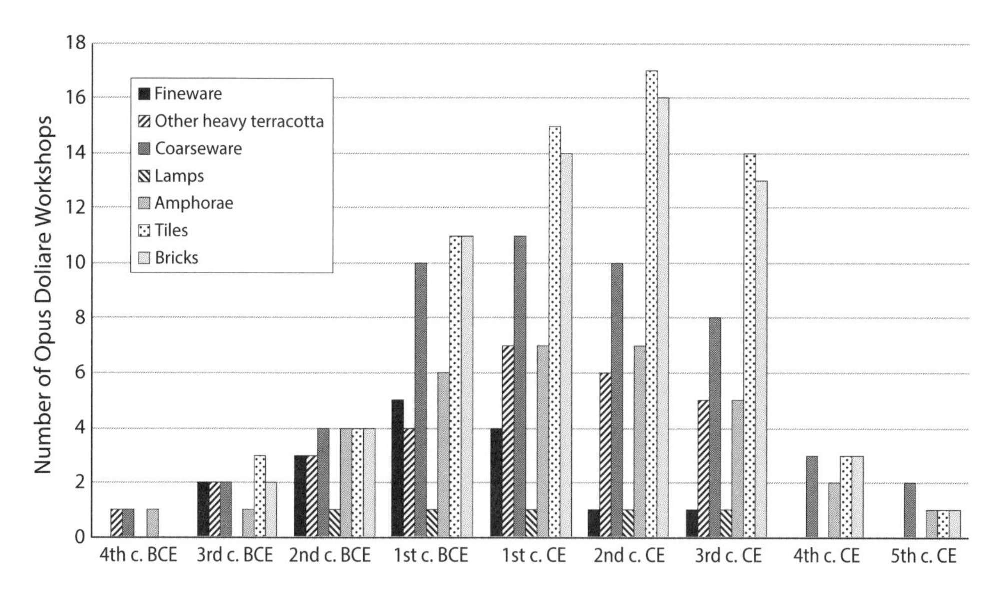

FIGURE 2.15. Other products made alongside dolia in workshops in Tuscany, Latium, and Campania.

From the first century BCE onward, brick and tile production rose significantly, at the same time that concrete-based villa construction proliferated in Latium.[62] Over time, dolium production overlapped more and more with brick and tile production, and less with other products, and the association between them near the capital became even more pronounced;[63] in fact, almost all the workshops that produced their dolia also manufactured bricks and tiles.

Marrying dolium production with brick and tile production offered several advantages. Dolia, bricks, tiles, and other opus doliare products share a similar ceramic makeup using groggy clay, and their materials were prepared the same way. Larger and more profitable workshops probably had the space and equipment necessary for dolium production, such as levigating tanks for processing clay and kilns for firing large quantities of bricks and tiles. Moreover, large and well-equipped workshops probably also facilitated specialization and division of labor, as workers could be assigned to certain areas of the workspace to perform their tasks. As mentioned earlier, the enslaved potter Corinthus gradually acquired different skills, from producing pottery on the wheel to working with groggy clay to make large tiles (which could distort easily) to making dolia; the production of tiles (and bricks, which used molds) entailed skills in working with groggy clay but was not as prone to failure or major losses as dolium production, since a warped tile, for example, was often still usable. The other crucial advantage of opus doliare workshops was that, if dolium production failed, the materials could be reused, cutting down on potential losses:[64] if

62. Mogetta 2021.

63. See Lazzeretti and Pallecchi 2005; entries in Olcese 2012.

64. Nicoletta (2007), Tol and Borgers (2016), and Olcese (2012) identify dolium production sites with wasters.

there were problems before firing, the dolium could be slaked back down into a clay mixture again; if the pot had already been fired, it could be recycled as refractory material to build up strength and heat resistance in bricks, tiles, other opus doliare products, and even the workshop's kilns.[65] By merging dolium, brick, and tile production, then, workshops balanced the lucrative yet risky manufacture of dolium with revenue from the stable but low-profit bricks and tiles for Rome's many architectural projects.[66]

The frequent overlap between *domini* and *officinatores* mentioned on dolium stamps and on brick stamps documents numerous clay lands or districts diversifying their production, which seems to have been a successful strategy as well as establishing a complex organization of workers (Table 2.2). Alessandra Lazzeretti and Silvia Pallecchi's study of opus doliare workshops and stamps found that many operated as "polyvalent" workshops producing multiple types of products, primarily bricks and tiles coupled with at least one other type of object.[67] Through careful study of a large corpus of material, they show that all opus doliare workshops were organized in a complex way, with various sectors of production entrusted to different positions. Often a workshop would have several workers in charge of bricks and tiles, a set overseeing dolium production, and another group in charge of *mortaria*. Many of these individuals were from prominent families who had a hand in industries that supplied various opus doliare products to the greater region of Rome as early as the late Republican period, such as the Cornelii, Domitii, Fulvii, Tossii, and Vibii, as well as the imperial family.[68] By the first or second century CE, dolia were firmly established products of opus doliare workshops in the Tiber River Valley that also produced the bricks and tiles for Rome's many construction projects.[69] A certain C. Cornelius Felix, for example, has been attested on dolia in Rome, while his slaves, Cimber and Calateus, have been attested on both bricks and dolia in Rome.[70] A M. Fulvius and his dozen enslaved workers also dominated the opus doliare scene, producing bricks, tiles, terracotta sarcophagi, and dolia for Rome and Ostia, with some products appearing also in Pompeii.[71] Some families, such as the Sulpicianae, exported products as far afield as Spain, and some had particularly long-lasting enterprises, such as the Calpetanii, who produced opus doliare products for almost two centuries.[72]

The widespread appearance of certain families on stamped ceramic and terracotta products suggests that dolium production benefited from the resources and networks of large opus doliare enterprises. Because dolium production was such a difficult and risky activity

65. See Chapter 8 for examples of dolium rims repurposed in a kiln.

66. For discussion of failure and a case at Marzuolo, see Van Oyen 2023.

67. Lazzeretti and Pallecchi 2005. On workspace and organization of production, see Murphy 2016.

68. Taglietti 2015: the Tossius family was active in the opus doliare industry for several generations. Carrato 2017, 619: a dolium with stamp of Q. Tossius Priscus was found in Gaul. For the extent of certain opus doliare workshops, see Chausson 2005; Chausson and Buonopane 2010; Graham 2009.

69. Lazzeretti and Pallecchi 2005. The literature on opus doliare workshops in the Tiber River Valley is vast; see papers in Bruun 2005; Bodel 1983; Steinby 1987; Graham 2006; Bergamini 2007; Comodi et al. 2007; Manca et al. 2016; papers in Spanu 2015; Gliozzo 2007.

70. C. Cornelius Felix: *CIL* 15.2430–2431. Cimber: *CIL* 11.2.2 8114.2 (dolium); Bloch 1947, 98n480 (brick).

71. Bloch 1947, 68–70 nos. 299–310.

72. See Bloch 1947, 96–97: their products span 160–170 years.

TABLE 2.2. Hypothesized opus doliare workshops thought to be producing multiple products, based on opus doliare stamps (from Lazzeretti and Pallecchi 2005).

Opus doliare products	Attested *figlinae*
Construction material and dolia	Fabianae, Naevianae, Publilianae
Construction material and *mortaria*	Ab Appollini, Caepionianae, Genianae, Terentianae
Construction material, dolia, and *mortaria*	Caninianae/Portus Licini, Camillianae, Sulpicianae, Tempesinae, Viccianae
Construction material, dolia, *mortaria*, other heavy terracotta objects	Castricianae, Marcianae, Oceanae

that required substantial investments, it became an operation that, over time, large workshops with access to capital could develop more successfully. Workshops increased their profits through new economies of scale by producing greater quantities of bricks, tiles, dolia, and other heavy ceramic and terracotta products at a lower cost. These families, often senatorial or even the imperial family itself (which became a major producer in the second century and the dominant producer by the third century), would have also had access to fluvial ports, boats, and other resources for transporting their products.[73] Domitia Lucilla (mother of Marcus Aurelius), for example, owned several estates with multiple clay lands that not only supplied a vast territory but also boosted her ability to commission building projects, supply her own villas, and perhaps even produce her own wine, all of which surely contributed to the accumulation of wealth and social and political power.[74] As Shawn Graham notes, over two hundred people can be connected to Domitia Lucilla, positioning her to "control the flow of information" and providing her access to skilled slaves, lucrative opportunities, and professional networks.[75] The increasing capital of various *figlinae* and *praedia* (estates) probably contributed to the concentration of ceramic industries, as well as dolium production, in the hands of fewer, larger workshops owned by wealthy elites.

Opportunity and Profit in the Ceramic Valley

The production of dolia was unlike the production of other types of packaging containers, requiring time, specialized knowledge, and substantial resources and investment. Dolium production was particularly challenging and prone to failure, and dolium potters worked with mixtures of groggy clay that would reduce shrinkage, warpage, drying times, and thermal expansion. Dolium production needed not only a *doliarius*' skill but also significant capital and funding before the dolia could turn a profit, and only if production was successful, since the vessels would not be finished until months after the work began. Making dolia therefore relied on a stable and capable labor force, materials, and equipment, upfront

73. See Graham 2006, 2009, for discussion on networks among opus doliare workshops.

74. Graham 2006, 2009; Chausson and Buonopane 2010. On Villa Magna, where many of the bricks were produced by Domitia Lucilla's workshops, see Fentress et al. 2017. On women in the opus doliare industry, see Braito 2020.

75. Graham 2009, 681.

investments that well-financed workshops could often provide.[76] The dolium stamps show that these opus doliare workshops produced not only dolia but also bricks and tiles. Because these opus doliare workshops were larger, functioned in an expansive network, and likely had access to more resources, including transport and port facilities along the Tiber River, they had the space, equipment, and materials to manufacture better and grander vessels and the means to transport the bulky vessels.

The challenges and risks of dolium production were surely off-putting for many, but the potential payoff enticed wealthy investors. With the long production time of several months and the need for highly specialized craftspeople, special materials, and large-scale equipment, most people could not afford the upfront costs of funding and pursuing such ventures. Dolium stamps point to wealthy and powerful families as owners of these successful opus doliare workshops by the first and second centuries CE. Operating and financing an opus doliare workshop did not depend solely on monetary resources, though—one needed access to natural resources, specialist potters, and a stable customer base. Moreover, the benefits of successful dolium production were not just about financial gains. Workshop owners could fulfill orders for friends, allies, or other wealthy, elite landowners, currying social and political favors. Many of these opus doliare workshop owners practiced viticulture themselves, such as Domitia Lucilla, and added dolium production to their portfolios to guarantee their own supplies; this form of "portfolio capitalism," whereby groups trying to mobilize agricultural surplus engage in commercial services and credit operations and diversify their portfolios to command as wide a range of resources as possible, is something we will continue to see with different investments in dolia.[77] In fact, it is likely that early investors in dolium production were elite landowners producing wine who wanted dolia; by funding dolium production, they were able to supply and install dolia in their own wine cellars and expand their operations while cutting costs as a type of vertical integration (a form of organization that controls the supply chain—e.g., producing the building materials and dolia to build and operate one's own vineyard). Dolium production could lead to financial gain, guaranteed equipment, control over resources, and even social prestige and influence.

Because it was such an expensive and risky endeavor, the production of dolia increasingly became entangled with the manufacture of architectural materials. Opus doliare workshops were so successful in balancing the risky yet profitable production of dolia with the manufacture of architectural materials that they were able to increase new economies of scale and cultivate a specialized workforce; over time, these opus doliare workshops dominated the Tiber River Valley. Dolium production became more sophisticated and successful near the capital, where resources for workshops abounded and demand for building materials soared; the dolia there were larger, more robust, and better fired because they were the products of workshops with the means, and incentive, to develop a sophisticated and lucrative craft. The dense array of the Tiber River Valley's large-scale opus doliare

76. Itinerant potters might have made pithoi in earlier periods—e.g., pithoi that fit in a unique curvilinear bulge in the wall at the House of Ganymede, Morgantina, and would not have fit through the doorways of the house, suggesting the pithoi were made on-site; Tsakirgis 1984, 78–79.

77. Bang 2007, 31–39.

workshop setup enabled an unprecedented level of dolium production, as well as the growth of a craft community. Artisans were able to share not only materials and equipment, such as clay, water, wood, and large kilns, but also techniques and working knowledge. At production sites near the capital, dolium production became folded into large workshops that also produced bricks and tiles, a low-profit but stable commodity. By the second century CE, members of the sophisticated urban opus doliare workshops that produced dolia, bricks, and tile developed their methods to build massive, standardized vessels; the proliferation and success of these major workshops were so great that the Tiber River Valley, in addition to supplying Rome with wine, olive oil, and building stones, had been transformed into a ceramic valley, a hub of ceramic and terracotta production. Building such large vessels, however, was a task fraught with risk, and craftspeople also developed new methods in their routines. Knowing well the material properties of dolia and their ideal firing temperatures could lead to increased efficiency and more successful production. Dolium makers made several significant changes to construct such large-scale dolia for the fermentation and storage of liquid products, and the pride they took in their skills, craft knowledge, and products explains why someone such as Lucius Aurelius Sabinus would choose to highlight his occupation as a *doliarius*. His skills, however, were probably not just limited to making dolia (and perhaps amphorae as part of his training, which might be why two amphorae are depicted on the altar), but could have also translated into managerial knowledge in a workshop, perhaps even his own workshop.

3

Dolia on the Farm

CONSPICUOUS PRODUCTION AND STORAGE

> It is advantageous for the paterfamilias to manage a well-built country farm—an oil press room, a winery, plenty of dolia, so that one may have the pleasure of waiting for high profit, which will be better for income, better for self-worth, better for reputation.
>
> patrem familiae villam rusticam bene aedificatam habere expedit, cellam oleariam, vinariam, dolia multa, uti lubeat caritatem exspectare: et rei et virtuti et gloriae erit.
>
> CATO *DE AGRI CULTURA* 3.2

IN THE EARLIEST surviving text of Latin prose, Cato the Elder in the second century BCE prescribed dolia on the farm, underscoring their role as essential equipment for viticulture and olive oil production. He also noted not only that farmhouses with more dolia were more productive, but also that having more dolia allowed farmers to sell their products at favorable prices, advice that was continuously echoed in later periods.[1] Owing to their size, dolia made it possible to produce and store extremely large quantities of valuable agricultural goods. From the late third century BCE onward, estate owners installed dolia in great numbers in farmhouses and villas throughout central Italy, massively increasing their productive capabilities. Large villas in the Italian countryside, such as the Villa della Pisanella in Boscoreale, could produce and store nearly seventy thousand liters in their large wine cellars.[2] Even more modest farmhouses, such as the Villa Regina in Boscoreale, could potentially produce and store over ten thousand liters of wine, well beyond what a single family needed for its own annual consumption. Bolstered by extensive storage, Italian estate owners and merchants increasingly traded and sold wine in large quantities, often to faraway places, and could turn quite a profit. This chapter begins with a general overview of the installation and use of dolia before it surveys the various ways dolia were used on farms and villas across central Italy.

As "specialized" hardware, dolia were installed, used, and maintained according to certain standards, requiring workers with the right training and know-how. Special care had

1. Varro *Rust.* 1.2.8, 1.22.4; *Dig.* 33.7.7.

2. For comparison of different villas, see De Simone 2017, 35–40, table 1.5.

to be taken to move the bulky dolia from their place of manufacture to their destination. Once on site, workers set up dolia in special rooms and in particular ways depending on how they would be used. Although dolia were designed for wine production and storage, they became useful vessels for other agricultural goods, especially olive oil. Villas had designated wine and oil cellars in different locations, and workers were expected to maintain the vessels on a regular basis. Dolia required substantial resources, but they greatly increased an estate's ability to produce and store profitable products. From the second century BCE onward, central Italian villas invested in the bulk ceramic storage technology to specialize, diversify, and expand wine production. Dolium installation went hand in hand with the development of villas across central Italy, and dolia became symbols of expansive production, abundance, and wealth.

Investing in Storage

Adopting the dolium-based storage technology was no arbitrary decision or undertaking. It required planning and enormous inputs of resources every step of the way. This could begin before the dolium was even purchased. As Chapter 2 discussed, certain workshops and areas had better reputations than others, and some buyers even visited the workshops and personally selected their valuable investments with great care. Anatolios (*Geoponika* 6.3) advised that these visits should also include inspecting the clay source and the vessel itself, but this might not have been feasible for every buyer. Ethnographic work on pithoi in the Messenian Gulf during the nineteenth and early twentieth centuries showed that customers had several different options. They could purchase pithoi directly at the workshop or markets from the pithoi potters themselves or through merchants who purchased the vessels from the potters.[3] In antiquity, customers could probably acquire dolia at markets or from merchants, and they could trace the stamped vessels to their workshops; without visiting and evaluating the workshop, they might have preferred dolia from a reputable workshop.

After purchase, the vessels had to be carefully transported from the workshop or market to their destination. Archaeological evidence (petrography, dolium stamps, and shipwrecked dolia) shows that dolia were sometimes transported on boats and ships. Cranes installed at ports presumably lifted these cumbersome jars in and out of boats and ships. Due to the massive size and weight of dolia, maritime and fluvial transport was probably the most cost-effective way to transport the large storage pots.[4] Some villa owners strategically situated their opus doliare workshops and *figlinae* along rivers, such as the Tiber River Valley, and by the sea to access conducive shipping routes. Recent work has revealed that dolia were being shipped from west-central Italy (from the Tiber River Valley or Minturnae) to more faraway places such as the imperial estate at Vagnari in Puglia and a villa on the island of Elba.[5] Given how precious and bulky the cargo was, shipping dolia, especially to more distant destinations, probably occurred primarily during the Mediterranean sailing season of April through October. Although some dolia were installed near rivers or the sea at port

3. Blitzer 1990, 698–707.

4. Vitr. *Arch.* 10.2.10; A.I. Wilson 2011, 51.

5. Carroll 2022a, 2022b; Carroll et al. 2022; Montana et al. 2021; Manca et al. 2016.

warehouses, the majority of dolia were transported at least in part on carts or wagons to places farther inland by the customer, contract drivers, or off-season farmers.[6]

Overland transportation was generally a more expensive way of moving goods, and this could be a significant factor, in addition to the cost, size, and quality of a dolium, in one's decision to acquire a vessel.[7] Cato's (*Agr.* 22.3) discussion of purchasing an olive mill from Suessa or Pompeii provides useful points on the topic of moving heavy farm equipment. A buyer took into consideration not only the price of the mill itself but also the expense and mode of transportation. Moving an olive mill from a local site, Suessa in this case, as opposed to a more distant site, Pompeii, incurred only a fraction of the price as transportation costs (1.44 sesterces per mile versus 3.73 sesterces per mile);[8] moving a mill from Suessa could form up to 40 percent of the total purchase price, while transport from Pompeii made up an astounding 70 percent of the total cost.[9] But there was another factor. Eric Poehler brings up several important points in his discussion of a household or farm transporting their goods themselves (household mode) or contracting drivers (commercial mode):

> At 25 miles distant from his villa, Cato reports the transportation cost for bringing the mill from Suessa was 72 sesterces using six men for six days, while the cost for delivery from Pompeii, 75 miles away, was nearly four times that amount, 280 sesterces. These figures seem to show transportation costs growing dramatically with distance. There is however, an important difference. While the trip to Suessa was a round trip as Cato's own carts were used, making the trip a total of 50 miles, the mill from Pompeii was being delivered. Comparing these costs per mile, the trip to and from Suessa was 1.44 sesterces per mile while the trip from Pompeii was 3.73 sesterces per mile. The difference in price is striking; transportation costs are two and a half times less when one owns the means of transport. Such variance in cost is the economic advantage behind the Household Mode of transport. On the other hand, this same example also demonstrates that the Household Mode was restricted by distance. If Cato had sent his own vehicles to Pompeii, the return trip of 150 miles would have cost 216 sesterces by his figures, or 77 per cent of the cost to have it delivered. More importantly, the time necessary for the trip would have tripled as well, taking away six men and six oxen from other work for eighteen days and delaying the arrival of the mill by nine days. The value of lost production from men and machinery, though not discussed by Cato, would have made the Household Mode of transport less efficient than the Commercial Mode at this distance.[10]

The acquisition of a dolium could be costly (72 or 280 sesterces in Cato's example, whereas a craftsperson earned 8 sesterces for assembling an olive mill axle, and a *modius* [ca. nine liters] of olives was valued at 5 sesterces) and depended on several factors: the distance between the place of manufacture or sale (the workshop or market) and the place of

6. Erdkamp 1999.

7. Duncan-Jones 1974, 368: according to Diocletian's *Price Edict*, fluvial transport was 4.9 times more and overland transport was 28–56 times more than maritime transport. James 2020, 28–33.

8. Poehler 2011, 205.

9. McCallum 2010, 85.

10. Poehler 2011, 205–206.

TABLE 3.1. Legal maximum vehicle load weights based on the *Codex Theodosianus* 8.5.8, as converted to metric weights by Weller (1999).

Vehicle term in Latin	Load weight in Roman pounds	Load weight in kilograms
Angaria	1,500	492
Raeda	1,000	330
Currus	600	198
Vereda	300	99
Birota	200	66

use (the vessel's destination); whether the dolium buyer had the resources (vehicle, draft animals, labor) to transport the dolium; and, if so, whether the dolium buyer wanted to use them to move the vessel rather than contract a driver. For buyers who did not have the means to do it themselves, or chose not to, transporting a dolium depended on the availability of contract drivers and well-connected roads.[11]

In addition to draft animals and a team of laborers, it was also important to have the proper vehicle to deliver a dolium, which could be pricey. The legal weight limits of carriages and carts outlined by the *Codex Theodosianus* suggest that the very large and heavy dolia were likely moved in four-wheeled wagons known as *raedae*, and not two-wheeled carts (Table 3.1): "We establish that only one thousand pounds of weight may be placed on a four-wheeled carriage, two hundred on a two-wheeled vehicle, and thirty on a courier's horse, for they appear not to support heavier burdens" (8.5.8: *statuimus raedae mille pondo tantummodo superponi, birotae ducenta, veredo triginta; non enim ampliora onera perpeti videntur*). Yet Poehler notes that these figures were likely the legal maximum weight limits for different means of overland transport, probably as an attempt to reduce wear on roads, but should be considered the average, or even minimum, vehicle loads.[12] Two-wheeled carts were likely able to support the weight of some dolia, but large dolia probably required large four-wheeled wagons. Ethnographic work on pithoi and *tinajas*, too, demonstrates that two-wheeled carts, pulled by two mules, were sufficient to transport a single large ceramic vessel, which workers carefully moved, possibly with ropes and mats, loaded onto the cart, and packed with straw or other supportive material.[13] Drivers probably used carts to move smaller dolia, reserving heavy-duty wagons for larger or multiple dolia and more arduous or lengthy journeys. Diocletian's *Price Edict* tells us that, depending on the size and weight limits, different types of freight wagons could cost 800, 3,500, and up to 6,000 denarii (compared with the *Price Edict*'s prices of 4 denarii for a *sextarius* of olives and almost 6 denarii for a liter of wheat);[14] moving an especially large dolium could require a larger freight wagon that could cost seven

11. Poehler 2011, 206: Varro (*Rust.* 1.2.23) also notes the importance of roads and infrastructure of transport for the household mode of transportation, and hence the success of a villa.

12. Poehler 2017, 108–109.

13. Carrato 2017, 147–152.

14. Diocletian's *Price Edict* 15.1, 15.31a, 15.32.

and a half times more for just the vehicle, not counting draft animals, drivers, or other fees. Delivering a dolium, especially over long distances, could add significant costs to the already pricey investment and might be why some people continued to use old dolia, even if they affected the taste of the wine.[15]

Dolia for Wine and Olive Oil

Dolia offered several advantages for storing liquids, but they occasionally served as general-purpose receptacles for dry goods, usually after they were no longer suitable for liquids. The dolium's strawberry shape, with wide shoulders that tapered to the base, facilitated removing the vessel's contents by concentrating any residual liquid into a narrow area where it could be effectively removed by a ladle or submerged pitcher. To further insulate and protect wine and oil, farmers sealed dolia using a ceramic double-lid system (Figure 2.10).[16] First, they placed a flat, disk-shaped lid (*operculum*) on top of the dolium rim. For wine storage, vintners smeared mastic or pitch on the inner surface of the internal lid to help create a tight seal between the lid and dolium. Next, they would place a larger, convex external lid (*tectorium*) on top of the inner lid. The external lid's three ceramic feet sat on top of the inner lid and provided a small air chamber between the *operculum* and *tectorium*. Each lid was usually about three centimeters thick, and the outer lid further insulated the dolium and its contents from fluctuating ambient temperatures. Beyond general storage, though, workers had to follow different protocols depending on how they planned to use the dolia, starting with installation.

Once the dolium arrived, it had to be installed properly. Setting up a dolium followed certain standards and depended on its use. Owners or managers of properties planning to install storage vessels often procured the right number, type, and size of jars before they installed and incorporated them into the property; some owners, however, were limited to whatever vessels were available at different workshops, found a secondhand jar, or added vessels after the initial setup.[17] Workers spaced out the vessels so that there would be adequate space to navigate between them. Some dolia might have been placed in certain areas for specific functions such as collecting freshly pressed liquids, providing long-term storage, or storing certain contents:

> And it should have a connected wine cellar that has windows facing north. For if on any other side there is an opening through which the sun could warm, the wine in this room, disturbed by heat, becomes weak. The oil room ought to be placed so that there is light from the southern and the warm regions, for the oil should not be chilled, but rather kept thinned by the warmth of heat. The extent of these rooms should be made according to the amount of harvest collected and the number of dolia, which, if they are [the size] of *cullei* [ox-hide containers], should occupy four feet in diameter.

15. *Geoponika* 6.3. For ways sigillata pottery was distributed to settlements, see Van Oyen 2015b.

16. The second-century CE funerary relief depicting a wine cellar in the Ince Blundell collection in Liverpool shows some of these flat lids as possibly wooden lids; see Figure 6.1.

17. See Peña 2007b, 46–47, 194–197. Pecci 2020 identifies wine dolia reused for garum production.

> habeatque coniunctam vinariam cellam habentem ab septentrione lumina fenestrarum. cum enim alia parte habuerit qua sol calefacere possit, vinum quod erit in ea cella confusum ab calore efficietur inbecillum. olearia autem ita est conlocanda ut habeat a meridie calidisque regionibus lumen. non enim debet oleum congelari sed tepore caloris extenuari. magnitudines autem earum ad fructuum rationem et numerum doliorum sunt faciundae, quae cum sint cullearia, per medium occupare debent pedes quaternos. (Vitr. *Arch.* 6.6.2–3)[18]

Because of their varied storage needs, villa and farm sites engaging in both viticulture and olive oil production often had separate equipment and rooms, facilitating identification of the archaeological remains.[19] Some large villas even had the production facilities for wine and olive oil located on opposite ends of the villa estate, as Vitruvius recommended, though poorer, smaller farms did not necessarily have the resources for separate wine and oil presses.[20]

Winemaking included the cultivation of grapes and their harvesting, processing, and fermentation.[21] Workers usually extracted juice from the grapes by treading or pressing, though producing higher-quality wines might have entailed drying, boiling, or resting the grapes to obtain a sweeter juice.[22] After treading or pressing grapes, vintners collected and placed the liquid known as "must" into vats or dolia for fermentation. Rooms with treading vats or presses were usually connected via channels or pipes to waterproof collection basins or dolia, where wine underwent primary fermentation over the course of several days (Figure 3.1A). Vintners could choose to include additives such as herbs, spices, and even seawater to impart certain flavors or mask the vinegary taste of the wine.

Dolium potters designed dolia for not only wine storage but also fermentation. The dolia's shape promoted free circulation during wine fermentation, which enhanced the flavor and texture of wine while regulating the temperature.[23] Coupled with the flavor enhancements from the resin coating the vessel walls, the dolium's unique features contributed to flavors and textures particular to Roman wine.[24] More importantly, however, their wide shoulders and small base accommodated the expansion of gas during the fermentation process, a potentially risky and catastrophic stage, as ancient authors noted and broken and repaired dolium lids, both *opercula* and *tectoria,* in Pompeii confirm.[25] Varro, for

18. Palladius *Opus Agriculturae* 1.18 also recommended that wine cellars face north.

19. Archaeologists often identify rooms with embedded dolia as wine cellars; see Rossiter 1981; Marzano 2007. The example of the Villa della Pisanella, which featured a wine cellar with a few unburied dolia used for grains and olive oil, demonstrates that we should be cautious about overgeneralization. There was another, separate room for oil storage, so the mixed use of the dolia in the wine cellar could show deviation from Vitruvius' advice or reflect the post-earthquake habits and disarray of the villa.

20. On large estates with separate facilities, see Teichner 2011/2012. On wine and olive oil production sites at smaller and less affluent farms, see Vaccaro et al. 2013; Bowes 2021a.

21. Dodd 2022, 2020; Brun 2003, 2004; Thurmond 2006, 111–164.

22. On the technology of winemaking, see Dodd 2022; see Lewit 2020 on wine and oil presses.

23. Godyn 2017; Carey 2017; Jarvis 2019.

24. See Plin. *NH* 14.27 on the differences between wine stored in dolia and barrels.

25. Dolium lids were probably cheap and relatively easy to replace once dolium rim sizes and lids were standardized. Repaired lids could indicate inaccessibility or shortages.

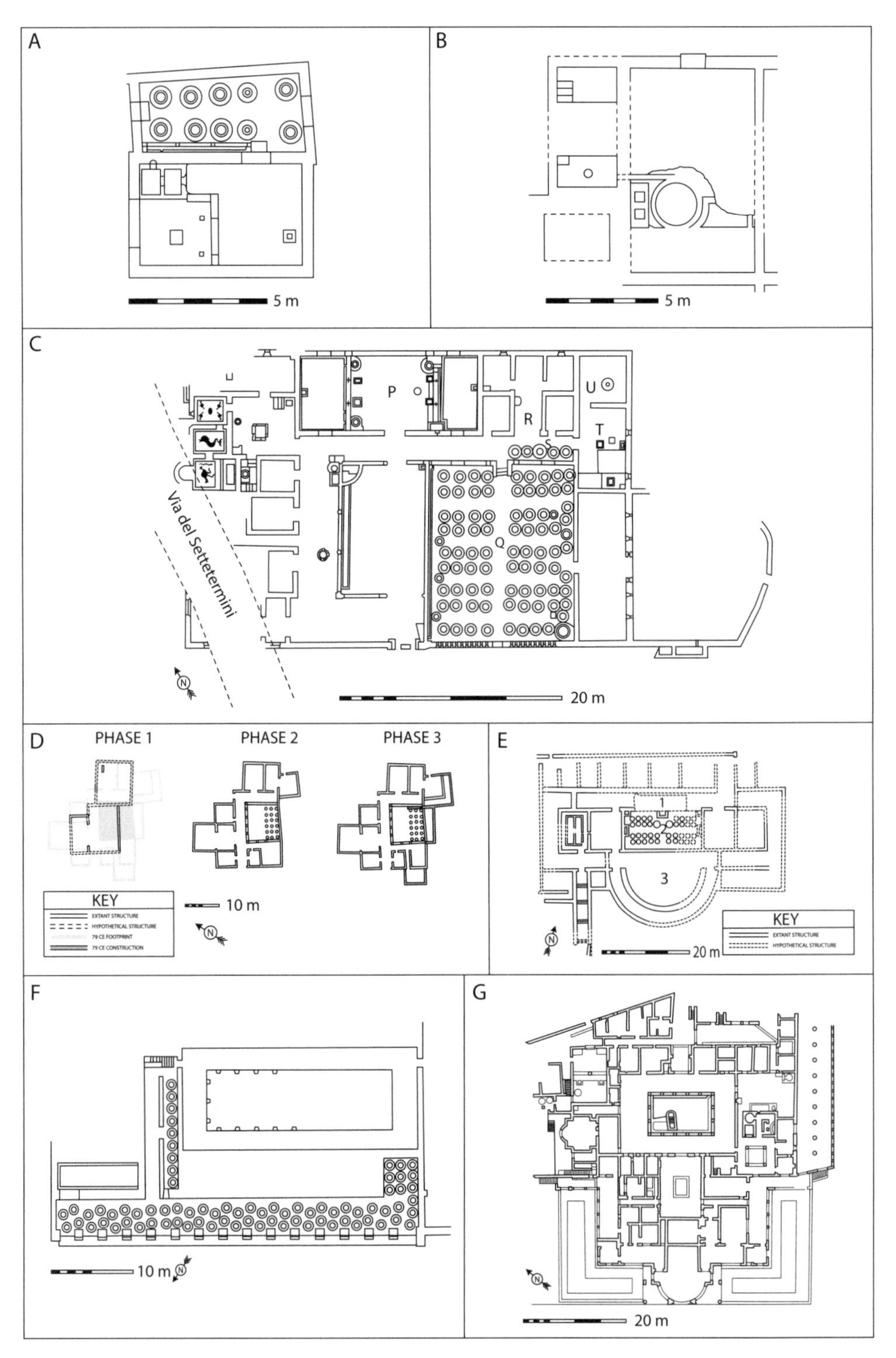

FIGURE 3.1. Plans of (A) winery, Pompeii II.5, after Rossiter (1981); (B) olive oil cellar, villa at Valle Lungha, after Rossiter (1981); (C) Villa della Pisanella, Boscoreale, after Pasqui (1897), P = wine press room, Q = wine cellar, R = rooms of the wine press room, S = olive oil cellar, T = olive oil press room, U = *trapetum*; (D) Villa Regina, Boscoreale, after De Caro (1994); (E) winery of Villa Magna, San Pietro di Villamagna, after Feige (2021), 1 = treading area, 2 = wine cellar, 3 = exedra; (F) winery of Villa Settefinestre, Ansedonia, after Carandini (1985); (G) Villa of the Mysteries, after Clarke and Muntasser (2014). Illustrated by Gina Tibbott.

example, mentions that farmhouses had reservoirs (*lacus*) and jars (*dolia, orcae*) for wine production:

> There a farm was praised if it had a good rural kitchen, spacious stables, and wine and oil cellars in proportion to the size of the farm and with a floor sloping to a vat, because often, after the new wine is stored, not only the large vessels used in Spain but also dolia used in Italy are burst by the fermentation of the must.
>
> illic laudabatur villa, si habebat culinam rusticam bonam, praesepis laxas, cellam vinariam et oleariam ad modum agri aptam et pavimento proclivi in lacum, quod saepe, ubi conditum novum vinum, orcae in Hispania fervore musti ruptae neque non dolea in Italia. (*Rust.* 1.13.6–7)

Varro's passage also tells us that the fermentation process was violent enough to break a dolium and would explain why some vintners might have chosen to ferment must in vats first before placing it in dolia. Anatolios also advises on the types of dolia appropriate for different stages of the fermentation process and varieties of wine:

> Pithoi must not be large. For the wine does not bubble up too much in [smaller] ones, seeing that it isn't confined, but the very excess rises up against itself and ejects not only the smell but also the surface yeast. Small vessels help greatly with protection and with wine quality. Because of this it is necessary to set up small pithoi. If we already have some large old pithoi, we should put the weaker, poorer wine into them; the more potent should go into the smaller pithoi.
>
> ἔστωσαν δὲ οἱ πίθοι μὴ μεγάλοι· οὐκ ἐπὶ πλέον γὰρ ἐν τοῖς τοιούτοις ἀναζεῖ, ἅτε μὴ στενοχωρούμενον, πλεονάζον δὲ αὐτὸ ἑαυτῷ ἐπανίσταται, καὶ οὐ μόνον τὴν ὀσμὴν ἀλλὰ καὶ τὸ ἄνθος ἀποπτύει. τὰ δὲ μικρὰ ἀγγεῖα πολὺ καὶ πρὸς φυλακὴν καὶ καλλιοινίαν συμβάλλεται. διὰ τοῦτο μικροὺς χρὴ κατασκευάζειν τοὺς πίθους. εἰ δὲ φθάσαιμεν παλαιοὺς πίθους ἔχειν μεγάλους, τὸν ἀσθενέστερον καὶ χείρονα ἐγχέωμεν εἰς αὐτούς, εἰς δὲ τοὺς μικροὺς τὸν κρείττω. (*Geoponika* 6.3.9–11)

In traditional winemaking in the Republic of Georgia, vintners preferred large *qvevri* for wine fermentation and smaller *qvevri* for aging wine, and Anatolios' advice indicated that ancient winemaking also had its preferences.[26] His discussion also brings up the importance of using dolia of a certain girth during fermentation, a process during which the wine would swirl and froth. The fermentation process was so potentially volatile that, in a letter to his teacher Fronto, Marcus Aurelius compares his effusive affection to wine fermentation in a dolium: "When you see the must fizzing in the dolium, let it remind you that, for me, this is the way the longing for you bubbles up and overflows and foams in my heart" (*Ad M. Cae.* 4.4: *quom videbis in dolio mustum fervere, in mentem tibi veniat mihi sic in pectore tuum desiderium scatere et abundare et spumas facere*).[27] Potters shaped dolia for wine fermentation and storage, and in the process, the dolia accommodated and even encouraged the swirling, bubbling wine and shaped its flavor profile.

26. Caillaud 2020; Brun 2003, 79–80.

27. Also Prop. *Elegies* 3.17 l. 17; Verg. *G.* 2 l. 6.

FIGURE 3.2. Wine fermentation and circulation within dolia. Illustrated by Gina Tibbott.

The unique shape and material of dolia benefited wine fermentation and storage, as well as contributing particular qualities to wines. Vintners today employing a similar fermentation vessel, commonly referred to as an egg, extol the design because of several advantages it offers.[28] The egg's shape promotes free circulation, whereby the fermenting juice is in constant contact with the skins, so vintners are not required to stir the wines (Figure 3.2). The continuous circulation provides a thermally stable environment for biodynamic fermentation, resulting in wine with more robust texture and flavor. The material of these egg-shaped vessels and their thick walls further contribute to low temperatures, and the porosity of the ceramic material is ideal for fermentation.[29]

Most wine requires time and cool temperatures for proper fermentation, whereas lower-quality wine beverages (e.g., *posca*), such as those for the (enslaved) workforce, were fermented for shorter periods of time, or practically not at all.[30] Vintners today recommend keeping temperatures cool—20°C–30°C for red wine and 7°C–16°C for white—to promote a lengthy fermentation process for quality wine; cool storage temperatures were important too, ideally at 12°C or 13°C (though a maximum of 20°C is acceptable by today's standards).[31] High temperatures could "cook" and even kill the yeast during fermentation and strongly affect the taste, texture, and quality of the wine, giving the wine a boiled taste, or even turning it into vinegar. Too cold, though, and the wine could freeze and expand,

28. Carey 2017; Godyn 2017.

29. Caillaud 2020, 148: ceramic jars kept contents on average one degree cooler than steel tanks.

30. Broekaert 2019; Curtis 2001; Varro *Rust.* 1.54.3.

31. Cheung 2020.

compromising its quality and even breaking its container. Dolia protected their contents from oxygen, light, and high temperatures, the culprits of prematurely aged wine and flattened textures and aroma. Studies have also shown that they swirled the wine and exposed it to higher oxygen transmission rates, lower temperatures, and a longer fermentation period, resulting in wines with different levels of acidity, more minerality, and higher levels of antioxidants; in other words, wines with more fruity, spicy, and astringent flavors, tastes and textures particular to Roman wines.[32]

To provide and ensure cooler temperatures, wine dolia were positioned in particular ways to maximize their storage capabilities.[33] The first major difference between oil and wine dolia is that wine dolia were buried, for which ancient authors provide several reasons. Burying the dolia supposedly minimized the exposure of their contents to air, reducing the risk of over-aeration and wine turning into vinegar.[34] According to Vitruvius, as noted earlier, and Pliny the Elder here, storing wine in buried *dolia* was also optimal and distinctive for Mediterranean climates:

> More mild districts store [wine] in dolia and bury them in the ground entirely, or else up to a part of their position. In this way they protect them against the climate; elsewhere they keep off the weather by building roofs over them. . . . But to speak also about the shapes of the vessels: pot-bellied and broad ones are less useful. . . . Weak vintages should be preserved in dolia sunk in the ground, but strong wines should be in dolia exposed to air. The dolia must never be full, and the area that remains must be smeared with raisin-wine or boiled-down must mixed with saffron or iris pounded up with must.

> mitiores plagae doliis condunt infodiuntque terrae tota aut ad portionem situs. ita caelum prohibent; alibi vero inpositis tectis arcent. . . . quin et figuras referre; ventriosa ac patula minus utilia. . . . inbecilla vina demissis in terram doliis servanda, valida expositis. numquam inplenda, et quod supersit passo aut defruto perunguendum admixto croco pistave iri cum sapa. (Plin. *NH* 14.133–135)

Usually the climate was mild enough that these cellae vinariae were unroofed courtyards, though additional structures might have been set up so a shade cloth could be drawn to protect the dolia defossa during hotter days.[35] Pliny noted that wine dolia in regions with warmer climates were buried. In Italy, farmers usually clustered dolia and buried them up to their shoulders in north-facing rooms to keep their contents cool, harnessing the stability of soil temperatures to reduce temperature fluctuations. Workers removed hundreds of liters of earth, or placed earth around the vessels, and sometimes lined the recess with mortar, to

32. For traditional Georgian wine, see Caillaud 2020; Díaz et al. 2013; Bene et al. 2019. On experiments using ancient wine techniques, see Caillaud 2020; Indelicato 2020. Indelicato 2020: the experimental dolium wine featured a smoky taste from the resin, and aging wine was difficult.

33. For in-depth discussion of wine cellars, see Dodd 2022, 469–472.

34. Macrob. *Sat.* 7.12–15: farmers buried dolia to minimize contact with air.

35. Dodd 2022, 470. On the structure at the Villa Regina in Boscoreale, see De Caro 1994; on the structure at the Villa Magna, see Fentress and Maiuro 2011, 347.

install each vessel.[36] Because soil has a much higher heat capacity than air and provides thermal insulation, soil temperatures remain relatively stable throughout the day and their seasonal temperature changes lag behind ambient temperatures. In other words, soil temperatures were much cooler than ambient temperatures during the summer, and warmer in the winter.[37] The thick ceramic walls of the dolium further insulated and protected the vessel's contents from temperatures distinctive to the arid Mediterranean climate.

Making and storing olive oil followed parameters similar to those for wine production.[38] After harvesting olives, workers crushed them using a stone olive press known as a *trapetum* to prepare the olives to form a paste for pressing. Farmers would ideally press olives to extract the oil in temperatures no higher than 27°C in order to preserve the olives' freshness and nutrients.[39] The last stage was settling in order to separate the oil from the aqueous part of the olive extraction known as *amurca* (olive oil lees), and olive press rooms usually contained tanks for the settling process (Figure 3.1B). Olive oil usually lasts up to two years if processed and stored properly, which entails keeping it away from oxygen, light, and heat, elements that would degrade and turn the oil rancid. Long-term storage was especially important for olive oil because olive trees produced a significant crop of fruit once every two years and because olive oil was often transported over great distances.[40] The olive tree, though hardy, could only grow in a limited geographical range, so production was confined to the Mediterranean basin, while demand extended well beyond it. Most farmers today would recommend storing olive oil at 13°C, while olive oil stored at very low temperatures will crystallize and even solidify starting at 10°C, separating into two different substances and eventually freezing into a thick butter that would be difficult to remove and use. Although cold temperatures would not affect the quality of the oil, one would have to warm frozen olive oil to bring it back to its liquid state in order to remove and use it, a process that would take much more time with large dolium containers. For olive oil, then, dolia would ideally maintain a temperature within the range of 10°C–20°C, but settling tanks and smaller, more portable containers were more commonly found in olive oil cellars than dolia.

For oil storage, in order to ensure that oil did not reach low temperatures in the winter, dolia were only minimally installed in the ground to stabilize the large container, just around the base and occasionally the lower walls. The few oil dolia at Italian farmhouses were usually only partly buried to anchor the vessel in place in a south-facing room to keep the oil in moderate temperatures. At farms at Stabiae and Camerelle, for example, large dolia were placed next to olive presses to collect runoff from press beds, and at other olive oil

36. Barisashvili 2011: *qvevri* installation required lining the outer surfaces of the vessels with lime. See Carroll 2022a, 2022b, on mortar-lined recesses. At the Villa della Pisanella in Boscoreale, the courtyard was at a higher level than the rest of the property; the dolia were installed and then earth was placed around the vessels (rather than digging recesses for the dolia).

37. Martín Ocaña and Cañas Guerrero 2006; Mazarron and Canas 2008: temperatures of wine cellars three meters underground were 5°C–15°C when ambient temperatures were −14°C–37.6°C.

38. Curtis 2001.

39. On the technology of oil presses, see Curtis 2001, 380–394; Brun 1993, 2003, 2004.

40. Waliszewski 2014: Roman olive growers experimented with the synchronization of olive fruit production to promote annual crops.

production sites in southern Italy, dolia for oil storage were not embedded in the ground.[41] The dolium's terracotta material was enough of an insulator to keep olive oil cool.

To keep oil and wine dolia in working condition, workers provided routine maintenance, which comprised tasks regularly featured in agricultural calendars.[42] Cleaning and preparing dolia for wine and olive oil storage entailed different processes and materials, but with the same goal: to remove traces of previous contents and to avoid contamination. This began before the vessel was ever even used. Cato recommended, for a new oil dolium, pretreating the vessel by filling it with *amurca* for seven days, removing the lees and drying the vessel, and then heating the vessel and rubbing gum on the dolium walls.[43] To clean and reline the oil dolia, ancient authors advised removing any remaining sediments from the dolium first and then using one's hands to rub warm lye on the vessel to clean it. Workers then had to rinse the vessel and allow it to dry thoroughly before adding another layer of gum, a process Columella recommended once every few years.[44]

Wine dolia, on the other hand, needed much more frequent upkeep and involved different materials. Latin agronomists such as Columella and Pliny the Elder advised coating the walls of the dolium with pitch, usually pine resin, to prevent the dolia from absorbing residual old wine that would contaminate the new batch of wine. Newly made dolia were to be lined with pitch immediately after their production, and the vessels frequently cleaned and relined for use, ideally each year before the next grape harvest.[45] The latter required removing any leftover wine, melting off the previous layer of pitch, scrubbing and rinsing the vessels, allowing them to dry, and then spreading a new batch of pitch on the interior vessel walls. One way to do this was to turn dolia on their side and spread the pitch, as depicted in a mosaic from a villa in Gaul, where dolia that were not buried could be maneuvered and manipulated to clean and spread these substances along the vessel walls (Figure 3.3);[46] in Gaul, due to colder temperatures, burial in the ground may not have been common. But in Italy, where dolia were buried in the ground up to their shoulders, farmhands who could squeeze inside the small space to climb into the dolium, petite adults and perhaps women and even children, faced a difficult and even hazardous task.[47] Pliny the Elder warns that "therefore so potent are the lees of wine that they kill any who go down into the vats. A lamp sent down offers a good test; so long as it is extinguished it announces danger" (*NH* 23.63.1: *ergo vini faecibus tanta vis est ut descendentes in cupas enecet. experimentum demissa praebet lucerna quamdiu extinguatur periculum denuntians*).[48] The leftover

41. Rossiter 1981.

42. For an overview of cleaning and resurfacing dolia, see Peña 2007b, 211–213.

43. Cato *Agr.* 69.1–2.

44. Columella *Rust.* 12.52.14–17.

45. The Menelogium Colotianum and the calendar from Santa Maria Maggiore indicate that dolia for new wine were pitched during the month of September; Degrassi 1963, 284–298; Magi 1972, 1ff.

46. Plin. *NH* 16.22: types of pitch; Columella *Rust.* 12.52.14–17 (oil dolia), 12.18.5–7 (wine dolia).

47. Barisashvili 2011, 17–19; Diggory 2018. Young men in the village, who could fit through the vessels' narrow openings, cleaned *qvevri* over the course of several hours to two days, soaking the vessels with a lime mixture before scrubbing them with a brush.

48. *Cupa* most likely means "vat" in this context, though it could refer to dolia. Regardless, the potency of lees would have been dangerous to workers cleaning vats, dolia, and other vessels in which wine was kept.

FIGURE 3.3. (*L*) Moving dolia and (*R*) applying pitch to a dolium from Rustic Calendar mosaic, Saint-Romain-en-Gal.

sediments from winemaking could create an anoxic environment in which it was difficult for workers to breathe. In melting off old pitch too, workers had to be careful not to place the torch too close to the vessel, otherwise it could damage the dolium, or even cause it to burst.[49] This intense, long (several days), and frequent (once a year) hazardous work was only one component of many in viticulture (e.g., pruning, trimming, defoliation, staking, and harvesting), a type of cultivation that relied on enormous investments, as well as technical and specialized knowledge and labor.

Significant maintenance went into dolium care to ensure proper use when it came time for the grape or olive harvest. Ancient sources do not mention a dolium's "life expectancy," but these valuable vessels were probably used for several decades, perhaps even well over a century with proper upkeep. Modern vintners also note their *tinajas*, *talhas*, and *qvevri* are used for more than a century, and some even claim their vessels today date back to the seventeenth century, while craft vintners using dolium-like vessels tout a life expectancy of at least fifty years. Across central Italy, landowners allocated resources, space, and labor for what became such an integral part of viticulture and oleoculture that agronomists regularly featured advice on buying, installing, cleaning, and using dolia in their general discussion of running a farm. The costs for such an enormous scale of production were surely prohibitive for some, though. Presses and dolia were expensive equipment that also drew on specialized labor and knowledge for their operation, repair, and maintenance, and some farmers might not have had the budget to support them, opting instead to produce wine and olive oil with only one press (or none at all) and to store foods not in dolia but in other, smaller types of containers such as amphorae, which was common in Phoenician-Punic settlements.[50] Recent work on more modest peasant sites in central Italy has revealed smaller and less costly sets of equipment; instead of a large cella vinaria with dozens of dolia defossa, a small farm such as the one at Marzuolo might only have a couple of tanks dug into the ground.[51] Although more modest or impoverished farms could use tanks or vats rather

49. Columella *Rust.* 12.18.5–7.

50. Marzano 2013a; Dodd 2022; Lewit 2020.

51. Van Oyen 2023; Vennarucci et al. forthcoming; Van Oyen et al. 2019.

than dolia to store their wine, they did not offer the same protections from oxidation or temperature control as dolia. Due to a dolium's high costs, regular maintenance, and large footprint, it might not have been a practical investment for some people, but implementing dolia could bring potentially enormous profits and cultivate other economic ventures. Those who could afford them participated in an expanded type of production, in which they could boast not only their wealth, agricultural surplus, and resources but also their values and status.

Dolia and the Development of Villas in Central Italy

This section presents a survey of dolia at production facilities in central Italy. Dolia began to appear in central Italy in significant numbers as villas were being developed (Figure 3.4; Table A1.9). Although a survey of dolia at villas hinges on (often uneven) archaeological preservation of both dolia and where they were installed, as well as the frequency and focus of archaeological work in particular areas and the quality of publications, it can still provide a ballpark estimate of trends in the central Italian countryside.[52] Most of the dolia at production facilities come from elite villas, which are more archaeologically visible and studied more often than humble farmsteads. Owners of elite villas were also more likely to invest in dolia given their wealth and resources, as well as their investment in large-scale wine production. A survey of the published material shows that central Italian villas began to install dolia at a scale visible in the archaeological record starting in the second century BCE, when at least fourteen villas were identified as having purpose-built cellars installed.[53] The number of villas and farms with dolia substantially increases over time. From the second to first century BCE, the number of villas and farms with cellae in operation doubled, and almost doubled again in the following century: fourteen villas were equipped with cellae in the second century BCE, at least thirty in the first century BCE, and a peak of sixty by the first century CE. Almost all cellae vinariae and *cellae oleariae* in central Italian farms and villas were constructed between the second century BCE and the first century CE, a period of intense villa construction and profit-driven viticulture in central Italy. The number of central Italian villas with cellae begins to decrease gradually starting in the second century CE, with nearly half of cellae from the first century still in use, and the *new* construction of only a handful of cellae are dated to the second century.

Legal sources designated these vessels as architectural elements that belonged on the premises.[54] Most of the dolia were dolia defossa, and cellae were usually associated with winemaking. This is not surprising considering our estimates for the production and consumption of the main commodities (ca. 250 liters of wine versus ca. 20 liters of olive oil per person per annum).[55] Only larger estates seem to have invested in dolia for olive oil production, usually with separate sets of dolia for wine and olive oil production. The size

52. Included in this are data compiled by Marzano 2007; Van Oyen 2020b; also the recent studies of rural sites, notably those studied as part of the Roman Peasant Project (Bowes 2021a).

53. For dolia and cellars in the context of the monumentalization and centralization of storage, see Van Oyen 2015a, 2020b; for food storage and distribution within the climate history of Roman expansion, see Bernard et al. 2023, esp. 35–39.

54. *Dig.* 32.93.4, 33.6.3, 33.6.15, 33.7.8, 33.9.11, 33.21, 19.1.17.6.

55. For olive oil consumption, see Mattingly 1988.

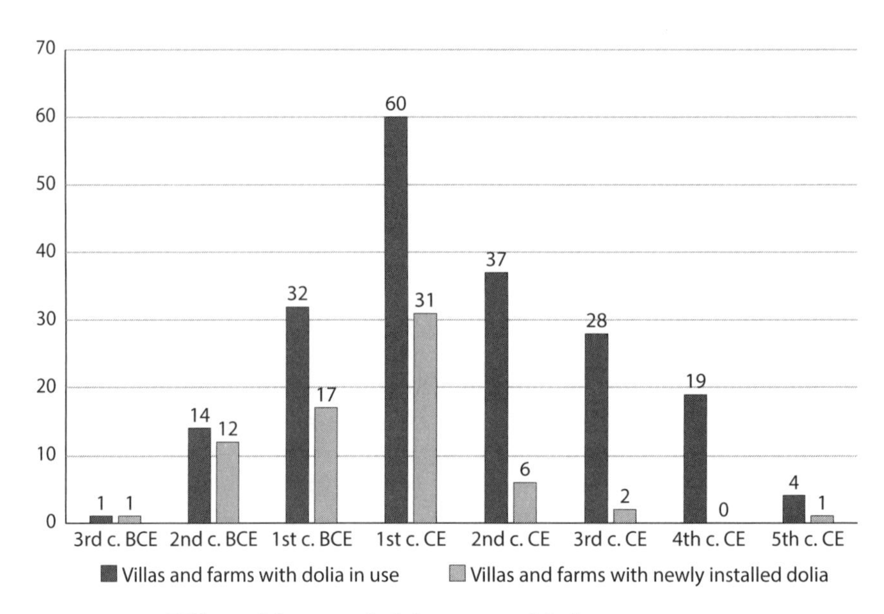

FIGURE 3.4. Villas and farms with dolia in central Italy.

of these cellae varies from very small rooms to several enormous ones, from an alcove with just a couple of dolia defossa to a massive cellar of several hundred square meters. The range in cellae vinariae's size likely reflects the range of estate sizes, the estate owner's investments, and perhaps cultivation strategies more broadly in central Italy. Although many elite villas are clustered around Latium, the overall discrepancy in cella size was not restricted to any region. In Boscoreale, for example, villas with as few as four dolia (Giuliana) and cellae as small as seventeen square meters (Stazione Ferrovia) were located in the vicinity of Villa della Pisanella, which had a total of ninety dolia in a cella vinaria of nearly two hundred square meters. Lest we assume that the number of dolia directly correlated to the estate owner's wealth or level of production, we should keep in mind that one could own multiple properties.[56] Various ancient authors, such as Pliny the Younger, noted that investing in several estates in different microclimates was much more prudent than purchasing a single expansive estate.[57] "I fear it is imprudent to expose so much property to the same climate and the same calamities. It seems safer to endure the uncertainties of fortune by distributing one's possessions into different situations" (Plin. *Ep.* 3.19: *contra vereor ne sit incautum, rem tam magnam isdem tempestatibus isdem casibus subdere; tutius videtur incerta fortunae possessionum varietatibus experiri*). If bad weather were to strike, one could lose an entire year's harvest in an instant. Nonetheless, the level of investment could be remarkably high. At Villa della Pisanella, a large open courtyard outfitted with eighty-four dolia served principally as a wine cellar while the adjacent room, next to the *trapetum*, contained a row of five olive oil dolia along the wall (Figure 3.1C); the agricultural equipment reflected high levels of production, while the residential part of the estate and

56. On difficulties in identifying the occupants and owners of farms, see Foxhall 1990.

57. On economic thought behind Pliny's letters, see Kehoe 1988.

the treasure trove of silverware and coins discovered further underscored the owner's wealth and status. This was a villa that belonged to an especially affluent person, who prioritized both urban amenities and agricultural production.

The installation of dolia, and in such large numbers, not only allowed farms and villas to produce much more wine and more efficiently but also opened new possibilities. One obvious advantage of using dolia for agriculture was sheer quantity. As a simplified case, with one dolium capable of holding one thousand liters, a farm could process and store one thousand liters of wine. With two, a farmer could double the annual product and consider regularly selling surplus. Most of the excavated cellae vinariae contained under a dozen dolia, but several were massive. By the first century BCE, some villa owners in central Italy were installing large sets of dolia, with some of the larger estates capable of processing and storing upward of one hundred thousand liters of wine. Dolia squeezed an incredible amount of wine into small spaces and, when buried to their shoulders, allowed not only greater concentration but also organization. Workers could navigate the storeroom more easily between vessels and keep track of their contents. The Villa Regina's eighteen dolia were arranged in three rows of six dolia, and their volumes were marked on the rim or shoulder (Figure 3.1D). Villa della Pisanella grouped dolia in the cella vinaria so that a wide path divided the room in half, and small spaces separated several clusters of eight or nine dolia on one side and several groupings of ten or eleven dolia on the other (Figure 3.1C). Other cellae vinariae might have been assembled into different areas, or even separate rooms. The cella vinaria of the Villa Magna set the thirty-eight dolia into four groups, whereas the Villa i Medici in Stabiae placed the wine storage jars in two separate rooms, one for large dolia, presumably for more common wines, and one for smaller storage jars, presumably to store higher-quality wines or create batches of wines using different recipes (Figure 3.1E). Distributing wine in two distinctive rooms, with at least two types of storage jars, also pointed to calculated strategies in wine production and sale.

These vessels also offered opportunities to vintners to add qualitative touches to their wine, such as creating different batches of wines modified by additives like marble dust and resins. Ancient consumers had different preferences for wine, often associated with place. At the same time, agronomists such as Cato, Columella, Palladius, and Pliny provided numerous recipes to mimic particular wines. One of Cato's recipes re-created the popular Coan wines by using seawater so a Roman farmer could imitate the famed wines of the east.[58] Pliny the Elder advised plunging beet leaves into spoiled wine to restore flavor, and other authors noted ways to sweeten the wine or remove unwanted odors.[59] Chemical analyses of wine jars have identified aromatics such as rosemary, thyme, and basil added to wines.[60] Immersing different herbs, spices, honey, and other substances masked any vinegary flavors and smells, which could easily develop in aged wine.[61]

Using dolia offered not only more control over the fermentation of wine but also strategic storage. By installing dolia in particular ways, farmers harnessed natural resources

58. Cato *Agr.* 104–115.

59. Columella *Rust.* 12.27; Palladius *Opus agriculturae* 11.14–19; Plin. *NH* 14.19.

60. McGovern et al. 2013. See also discussion in Dodd 2022, 468–469.

61. Dodd 2020, 56–59, 115–116; Van Limbergen 2020.

and available craft technology to optimize storage conditions for wine and oil, sometimes for long-term storage and even aging. After the process of fermentation, vintners, merchants, and consumers still needed to keep wine storage conditions stable and cool. Vintners probably sold most of their wines after fermentation was complete, and wine sale arrangements stipulated emptying dolia before the next harvest. Wine prices could fluctuate wildly, however, and vintners with more dolia in their cellars could opt to hold on to a portion of their wine in order to fetch higher prices, whether it was through aging a nicer vintage or selling when prices were climbing while supply was low.[62] Falernian wine, for example, was known as an exceptionally high-quality wine and for its value, which would increase every year, aged up to twenty years.[63] A passage by Pliny the Elder reveals how having the right number of vessels, and having prepared them at the right time, could affect vintners' profits:

> The right time of the vintage is the period of forty-four days from the equinox to the Pleiades [late September to October]; a saying of the organizers goes that from that day to treat with pitch a cold vessel is for nothing. But now I have seen even on the Kalends of January vintners at work due to a shortage of vessels, and must being stored in tanks, or previous vintages being poured out to make room for new wine of dubious quality. This happens not so often due not to an abundant crop as to sluggishness, or else to the greed of those lying in wait for rising prices.
>
> iustum vindemiae tempus ab aequinoctio ad vergiliarum occasum dies xliv; ab eo die oraculum occurrit frigidum picari pro nihilo ducentium. sed iam et kal. Ian. defectu vasorum vindemiantes vidi piscinisque musta condi aut vina effundi priora ut dubia reciperentur. hoc non tam saepe proventu nimio evenit quam segnitia aut avaritia insidiantium caritati. (*NH* 18.319–320)

As other agronomists have said as well, dolia and other wine vessels had to be prepared in time for the vintage. Pliny here recounts an episode where, due to either procrastination or avarice, vintners ran out of vessels and resorted to storing the must in tanks and perhaps even pouring *out* (*effundi*) the previous year's harvest that they had been holding on to for prices to rise. These were desperate measures, as tanks offered little protection for wine and pouring out old vintages jettisoned potential profits. Pliny ends with a piece of advice—to use the annual harvest each year—suggesting that this practice of holding on to wine for profitable prices was common, but also risky, in line with Roman advice against taking risks.

Long-term storage could be problematic because unfavorable conditions accelerated the process of wine turning into vinegar, which seems to have been a common problem especially with ancient wines that tended to be more vinegary than wines today.[64] With the right storage facilities, vintners could ferment and preserve the wine. Vintners had the option of aging their wines by placing them in special rooms that warmed or "smoked" the wines,

62. Plin. *Ep.* 8.2; *Dig.* 33.6.8, 33.17.1; Van Oyen 2015a, 2019, 2020b; Purcell 1985; Horden and Purcell 2000; Garnsey 1988.

63. Varro *Rust.* 65; Plin. *NH* 14.57.

64. Frier 1983.

lighting incense, or setting aside the wine in dolia of the cella vinaria.[65] A cella vinaria was not necessarily built for a single harvest. Astrid Van Oyen has argued that the somewhat limited capacity of the fermentation vats and the elongated layout and narrow corridor of the cella vinaria at the Villa Settefinestre could not have accommodated processing and storing an entirely new harvest every year; instead, some dolia were probably extras, installed for wine beyond the annual harvest and to age vintages (Figure 3.1F):[66]

> Settefinestre again provides a classic example, with a narrow, elongated semi-subterranean space measuring 44 × 4 m and containing no fewer than 56 sunken dolia. At Settefinestre, the maneuver space in between the vessels did not exceed 25–50 cm. As a result, transferring wine in and out of the dolia would have been even more laborious than usual, and was probably not a regular activity. The layout and organization of the storeroom promoted long(er)-term storage, in particular for the least accessible groups of dolia. This would have underwritten the creation of qualitative differences through the segmented process of production and storage: some dolia would have held their wine for a longer period than others, creating aged wines of high value and commensurate prices.[67]

Larger cellae vinariae and a greater number of dolia gave the vintner more options, but it was crucial to have high-quality storage equipment and facilities. With enough well-made dolia, vintners could seize opportunities not only to produce large quantities of wine but also to make a profit by strategically saving and aging batches of wine and waiting to sell when prices were favorable. The Roman villa as an architectural type thus seems to have featured dolia early on, and the dolia grew to feed into a landowner's arsenal of symbols of wealth and status.

Celebrating Surplus

Central Italy in the final three centuries BCE into the first century CE underwent several changes, including improved food storage. Bigger sets of larger dolia were installed across central Italy, increasing not only the quantity of wine production but even the range of wines vintners could produce, cultivating qualitative distinctions. Dolia were large storage vessels specially designed for wine production and storage, but they became associated with productive agriculture and emerged as the defining feature of not only wine and oil cellars but villas more broadly. Installed in large numbers in villas across central Italy, dolia worked as the engine of agricultural growth and surplus. Their use entailed a specialized setup, taking over and defining entire rooms and areas of an estate. Yet dolia significantly expanded and could potentially diversify an estate's viticulture. Their utility and value contributed to a growing trend of conspicuous production and storage across central Italy.

The archaeological evidence of dolium use for viticulture coincides with other advancements in agricultural technology, especially food storage. As Van Oyen has noted, during this time, storerooms for dry goods, especially grain, became more distinct and articulated,

65. Columella *Rust.* 1.6.19; Plin. *NH* 14.27; Mart. *Ep.* 10.36. See also Dodd 2022, 468.

66. Van Oyen 2020b, 52–53.

67. Van Oyen 2020b, 51–52.

and some were built as separate, stand-alone buildings in urban areas and near areas of production in the case of villas. Storage facilities became larger and better constructed, fitting the patterns of trade and consumption. Utility and luxury in villas were not necessarily exclusive to each other, and dolia came to represent different cultural values. While some grand villas certainly kept a more opulent area that reflected urban amenities (*pars urbana*), the productive part of a villa (*pars rustica*) was usually not isolated or far from the rest of the living quarters. The Villa of the Mysteries, for example, is known for its wall paintings and fine decoration, yet just off the peristyle courtyard in the center of the villa was the press room, where grapes were trodden and pressed before undergoing fermentation and storage in an adjacent cella vinaria (Figure 3.1G).[68] Conspicuous production and storage, as Van Oyen has shown, played into Roman elite self-fashioning as *boni agricolae* with a moral bent, and took off with large, diversified, and specialized storage in the countryside.[69] This was not just limited to wine storage and to dolia. The proliferation of dolia-laden cellars worked in tandem with the monumentalization, diversification, and centralization of other storage structures such as granaries, storerooms, and cisterns. With these larger, distinct storage structures, villa owners were able to stock up on and display valuable foods and supplies. This all drew on considerable resources, not only the land and equipment (dolia included) for viticulture but also the (often specialized) labor and knowledge to operate them. It is probably no surprise, then, that wine industries of such an enormous scale in central Italy are often at elite villas, but even more modest estates had a hand in the game. The Villa Regina, for example, was a small-scale farm, likely leased out to tenant farmers; the owners chose not to invest in renovations after the farm was damaged by the earthquake of 62 CE but instead employed a workforce that could hardly maintain a clean space: workers urinated in a makeshift urinal composed of an old amphora with a hole cut into it next to the hearth of their kitchen.[70] Though the living quarters would not have been suitable for someone like Pliny the Younger, the modest farmhouse was still capable of producing ten thousand liters of wine per year, though the capacity of the eighteen dolia likely represents maximum storage for exceptional bumper years or attempts to reserve wine for long-term storage. At ten thousand liters of wine, this could have been enough to supply at least forty people for an entire year.

On the other end of the spectrum, these productive spaces and equipment were celebrated and decorated. The Villa della Muracciola in the *suburbium* of Rome, for example, presented the dozen dolia defossa to the winery's visitors almost in a theatrical way, in rows along the back wall of a semicircular room.[71] Some spaces were even more extravagant. Villa Magna, an enormous imperial villa in the countryside that passed through the hands of the emperors Hadrian, Antoninus Pius, and Marcus Aurelius, featured a winery with a marble *opus spicatum* (masonry technique where bricks or colored stone are arranged in a herringbone pattern) floor and had thirty-eight dolia festooned with serpentine

68. See Wallace-Hadrill 1998 for the intersection of morals of the *pars rustica* and *pars urbana* of the Villa of the Mysteries.

69. Van Oyen 2015a, 2020b, esp. 50–53.

70. De Caro 1994, 50.

71. Purcell 1995, 157–173; Feige 2021, 36–37.

(Figure 3.1E).[72] Just above the cella vinaria was a raised platform containing the treading vat, while on the opposite side of the winery was a semicircular courtyard with a direct view of the cella vinaria and the treading vat. The winery, perched on a hill, celebrated viticulture and projected the emperor's annual sacrifices to Jupiter at the beginning of the Vinalia (grape harvest), with the productive space as a backdrop. Another recently excavated cella vinaria of an imperial villa, the Villa of the Quintilii just outside the ancient capital and on the Via Appia, was found to have an ornate and highly visible cella vinaria.[73] As Emlyn Dodd recently noted,

> Performative and conspicuous production occurred at the Villa dei Quintili in a slightly different sense from that at Villa Magna. Here, must flowed from a *calcatorium* [press room] and two presses (via a settling tank) into the *cella vinaria* and dolia through a series of marble-faced *canali* [channels] and a facade with fountains in a quasi-nymphaeum-like arrangement. Three luxurious rooms paved in multicolored *opus sectile* [pavement inlaid with colored stone] surround the *cella* and may have enabled residents and guests of the emperor to watch the spectacle of production. Even the treading floor is partially clad in red breccia marble. Indeed, the *cella vinaria* itself is a luxurious space commensurate with the broader context of the Villa dei Quintili, not in terms of scale but in architectural quality, where raised walkways separate rows of dolia (also at Villa Magna) and afford winemakers, or casual viewers, a pleasurable and opulent experience generally unheard of in *pars rustica*.[74]

Production and storage, especially at a large scale, was commended and even decorated. At the Villa of the Quintilii, the entire wine production area was embellished with colorful, polished stone and luxurious architectural features. These villas combined the luxury typically associated with posh estates and the productive, utilitarian areas of the farm, presenting the viewer with an array of symbols that praised the good (and profitable) farmer. Both the Villa Magna and Villa of the Quintilii demonstrate that not only local elites participated in showing off conspicuous production and storage; even the imperial family did too.

Some of these spaces found ways for the luxury typically associated with the *pars urbana* to embrace the functionality of the *pars rustica*. As Michael Feige observes in his survey of decorated Roman villas, some productive rooms also advertised the cleanliness of the viticultural equipment, a major source of pride for Roman agronomists. The productive spaces of the Villa of Russi of Ravenna did not feature the same theatrical architectural layout as the Villa Magna or Villa of the Quintilii, but the pressing floor and vat were paved with a surprising choice: marble.

> A pavement made of brightly shining marble provided an inventive way to visually underline and exaggerate the impression of a carefully cleaned surface; it thus attested to the good maintenance of the winery. Consequently, the *décor* here perhaps functioned as a means of advertisement. One possible occasion on which this might have occurred was a demonstration of the production facilities in the context of trade

72. Fentress et al. 2017, 203–210.

73. Dodd et al. 2023; Dodd 2022.

74. Dodd 2022, 472, figure 16.

FIGURE 3.5. Scene of cupids sampling wine, House of the Vettii (VI.15.1), Pompeii, by Gary Todd, 2000.

> negotiations with potential buyers. . . . The owner of the Villa of Russi, through the application of cost-intensive material *décor*, probably tried to demonstrate to the buyers his extraordinarily high degree of attention to the care of his equipment and goods.[75]

Good hygiene was paramount to winemaking, and ancient authors provided strict instructions. One was to keep the winery away from the dung heap, baths, and oven—any place with strong, and foul, smells.[76] While certain additives helped develop the aromas of wine, some were notoriously bad for wine.[77] Other substances could weaken the vintage. Strong smells would influence the flavor and aroma of the wine's developing tastes. Equipment maintenance, too, ranked high among the agronomists' priorities, and regular cleaning of dolia was crucial. To ancient agronomists and vintners, clean equipment formed the foundation of successful winemaking. In the case of the Villa of Russi, then, the choice to pave the winery with marble not only highlighted a valuable stone and the owner's wealth but also conveyed the cleanliness of the winery and good winemaking practices to visitors, including potential customers and merchants.

Most, if not all, of the farms and villas with dolia defossa that this chapter has discussed practiced viticulture beyond household consumption. Ancient textual and visual evidence shows that buyers could personally visit the winery and sample the wines. A fresco painting in the House of the Vettii in Pompeii, for example, shows cupids sharing wine (Figure 3.5). One interpretation is that two cupids, near rows of amphorae, exchange a patera containing

75. Feige 2021, 42–43.

76. Columella *Rust.* 1.6.9.

77. "The flavor of wine in a dolium is spoiled by cabbage" (Plin. *NH* 19.135: brassica corrumpatur in dolio vini sapor); see Cheung 2021a, 678–679.

wine from the amphorae. To their right, two other cupids work together to pour a small amount of wine from an amphora into another patera, presumably providing the potential customer a sample of the wine before concluding the sale.[78] Legal sources also stipulated that buyers had the right to sample wine (*degustatio*) before completing a transaction to ensure that the wine had not turned into vinegar or spoiled in some other way, and in fact, a number of sample vessels have been identified in the archaeological record.[79] The great investment and value in developing and maintaining these specialized cellae with dolia allowed farmers to produce and store at a much larger scale for the market. Local towns and markets were obvious targets, but Rome, with its large consumer base, was also a desirable market. In fact, accounts praising and criticizing vintages from various regions show that Romans had grown familiar with wines made across central Italy and beyond.

78. De Angelis (2011) and Monteix (2016) caution against a straightforward reading, instead suggesting this exchange was between fictive characters for a feast that embellished the triclinium where it featured.

79. On legal writings about *degustatio*, see Frier 1983. For a survey of sample jugs, see Djaoui and Tran 2014; Djaoui 2020.

4

Dolia Abroad

INNOVATIONS IN TRANSPORT

> First place is given to the Aminaean vines, on account of the body of that wine and its essence, which certainly improves with age. There are five varieties of these vines. . . . The next rank goes to the vines of Mentana, with red wood, on account of which some have called them the "ruby vines." These are less fertile, as they produce too much husk and lees, but they are most vigorous against frost. . . . Bees give their name to the "bee wine" because they are particularly fond of it. . . . The wines are at first sweet but acquire harshness over the years. Tuscany delights in no wine more than this one. So far the chief distinction is given to the vines particular and domestic to Italy.
>
> principatus datur Aminneis firmitatem propter senioque proficientem vini eius utique vitam. quinque earum genera. . . . proxima dignitas Nomentanis, rubente materia, quapropter quidam rubellas appellavere vineas. hae minus fertiles, vinaceis et faece nimiae, contra pruinas fortissimae. . . . apianis apes dedere cognomen, praecipue earum avidae. . . . vina primo dulcia austeritatem annis accipiunt. Etruria nulla magis vite gaudet. et hactenus potissima nobilitas datur peculiaribus atque vernaculis Italiae.
>
> PLINY THE ELDER *NATURALIS HISTORIA* 14.21–24

WHEN PLINY the Elder was writing in the first century CE, Italian vintners produced a range of wines, from the fine Setinum wine that Augustus favored to cheap Pompeian wine that could give one an excruciating headache the next morning. These wines became well known as they were traded and brought to different markets and consumers. Italian wines had earned a reputation not only among Romans but also abroad with customers extending across Italy and other areas of the Mediterranean. Long-distance trade had been practiced across the Mediterranean basin for centuries before the Romans, but the scale, distance traveled, and sheer variety of foods moved from the second century BCE onward were unparalleled.

To move so much wine, which was often made and stored in dolia, whether it was from a farm to local markets or around and across the Mediterranean, was no small undertaking. Ships loaded with wine crossed the seas and navigated rivers, even in precarious conditions, and were sometimes overtaken by storms to sink into the wine-dark sea. Carts and wagons groaning under the weight of the beverage traversed the countryside between farms, villas, and markets to reach more inland customers. Containers mattered too. Potters regularly manufactured large orders of amphorae and other jars so traders could bottle, move, and sell wine (wine dolia were too heavy and fixed to move). Porters working along the docks hoisted jar after jar of wine onto their shoulders as they moved the vessels in and off ships, boats, and carts. The thousands of amphorae found throughout the Mediterranean basin and beyond testify to the extraordinary apparatus of food shipment, one that lasted for centuries, if not millennia.[1]

An undercurrent of trials and errors, including a new way to use dolia, ran beneath these operations. This chapter both considers how dolia slotted into a traditional container system and charts an innovative and highly specialized way traders used dolia to deliver wine and connect markets in the western Mediterranean: in specially designed bulk transport ships where dolia were installed directly into the hull. This specialized ship, which could only transport one type of commodity (wine), emerged as the product of an ambitious group of entrepreneurs looking to move more wine quickly. To understand the significance and financial impact of this innovation, we first need to step back and situate it within the broader history and development of packaging and transporting wine (and olive oil) in the Roman world, which was becoming increasingly specialized and efficient. (Olive oil also required packaging, and often in the same types of containers as wine, but this chapter will focus on the transport of wine to highlight a new use for dolia.) The production and use of containers able to withstand long journeys had long been entangled with large-scale trade of agricultural goods, but they reached an unprecedented level of specialization under the Roman Empire. Within this specialized container system, dolia provided new opportunities for more massive and effective wine transport and profitable packaging.

Breaking into Markets

Cato's agricultural treatise advised estate owners to sell whatever they could after reserving provisions for the household, including wine.[2] The last grape pressings, for example, could be diluted with water and served as a cheap beverage to the farmhands, but the rest of the wines could be sold on the market. As the previous chapter discussed, farms and villas across central Italy installed dolia to practice more large-scale wine production starting in the second century BCE, if not earlier, the same time Cato was writing. Farmers could, and often did, sell their products locally. It was expected of them. But how did they conduct these sales? André Tchernia has pointed to wholesale as the most common way, though there was an exception:[3] farms and agricultural estates could sell smaller quantities of wine and other products directly to the consumer if the property had an adjoining or adjacent shop, tavern, bar, or inn.[4] Travelers stopping by an inn could have wine with

1. For a *longue durée* overview of containers in the Mediterranean, see Bevan 2014.

2. Cato *Agr.* 2.

3. Tchernia 2016, 140–149. For discussion of moving agricultural produce, see Morley 1996, 55–82. On wholesale retail, see Holleran 2012, 66–98.

4. Varro *Rust.* 1.2.23; *P.Oxy.* 59.3989; Vitr. *Arch.* 6.5.2; Tchernia 2016, 140ff.

their dinner before retreating for the night, while a local resident passing a shop or bar could pick up a jar to take home. The Villa i Medici in Stabiae, for example, had a roadside wine tavern connected to two cellae vinariae, in which the tavern keeper could draw the wine directly from dolia. Farmers could also cart and sell their wine and other goods at local, periodic markets such as the *nundinae* (scheduled markets that met every ninth day) in Campania and Latium. Packaging, moving, and selling produce was logistically demanding, requiring significant materials, equipment, and labor (both human and animal). Some vintners opted out of selling their products at markets entirely. Selling wine to merchants was the least risky option for the landowner, and was probably a common choice. Estate owners often sold their wines wholesale to merchants who bottled, moved, and sold the wine.[5] An agreement could be reached before the wine was even ready, as merchants bought the standing crop or vintage in advance and estate owners leased the vineyard to tenants who cultivated and processed the grapes.

Transporting food products depended on not only adequate vehicles but also the coordination of labor and the proper containers. Using dolia for storage relied on transfers from the dolia to other containers (Figure 1.6). Because dolia were fixed equipment, and often considered architectural elements, farmhands had to move their contents to other containers by manually pumping, transferring, or even ladling the contents.[6] Cato, for example, instructed estate owners to provide ladles in wine and oil rooms, while ladles and jugs have been found in storage rooms in Italy. Transferring liquids was a time-consuming and repetitive job given that a single dolium would hold several hundred, even over a thousand, liters and probably required hours, if not an entire day, to empty just a single large dolium. Transferring the wine from a large cella vinaria, such as the one with over eighty dolia at the Villa della Pisanella, must have been time consuming and labor intensive, but there was no other alternative for moving the wine for sale (or even general household consumption). The dolium-based storage system relied heavily on complementary containers. Choosing the receptacle into which the dolia's contents were placed hinged on several factors, such as their cost and availability, the contents, the volume, the mode of transport, and the degree and duration of protection needed.

People often turned to reusable animal-hide containers to move wine from dolia to local markets or bottling facilities.[7] Animal hides were widely available in antiquity since their production could be done at various scales: at home when animals were slaughtered or, at the other end of the spectrum, through more elaborate tanning processes in workshops.[8] Although producing animal-hide containers was originally a rural activity, leather was such

5. Plin. *Ep.* 8.2; Kehoe 1989; Tchernia 2016, 140–149; Shaw 2019. In Ptolemaic Egypt, estate owners themselves often produced the wine and sometimes also pruned vines; Kloppenborg 2006.

6. On ladles, see Cato *Agr.* 10.2, 11.3, 13.2, 66; Columella *Rust.* 12.52.8–12. Cato *Agr.* 66 advises placing a ladler (*capulator*) in the oil press room to skim off the *amurca* with a ladle. On jugs, see Pasqui 1897, 492. Varro *Ling.* 5.177: parallel between use of *multa* (fine) because magistrates impose one fine after another and how countrymen fill *dolium aut culleum* with wine.

7. Skin containers are rarely preserved, and much of what we know comes from textual and iconographic sources and ethnographic studies. See Marlière 2002; Churchill 1983; Borowski 1998; van Driel-Murray 2008.

8. Van Driel-Murray 2008, 485. On the tanner (*coriarius*), see Bond 2016, ch. 3. Tanning workshops have survived at Pompeii, Vindolanda, Saepinum, Timgad, and Vitudurum. On fulling, see Flohr 2013a, 2013b.

a widely used material during the Roman period that urban workshops often focused on its production, and certain regions became notable for their products.[9] Skin containers came in a range of sizes and costs. Small goat-hide containers known as *utres* were probably available universally, and their flexibility made them ideal for transport by cart and pack animals, as they could be secured and fastened to rest against the sides of animals, but their small size designated them as personal containers. Ox hides known as *cullei* were immensely useful for bulk overland trade and transport, capable of holding large volumes equivalent to twenty amphorae (20 amphorae × 26.1 liters = ca. 520 liters = 650 bottles of wine). Despite their massive size, cullei were both flexible and lightweight, making them the most efficient (ratio of volume to weight) containers in the ancient Mediterranean (Figure 4.1). Ox-hide containers were so bulky, however, that they were not portable and depended on vehicles, and sometimes specialized transporters known as *utriclarii*, to convey them;[10] because of their organic material (susceptible to getting eaten by vermin) and their propensity to burst when stacked on one another, cullei were not used for maritime trade. Still, ancient authors regularly listed ox-hide containers among farm equipment; the third-century jurist Ulpian, for example, includes cullei among pack animals, ships, vehicles, and barrels as equipment for exporting produce.[11] Cullei were expensive investments, though; according to Diocletian's *Price Edict*, a culleus cost up to 400 denarii, while the material for a goat hide cost, by comparison, a paltry 20–50 denarii.[12] With the cullei's high cost and legal status as farm equipment, merchants were expected to transfer the wine into other containers and return the ox-hide containers to the farm.[13] Cullei were thus used only for short-term purposes and for local, primarily overland, transport and trade.

Given how perishable animal-hide containers are, their absence from the archaeological record does not preclude that they were likely used by local estates to supply the city of Rome with wine and olive oil.[14] A slew of evidence suggests merchants packed wine into cullei and transported them to Rome. Excavations and surveys of the hinterland of Rome reveal an area of intensively cultivated agricultural estates, which potentially supplied as much as one-third of the wine (54 million liters) and a quarter of the olive oil (4.7 million liters) Rome consumed in a single year.[15] Texts by agronomists such as Cato and Columella as well as various iconographic representations suggest that agricultural estates

9. Pliny *NH* 9.5.14–15: Claudius killed a whale feasting on shipwrecked leather hides from Gaul.

10. Columella *Rust.* 3.3.10; Cato *Agr.* 105, 154; Ulp. *Dig.* 33.7.12.1. See the fresco from the Praedia of Julia Felix in Pompeii showing traders bringing in a wineskin on a cart. The term *utriclarius* has been found in several inscriptions in Gaul, but the occupation is unclear and could refer to boatmen who used inflated skin containers in transport or, more commonly accepted, professionals who transported liquid goods in skin containers. Cf. Kneissl 1981; Deman 2002; Marlière 2002, 18ff.; Leveau 2004; Liu 2009, 136–138.

11. Diocletian's *Price Edict* 8.6–12. Ulp. *Dig.* 33.7.12.1. *Mishnah Nezikin Abodah Zarah* 2.4 mentions both wine skins and jars. The parable of new wines in old wine skins (Matthew 9:14; Mark 2:22; Luke 5:37–38) suggests new or fermenting wine could burst older (brittle) wine skins.

12. Columella *Rust.* 3.3.10: a culleus equals twenty amphorae or forty *urnae*. Churchill 1983: goat-skin containers had a capacity-to-weight ratio of ca. 3 liters/kg and cowhide containers ca. 65 liters/kg; these were much higher than Dressel 1 (0.88 liters/kg) and Dressel 2–4 (1.09–2.04 liters/kg) amphorae.

13. Ulp. *Dig.* 33.6.3.

14. De Sena 2005; Marzano 2013a; Komar 2021.

15. Marzano 2013a.

FIGURE 4.1. Watercolor by Giuseppe Marsigli, 1828, depicting workers filling amphorae with wine brought in a culleus, Osteria della Via di Mercurio (VI.10.1), Pompeii.

transported their wine and oil in animal-hide containers en masse to local customers;[16] this explanation would at least partly reconcile the paradoxical lack of amphora finds and workshops with the presence of many wine and oil presses in Rome's hinterland, as well as demonstrating how the capital could be supplied.[17]

Some merchants needed different kinds of containers to bottle nicer vintages or to travel to more distant markets in hope of selling their goods at favorable prices. Prices for certain foods and beverages, especially wine, fluctuated, offering room for price speculation and high profits.[18] Merchants engaged in long-distance trade across the Mediterranean basin for centuries before the Roman period, but after the Second Punic War, around the same time dolia were used in central Italian villas, the exports of Italian goods intensified.[19] During the second century BCE, lucrative markets abroad offered many opportunities to Italian vintners who were beginning to mass-produce wine.[20] Moving wine to faraway markets incurred transportation costs and risks but could tap into unprecedented profits if competition was low. Gaius might have arrived with twenty jars to sell before he could proceed to his next destination, but Felix was also looking to unload and sell his wine, and the two could lower their prices in attempts to lure a customer. Communication

16. Columella *Rust.* 3.3.10; Cato *Agr.* 11, 148, 154.

17. Marzano 2013a; De Sena 2005. While most opus doliare workshops in Tuscany and Campania that produced dolia also produced amphorae, the overlap is nearly nonexistent in Latium; Olcese 2012. Cf. Marlière and Torres Costa 2007.

18. Purcell 1985.

19. Tchernia 1986; 2016, 286–296.

20. Purcell 1985. On Italian agriculture of this period, see Roselaar 2019, 2020.

in antiquity was slow and unreliable, though, and traders targeting a market they learned had favorable prices might arrive to find that the situation had already changed.[21] On the other hand, if consumers had a thirst for wine but had to purchase wines made abroad, they were at the mercy of visiting traders. Markets in areas with limited local viticulture but a demand for wine thus became popular destinations for traders who went to extraordinary lengths to package and move wines farther afield.

One of the most important practicalities of long-distance trade was to move and protect the goods. Selling wine at local markets could be time consuming and remove both labor and equipment from the farm, but moving wine so far afield posed different challenges. With overseas journeys lasting weeks, potentially months, traders had to protect and store wine properly while in transit.[22] Animal-hide containers played a vital role in the supply chain of packaging for wine for overland transport but were not effective for overseas trade. Because of their material, they were prone to wear and tear and vulnerable to hungry rats and vermin that could chew through them and destroy entire shipments of goods.[23] Furthermore, an ox-hide container, generally conveyed on a wagon, did not offer structural or ergonomic support for shipments. They were difficult to move and arrange since they did not have handles for porters to grab onto to adjust and position them. Porters could not stack them either, or they would burst from the pressure of excess weight. What merchants needed was another container that complemented the dolium-based storage system, one that porters could easily move and maneuver and that could also protect its contents from light, air, extreme temperatures, and pests.

For lengthier trips, and especially long-distance trade, merchants turned to another type of ceramic container: amphorae, the seemingly ubiquitous double-handled ceramic containers of the Mediterranean basin. As the most common and characteristically "Mediterranean" artifacts of the ancient world, scholars have studied them to quantify and trace long-distance trade. Merchants used amphorae from as early as the Bronze Age to transport goods in the Mediterranean, but variations and developments in their form and use took place over time.[24] Potters endowed amphorae with a distinct elongated shape, a narrow neck with two handles that connected to the shoulder, and a pointed base.[25] The standard wine amphora reached one meter tall with vessel walls about two centimeters thick, weighed eighteen kilograms when empty and over fifty when full, and even represented a standard unit of measurement (ca. 26.1 liters) (Figure 4.2).[26] With the thick vessel walls

21. Purcell 1985, esp. 3–5.

22. According to Stanford University's ORBIS project, a direct journey between Rome and Tarraco took eight days, while a journey between Antioch and Tarraco took twenty-eight days. Cabotage (shipping from port to port along coastal routes, rather than directed through large emporia) would add time.

23. In Carthage's olive oil storehouse, an inscription suggests the presence of a *coriarius* (leather worker) on-site who was responsible for the production and repair of leather packaging (*CIL* 8.24654 = *AE* 1890, 132). See Peña 1998; Marlière and Torres Costa 2007.

24. For a diachronic overview of Mediterranean amphorae, see Bevan 2014.

25. Moore 2011. Although customers often associated amphora shapes with certain contents and places, merchants also communicated key information to a buyer through labels, stamps, or tags. For amphora labeling, see Curtis 2015. For *tituli picti* (painted labels), see Liou 1987; Rodríguez Almeida 1989; Peña 2007b, 99–114. For lead amphora tags, cf. Lequément 1975; Ehmig and Haensch 2021.

26. Ulp. *Dig.* 33.6.3.

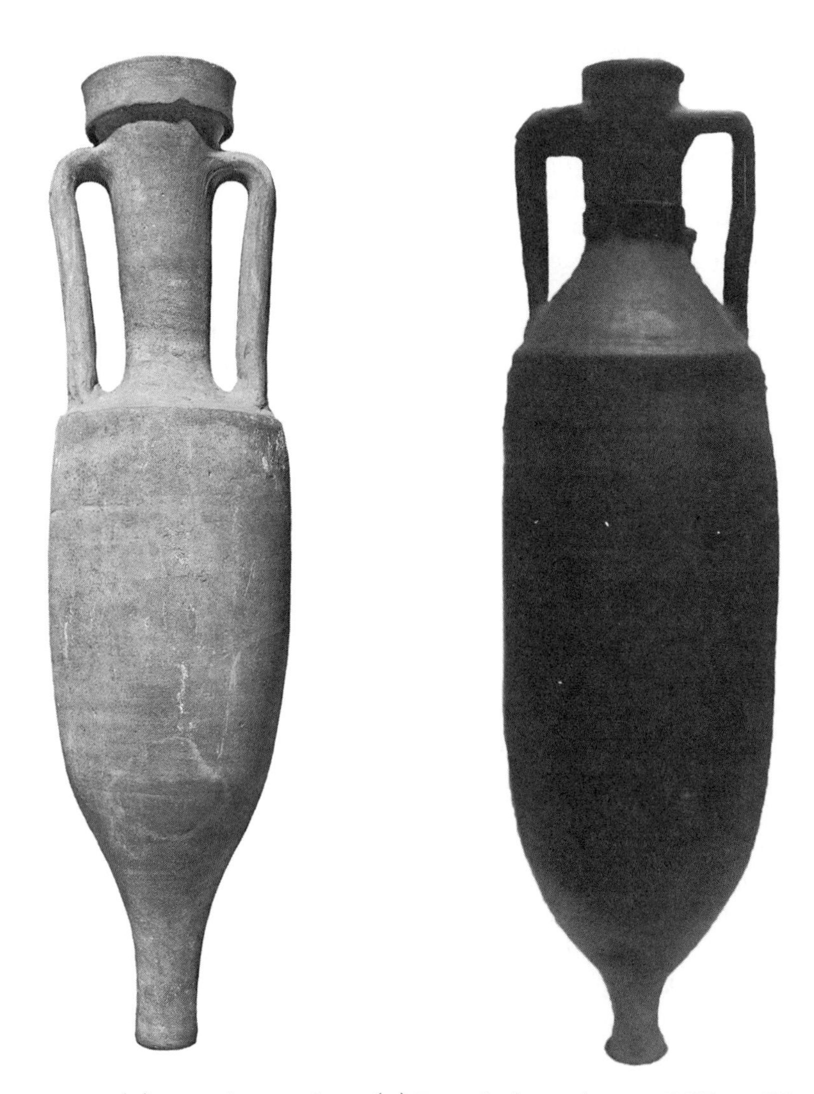

FIGURE 4.2. (*L*) Dressel 1B amphora; (*R*) Dressel 2/4 amphora, 25 BCE–79 CE, British Museum. © Trustees of the British Museum.

and pointed base, amphorae were designed to withstand long journeys and preserve their contents.[27] The ceramic material of amphorae insulated the product and blocked out light, though it was somewhat porous, so the inner surface was lined with pine resin to prevent the vessel from both leaking and tainting the wine.[28] When sealed properly with durable

27. Cheung 2020.

28. Plin. *NH* 16.21.52. *P.Oxy*. 3354 ll. 16–17: potters tested their wine jars. Olive oil amphorae usually remained unlined because the oil would dissolve the pitch and ruin the flavor of the oil, except for low-quality olive oil used for lighting.

plasters, it also kept out air and pests as well as guaranteed an untainted product.[29] Sealing the amphorae this way proved so effective that over two millennia later, underwater explorer Jacques Cousteau not only discovered the transport vessels intact at the bottom of the sea but also sampled the wine within.[30] The material and usage of amphorae therefore provided ideal conditions for storing wine (as well as other foods): lack of oxygen and light, low temperatures, and stable humidity levels. But amphorae were also designed for mobility. The pointed base aided in the transport and storage of the vessel. Porters could embed the vessel in soft ground when stored in warehouses or into slots built in ship hulls during shipment.[31] Stevedores carrying conventional amphorae generally placed the vessel on their shoulders, using one hand to hold on to one handle while the other could grasp the pointed base for extra stability;[32] treating the pointed base as a third handle also made pouring wine from the amphora easier.[33] With their design, availability, and low cost, amphorae dominated as the primary transport container in the ancient Mediterranean.[34]

Traders used great quantities of amphorae for millennia to transport wine and other foods.[35] From roughly 200 BCE to 200 CE, amphora potters produced an especially wide range of amphorae of varying shapes and sizes, many of which corresponded with certain places and products. The amphora type scholars call Dressel 10, for example, was known to have carried *garum* (fish sauce) from the region of Spain. Dressel 20, on the other hand, was a distinctive oversize globular amphora containing olive oil in bulk from the Guadalquivir Valley in Spain. Standing at sixty to eighty centimeters tall, the Dressel 20 amphora could carry seventy to eighty liters of olive oil, weighing eighty to one hundred kilograms when full, likely requiring at least two people to lift it. Potteries in the same region often featured their own containers that were so distinctive that ancient consumers often recognized and associated amphora shapes with different places and certain contents, and other potteries might even produce copycat amphorae to emulate containers of more valued vintages or refined olive oil.

The amphora's desirable qualities for transport extended into long-term storage. With their sturdy shape and material, tight closing and sealing, and ability to stabilize temperatures, they were often used to store wine. In fact, ancient authors discussed keeping wine amphorae in an *apotheca*, a storeroom on the upper floor, to age and store nice vintages of

29. Petron. *Sat.* 71.11.4: *amphoras gypsatas*. Egyptian merchants used mud clay to seal jars, but it degrades quickly. See Denecker and Vandorpe 2007; Vandorpe 2005; Nachtergael 2000, 2001, 2003; R.I. Thomas 2011, 2014; R.I. Thomas and Tomber 2006; Bos 2000, 2007; Mulder 2007; Davoli 2005; Bos and Helms 2000; Sundelin 1996; Minutoli 2014; O'Connell 2014. On (re)stoppering with clay and problems with the wine's status, see *Abodah Zarah* 5.3–5; I thank Dan-el Padilla Peralta for this reference.

30. Norton 1999, 228–229.

31. Amphorae found on the Madrague shipwreck were upright, leaning against each other in the hull.

32. McCormick 2012, 61–64.

33. See how the attendant pours from an amphora in the fresco in Cubiculum B, Villa Farnesina.

34. Brughmans and Pecci 2020: amphora (re)use was complex, varying regionally and chronologically. See also Bernal-Casasola 2015. For an example of difficulty in identifying amphora contents, even with epigraphic evidence, see Peña 2007a.

35. In the first century BCE, some ships held more than three thousand amphorae; Dion. Hal. *Ant. Rom.* 3.44.

wine. Columella outlines the rationale behind setting up an *apotheca* upstairs above areas with smoke to enhance the wines' flavors:

> Wine storerooms will be placed appropriately over these places from which ample smoke is usually rising, because wines age more rapidly when they are brought to a premature ripeness by smoke of a certain nature. On account of this there should be another floor to which they may be moved, to keep them from becoming tainted, on the other hand, by excessive smoking.
>
> apothecae recte superponentur his locis, unde plerumque fumus exoritur, quoniam uina celerius uetustescunt, quae fumo quodam genere praecoquem maturitatem trahunt. Propter quod et aliud tabulatum esse debebit, quo amoueantur, ne rursus nimia suffumatione medicata sint. (*Rust.* 1.6.20)

According to Columella, the smoke accelerated the aging process for finer wines, imparting certain flavors, but control remained vital to prevent tainting the wine. The *apotheca*, and the jars in that space, came to represent higher-quality and more aged wines: Horace, for example, mentions retrieving jars of nicer wine from the *apotheca* for dinner guests.[36] As the previous chapter discussed, cellaring and aging wine could greatly increase a vintage's value, and vintners could set aside those wines in dolia or, better yet, amphorae. Although they are generally considered transport containers, amphorae also offered insulation and blocked out light while their narrow necks (diameter of ca. 10 cm compared with a dolium's ca. 40–65 cm) minimized exposure to air. Since they were already packaged for transport, too, using amphorae for storage reduced the number of transfers and kept the wine sealed. Furthermore, the smaller quantities distributed the risk of wine spoiling. If a vintner placed wine into a five-hundred-liter dolium that would later leak or crack, that was a sudden and complete disaster as opposed to a couple of faulty or improperly sealed amphorae among twenty.

Merchants often treated amphorae as single-use vessels, and customers expected to keep the jars when they purchased the contents as consumers do today.[37] Some traders reused amphorae, but they could crack or break. Moreover, certain foods, such as olive oil, seeped into the porous ceramic walls of the vessels, eventually turning rancid, and made the vessels unreusable for food products.[38] Amphorae from Egypt and Villa B in Oplontis, Italy, also bear holes on their necks that some scholars have interpreted as punctures that merchants made to allow excess gas to escape as wine fermented on its journey to markets.[39] With the open sailing season in the ancient Mediterranean between the months of April and October and the grape harvest in August or September, merchants had only a narrow window to transport their products to markets farther afield before they would have to wait months until the next sailing season. (Justin Leidwanger and James Beresford have noted that, although during the nonwinter months "the seas are calmer, nights are shorter,

36. Hor. *Carm.* 3.21; see also Hor. *Carm.* 3.8, *Sat.* 2.5.

37. Proculus *Dig.* 33.6.15; Ulp. *Dig.* 33.6.3.

38. On reusing and discarding amphorae, see Peña 2007b, 69–192, 299–306; Lancaster 2005, 68–85.

39. On air holes, see Denecker and Vandorpe 2007. On tap holes, see Loughton and Alberghi 2015. On Oplontis amphorae, see Muslin 2019; Pecci et al. 2017.

visibility is more reliable, storms are milder and less frequent, and the weather is generally more predictable," this did not mean there was no sailing at all during the winter months.[40] Some traders might have attempted to transport wines when the sailing season was "limited" along localized routes via cabotage, but that was probably a risk few would take given wine's high values.) With the single usability of amphorae and the sheer volume of foods moved in the ancient world, farms and merchants therefore needed a steady supply of these containers.[41] Agricultural estates with access to clay beds, skilled potters, fuel, and kilns produced their own amphorae, with some able to operate at a scale large enough to supply neighboring estates.[42] Not all estates could support their own pottery production, though, and many procured amphorae elsewhere.[43] Long-distance wine trade went hand in hand with amphora manufacture, which offered its own economic opportunities and complemented the type of large-scale viticulture dolia supported.

Dressel 1 Amphorae and Trade: The Sestius Family's Amphora Enterprise

In the second century BCE, central Italy, with its large-scale viticulture and growing villa culture, was serviced by amphora workshops in regions producing reputable wines, such as Falernian and Caecuban. Thanks to felicitous evidence, both textual and archaeological, we know about one family with an amphora enterprise near Cosa, including not only how their amphorae were used and distributed but also the kinds of portfolio diversification they could pursue. The new colony and its hinterland provided ample space and opportunity for landowners to build large estates and invest in profitable production, often cultivated by enslaved workers. Farms and villas became more widespread in the *ager Cosanus*, especially in the areas south and west of Cosa from the early second century BCE onward.[44] Smaller farms and houses were probably already producing wine and other products for export in the first half of the second century BCE.[45] From the mid-second century onward, however, the landscape of production began to change as large villas with plenty of dolia began to dominate the countryside, some, such as the Villa Settefinestre, capable of producing as much as one hundred thousand liters of wine.[46] Cosa's port, a naturally small harbor on the promontory, also underwent numerous modifications, including the addition of wharves, piers, and a lighthouse, as more agricultural products were exported from the region.[47]

One family located just outside Cosa seized on the opportunity to produce amphorae for a growing wine market. Known as a wealthy family, the Sestii were also involved in

40. Leidwanger 2020, 63, 62–67; Beresford 2013.

41. For contracts of amphora production from Roman Egypt, see Gallimore 2010; Cockle 1981.

42. Varro *Rust.* 1.2.22: digging clay pits on an estate that benefited the farmer.

43. They procured them from independent potters, who had their own workshops, rented the space and equipment, or worked for a pottery workshop; see Peacock and Williams 1986, ch. 3; Gallimore 2010.

44. Fentress and Perkins 2016.

45. Celuzza 1985, for a house excavated at Giardino Vecchio.

46. Carandini 1985, 121, 173–177.

47. Gazda and McCann 1987, 137–155; Gazda 1987; Will and Slane 2019.

politics; prominent family members included Publius Sestius, a senator who was a friend and defendant of Cicero (*Pro Publio Sestio*), and Lucius Sestius, who sided with Brutus and was proscribed but later pardoned.[48] The Sestii were not the only ones in the amphora business, but their example is especially striking and visible in the archaeological record, highlighting both the role of amphorae in large-scale trade in the Mediterranean and the opportunities for wealthy families to diversify their portfolios, something we will continue to see with dolia in later periods. Although elite Romans spoke and wrote disparagingly about trade, they considered ceramic production a form of working one's property and extracting the property's natural resources (often almost parallel to agriculture), and hence a respectable activity.[49] The Sestius family's industry demonstrates just how large scale this type of economic activity could be, but we should keep in mind that this is just one family of many that profited from ceramic production, often in relation to viticulture.

During the heyday of the *ager Cosanus*, the Sestius family produced large quantities of wine amphorae, attested by stamps bearing their name, abbreviated as SES or SEST.[50] Their enterprise in large-scale pottery production is extraordinary, and our discussion will only scratch the surface of the abundant scholarship on this topic.[51] The Sestius amphorae were popular among vintners in the *ager Cosanus*. The makers of Sestius amphorae marked their products with stamps or pictorial symbols such as a trident, palm branch, or pointed star, perhaps to authenticate the amphora's high quality. Elizabeth Lyding Will's studies found that the amphorae share a remarkable homogeneity in their ceramic makeup, suggesting that the potters used a specific clay source in the vicinity and added particular materials to make the clay redder.[52] In early phases of the workshop, the potters developed and refined the forms of the amphorae as well, producing the Greco-Italic type, a type of amphora with an elongated form of earlier Greek amphorae that became prominent in the third to first half of the second century BCE.[53] The Sestius amphora workshop (as well as amphora workshops around Pompeii) later developed the Dressel 1 amphorae during the second half of the second century BCE, which quickly earned the title of the most commonly used amphorae for trade in the western Mediterranean (Figure 4.2L).[54] The large number of Dressel 1 amphorae have led scholars such as André Tchernia to suggest that central Italy had over one hundred workshops producing these amphorae for wine export; in fact, according to Greg Woolf, the Dressel 1 "represents the most widely distributed amphora type in the ancient world."[55]

48. For a full account of P. Sestius' life and Cicero's speech, see Kaster 2006, esp. 14–22. For Lucius Sestius and the reference to him in Horace *Odes* 1.4, see Will 1982a; Corbeill 1994.

49. See Varro *Rust.* 1.2.21–24.

50. Will 1979, 1982b; Manacorda 1978; O. Patterson 1982; Woolf 1992.

51. Fentress and Perkins 2016: at least fourteen individuals affiliated with the Sestii are attested on amphora stamps from the late second through the first centuries BCE. Vitali 2007: a large amphora workshop was north of Cosa in the modern town of Albinia.

52. Will 1979: mineralogical analyses also point to the purposeful addition of iron titanium oxides.

53. For an overview of Greco-Italic amphorae, see Will 1982b; for early studies, see Benôit 1961. Will 1982b: potters enlarged the Greco-Italic amphorae over time as they refined the form.

54. Dressel 1 amphorae were produced at kiln sites that produced Greco-Italic amphorae, and the *ager Cosanus* (and *ager Pompeianus*) was the primary site; Tchernia 1986, 42–48, 126–127.

55. Woolf 1992, 285.

In a span of about a century to a century and a half, thousands of Sestius wine amphorae reached markets across the Mediterranean, from Athens to Gaul. Central and southern Gaul was a particularly promising place to sell wine. With a long history of trade with Greek and Etruscan merchants from as early as the seventh century BCE, Gaul became a major market for wines, especially Italian wines, after the Second Punic War.[56] Although beer was the more common alcoholic beverage in Gaul, local elites consumed the exotic beverage at feasts that incorporated religious and ceremonious rituals to flaunt their social status. Matthieu Poux's survey of wine consumption in Gaul during the second and first centuries BCE reveals the important rituals associated with these feasts.[57] Many of these feast sites featured ditches or pits where ritual libations were made, as well as the scattered remains of wine amphorae that attendees smashed, chopped, and slashed. Poux observes that the Gauls' preference for red wine (while Italians consumed white) and their treatment of amphorae were probably due to an association of the red wine and vessels with communal ritual feasts, with red wine symbolizing blood and amphorae symbolizing bodies. Archaeological evidence shows a rise in the consumption of both wine and olive oil during the second century BCE.[58] Although some local viticulture was developed, possibly by the Greek colonists of Massalia, viticulture in the region was limited, yet the thirst for wine was growing. Consumers turned to and mostly relied on foreign imports. Seizing new opportunities, many of the Italian vineyards specialized in producing red wines for the lucrative Gallic market. Writers pointed to the high prices wine fetched in Gaul, which gained a reputation for its insatiable thirst for wine.[59]

> And because the temperate climate is destroyed by the excessive cold, the land bears neither wine nor oil. As a result, those Gauls who are deprived of these fruits prepare a drink out of barley they call beer, and they also drink the water with which they wash off their honeycombs. Being exceedingly addicted to wine, they fill themselves with the wine that is imported by merchants unmixed, and, greedily consuming this drink due to their craving for it, when they are drunk pass out or fall into a state of madness. On account of which, many of the Italian merchants, because of their customary love of money, believe that the Gauls' love of wine is their godsend. For they transport the wine on navigable rivers with boats, and through the flat fields on wagons, and receive for it an unbelievable price; for delivering a jar of wine they receive instead a youth, getting a servant in exchange for the drink.

διὰ δὲ τὴν ὑπερβολὴν τοῦ ψύχους διαφθειρομένης τῆς κατὰ τὸν ἀέρα κράσεως οὔτ᾽ οἶνον οὔτ᾽ ἔλαιον φέρει: διόπερ τῶν Γαλατῶν οἱ τούτων τῶν καρπῶν στερισκόμενοι πόμα κατασκευάζουσιν ἐκ τῆς κριθῆς τὸ προσαγορευόμενον ζῦθος, καὶ τὰ κηρία πλύνοντες τῷ τούτων ἀποπλύματι χρῶνται. κάτοινοι δ᾽ ὄντες καθ᾽ ὑπερβολὴν τὸν εἰσαγόμενον ὑπὸ τῶν ἐμπόρων οἶνον ἄκρατον ἐμφοροῦνται, καὶ διὰ τὴν ἐπιθυμίαν λάβρῳ χρώμενοι τῷ ποτῷ καὶ μεθυσθέντες εἰς ὕπνον ἢ μανιώδεις διαθέσεις τρέπονται. διὸ καὶ πολλοὶ τῶν Ἰταλικῶν

56. Dietler 1997, 2005, 2010a; Garcia 2004; Py 1993.

57. Poux 2004a, 2004b. For a summary, see Tchernia 2016, 281–286. See also Metzler 1991.

58. Tchernia 2016, 277–291; Morel 1998, 2004; Poux 2004b; Ferdière 2020.

59. Cic. *Quinct.* 6 (81 BCE): trader Lucius Publicius brought slaves from Gaul to Italy.

ἐμπόρων διὰ τὴν συνήθη φιλαργυρίαν ἕρμαιον ἡγοῦνται τὴν τῶν Γαλατῶν φιλοινίαν. Οὗτοι γὰρ διὰ μὲν τῶν πλωτῶν ποταμῶν πλοίοις, διὰ δὲ τῆς πεδιάδος χώρας ἁμάξαις κομίζοντες τὸν οἶνον, ἀντιλαμβάνουσι τιμῆς πλῆθος ἄπιστον: διδόντες γὰρ οἴνου κεράμιον ἀντιλαμβάνουσι παῖδα, τοῦ πόματος διάκονον ἀμειβόμενοι.

(DIOD. SIC. 5.26)

Diodorus Siculus' account suggests that climate was a major hindrance to local viticulture in Gaul, and that the local beverages were beer and mead. The desire for wine and different customs, however, led to excessive consumption of wine, unmixed, and bartering that overwhelmingly profited Italian traders. From these favorable markets, Italian traders brought back highly valued return cargo, such as slaves, some of whom were probably put to work on the vineyards producing wines that led to their enslavement. While Diodorus might have exaggerated aspects of this exchange, southern Gaul nonetheless seems to have been a promising market for Italian wines.

The lure of markets overseas and the fruits of the Sestius amphora enterprise are attested by thousands of vessels found at different sites and shipwrecks in the Mediterranean Sea, especially concentrated around southern and central Gaul. Numerous shipwrecks also show how closely merchants packed the ship hulls with wine amphorae. Many, if not all, of the ships that transported the Sestius wine amphorae from central Italy were filled with wine amphorae, and probably targeted specific markets along the coast and rivers in Gaul and elsewhere.[60] Later evidence shows that ships could feature a regional wine from different producers, suggesting that merchants collected regional goods for export at warehouses before shipment. A bottling and shipping facility at Villa B in Oplontis in the first century CE, for example, was stationed both on the ancient coastline and where local wines were assembled and packaged before being shipped for export, illustrating the region's massive wine export industry.[61] There was no evidence for viticulture on-site: no press, vats, or dolia. Instead, the warehouse contained nearly 1,400 previously used amphorae, which collectively could hold approximately 36,300 liters of wine, the equivalent of 45,325 bottles of wine today. At the time of the eruption, the amphorae had been washed and relined with pitch, ready to bottle new batches of wine for shipment overseas.

Furthermore, the shift from the traditional Greco-Italic amphorae, which were in use from the fourth to first half of the second century BCE, to Dressel 1 amphorae gave merchants a taller vessel with a more pronounced, spiked base, a vessel that was easier to stack and fit into commercial ships. Developed from Greek amphorae that originated in the Aegean (specifically the island of Cos), Italian amphorae became more distinctive and

60. Horden and Purcell (2000) favor the idea that most shipments over the course of history in the ancient Mediterranean operated under cabotage; A.I. Wilson et al. (2012) advocate for the idea of direct shipments to major ports and for single-commodity shipments, such as stones, ceramic building materials, and metals, during the Roman period. Rice 2016: shipwrecks reveal ancient ships loaded with wine amphorae heading to distant markets. On harbors and harbor infrastructure in the Roman world, see D. Robinson et al. 2020.

61. Muslin 2019; Rice 2016, 174–175; Carreras Monfort 2013; Martínez Ferreras et al. 2013; M.L. Thomas 2015; Pecci et al. 2017.

elongated over time.[62] Overall, potters modified amphora design to align better with contemporary ship design (or other modes of transport) for traders to pack more amphorae, and hence more wine, into the ships for each journey.

By the mid-first century BCE, however, the situation in both Cosa and Gaul had changed. The Sestii's activities in southern Tuscany ceased. The number of amphorae dropped off by 40 BCE, and instead, the Sestii later relocated business to Rome, where they developed other industries.[63] Near Cosa, some villas no longer invested in mass wine production or export.[64] The port stopped exporting agricultural products and instead imported goods from afar.[65] It is still unclear why this occurred, whether it was due to broader developments and changes in Italian agriculture and viticulture, disruptions specific to the *ager Cosanus* from the Social and Civil Wars (and to the Sestius family, such as Lucius' proscription and the confiscation of his pottery workshop), or other factors. Local hostility and violence toward Roman traders in Gaul might have been another reason.[66] After the Gallic Wars, the Italian amphora evidence in Gaul is not as striking, and scholars believe the incorporation of Gaul into the empire meant Gallic chieftains were no longer able to enslave and sell their own peoples for wine and that large, alcohol-driven feasts were banned due to fear of sedition.[67] The shifts could have also stemmed from changes to how wine was moved en masse or expanding agricultural production in provinces (Chapter 5). Although the exact reason is still unknown, what is clear is that amphora production in the *ager Cosanus* changed by the late first century BCE, but markets abroad still presented enriching opportunities, which enticed other wealthy families.

Bulk Transport: The Piranus Family's Dolium Tanker Ships

From the late first century BCE onward, wine consumption continued to rise and merchants operated in what scholars believe to have been a more integrated wine market. With mounting levels of urbanism in the Mediterranean from the second half of the first century BCE through the first two centuries CE, urban markets and large consumer bases continued to attract merchants. By the last quarter of the first century BCE, wine from different parts of the Mediterranean regularly reached consumers at faraway markets and destinations. Supporting this increased connectivity and heightened scale of distribution across the Mediterranean were innovations in ceramic technology.

Viticulture was booming in the Mediterranean, and potters modified their products to transport even more wine. One effective outcome was the Dressel 2/4 amphorae (Figure 4.2R). Originally associated with Coan wine in the east, Dressel 2/4 amphorae

62. Interestingly, an eastern counterpart, known as the Lamboglia 2 amphora, took off in the Aegean and the distinction between the eastern and western Mediterranean—in amphora types and trading routes—continued to persist for centuries.

63. Manacorda 1978; Will 1987, 2001; McCann 2002, 28–30.

64. Carandini 1985. On sources for the Roman slave population, see Scheidel 1997, 1999; Bradley 1987; O. Patterson 1982.

65. Will and Slane 2019.

66. Caes. *BGall.* 7.3: a local tribe massacred Roman traders who settled in Cenabum in 53 BCE.

67. Tchernia 2016, 292.

proliferated first in Italy during the first century BCE, especially for trade in the western Mediterranean.[68] Amphora potters in central Italy, and later in other regions, created a new type of amphora that could carry more while taking up less cargo space on ships. Although they were shorter than their predecessors, Dressel 2/4 amphorae were designed more efficiently by reducing the height of the vessel neck while elongating the vessel's body (Figure 4.2R). Endowed with thinner walls, the Dressel 2/4 amphorae were also lighter vessels. Overall, they were more capacious, able to hold one and a half to two times more than their weight compared with Dressel 1.[69] Their shape also enabled closer packing, with the result that a ship that carried 4,500 Dressel 1 amphorae in the previous generation could now hold a third more wine packed into 6,000 Dressel 2/4 amphorae.[70]

Some scholars have posited that amphora potters might have also developed the Dressel 2/4 as a more robust container for shipping. An innovation in their design was the addition of a "salt skin." Firing clay mixed with saltwater brought the salt to the surface, resulting in a hard, whitish surface that made the vessel more durable for shipping and handling, an appealing feature for overseas journeys.[71] One of the most distinctive, and identifying, features of Dressel 2/4 amphorae is the handle. Known as bifid handles, or *doppio bastoncino* in Italian, each handle consisted of two joined rods positioned vertically from the neck to the shoulder of the amphora. The Dressel 2/4 amphora also featured a smaller, more robust toe. The widespread finds of Dressel 2/4 amphorae across the Mediterranean and beyond, from Britain to as far as India, reveal the extent of the Italian wine trade, making it the most widely distributed container type in the ancient world. The vessel became a popular shape and type, with different regions eventually developing their own variations of the Dressel 2/4 amphora.

Attempts to improve and make more efficient amphorae fed into a grander process of increasing specialization. As Steven Ellis recently pointed out, the emergence of retail in urban areas was part of a wider phenomenon of specialized infrastructure to accommodate the rapid rise of urbanism in the Roman world, which also included bakeries, horrea, and even Roman concrete-based construction.[72] The development of Dressel 1 amphorae, and then Dressel 2/4, demonstrates the amphora potters' part in streamlining trade and improving the capabilities of these vessels in line with the contents and mode of delivery. But changes to shipments and attempts to specialize containers for trade were not limited to amphora design. People found a creative way to adapt an unexpected container for supersized bulk wine transport: the unwieldy, typically immovable dolium.

Several central Italian entrepreneurial families invested in innovative bulk wine shipping through dolium tanker ships, a unique and unprecedented ship designed specifically for wine transport.[73] Multiple unusual (and often very poorly preserved and difficult to interpret) shipwrecks along the coast of the western Mediterranean point to this new

68. For an overview, see Moore 1995; Hesnard 1981.

69. Corsi-Sciallano and Liou 1985, 167–168.

70. Hesnard 1977, 162n28; Panella 1981, 59; Corsi-Sciallano and Liou 1985, 168; Moore 2011, 95.

71. Peacock 1984, 263; Bonifay 2004, 41; Moore 2011, 104–105.

72. Ellis 2018, 145–147.

73. For information on shipwrecks, including dolia, dolium stamps, and cargo, see Heslin 2011; Rice 2016; Monticone 2017/2018; Carre and Cibecchini 2020.

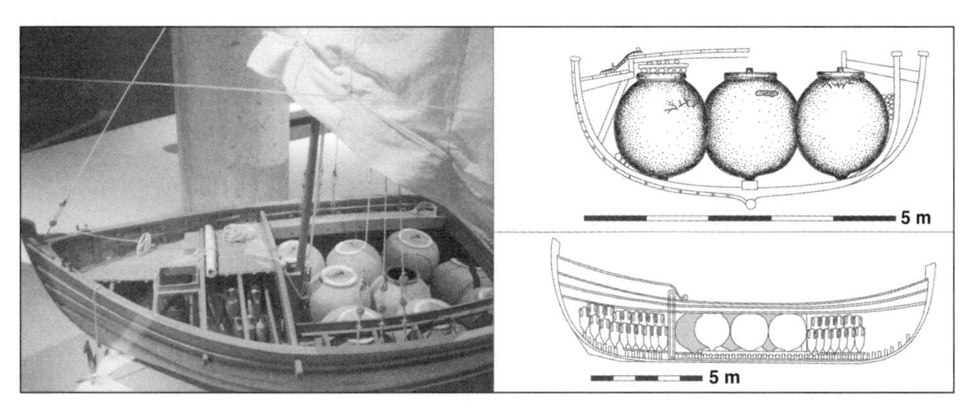

FIGURE 4.3. (*L*) Tanker ship model, Centre National de la Recherche Scientifique, Centre Camille Jullian, by Véronique Pagnier, 2010. (*R*) Drawings of tanker ships, illustrated by Gina Tibbott, after Carre and Roman (2008, figures 1 and 15).

method of bulk wine shipping between the late first century BCE and the mid- to late first century CE (Figure 4.3).[74] Several decades of underwater archaeology by international scholars primarily in France and Italy have supplied much more information about these peculiar ships.[75] The dolium tanker shared the same function as earlier ships used for bulk wine shipments, but they differed in form and arrangement. They were specialized wine transport ships, known as *bateaux citernes* (cistern or tanker ships). One major difference in their design is that they were smaller than other commercial ships, only about eighteen to twenty-two meters in length and about six to seven meters in width. Merchant ships, on the other hand, could reach fifty-five meters in length and fourteen meters in width, more than double the size of a dolium tanker ship. The other major distinction in their design is that dolium ships were constructed with dolia built and cemented directly into the hulls to hold bulk quantities of wine.[76] Found mostly along the coasts of Italy and France, with a few tentatively identified dolium tanker ships also located along the coasts of Spain and Croatia, a total of sixteen ships have been securely identified as dolium tanker ships and at least another sixteen ships have been posited as such (Figure 4.4).[77] Fourteen other discoveries of individual dolia, not obviously associated with shipwrecks due to their

74. Corsi-Sciallano and Liou 1985; Hesnard et al. 1988; Carre 2020. Cibecchini 2020: tanker ships span from the reign of Augustus through the mid-first century CE. See A.I. Wilson 2009 for difficulty in interpreting shipwreck evidence. Rice 2016, 167–168: low numbers of shipwrecks from North Africa and the eastern Mediterranean likely reflect "a scarcity of documentation and systematic research." See Leidwanger 2020 for maritime trade in the eastern Mediterranean.

75. E.g., Hesnard and Gianfrotta 1989; Gianfrotta and Hesnard 1987; Cibecchini 2020; Marlier 2008, 2020; Marlier and Sibella 2008.

76. Marlier 2008, 2020; Carre and Roman 2008; Cibecchini 2020; Rice 2016; Heslin 2011.

77. See Radić Rossi 2020: a shipwreck off the coast of Croatia originally had at least twelve dolia as well as other scattered dolia without context in the eastern Adriatic, including several (intentionally) perforated dolia. For dolium shipwrecks off Spain, see De Juan 2020. See Monticone 2017/2018 for a wreck off the coast of Sardinia.

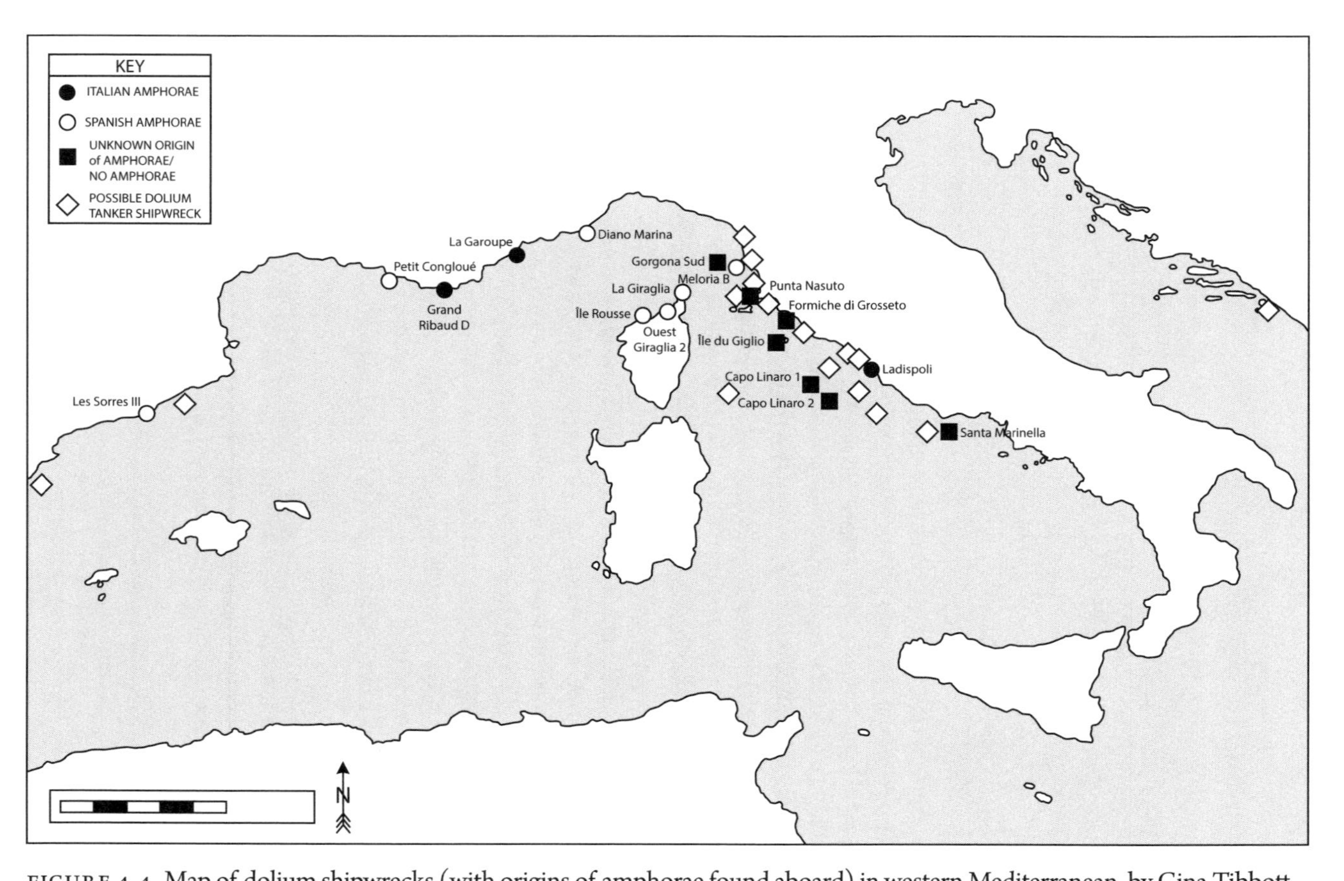

FIGURE 4.4. Map of dolium shipwrecks (with origins of amphorae found aboard) in western Mediterranean, by Gina Tibbott, after Rice (2016) and Cibecchini (2020).

state of preservation, were likely part of dolium tanker ships or from ships transporting dolia from workshop to warehouse;[78] as Jean-Pierre Brun and André Tchernia aptly note, people did not throw dolia into the sea.[79] The question is whether those dolia were part of a dolium tanker ship or whether they sank in the midst of being transported. Nonetheless, the present count points to these dolium tanker shipwrecks making up 6–13 percent of all known wrecks during this period.[80] Although the shipwrecks have been found at different places, both petrographic analysis and dolium stamps trace the production of these dolia, and likely the ships, to the Tyrrhenian volcanic zone and to families epigraphically attested in Minturnae.[81]

Minturnae was a coastal town on the border of Latium and Campania in an agriculturally fertile region, which included the *ager Falernus*, known not only for its wine but also its ceramic production. Situated on the Tyrrhenian coast and the banks of the Liris (Garigliano River), one of the largest rivers of Italy and recognized for its supreme navigability, Minturnae became a strategic harbor with at least fifteen ship sheds along roughly 200 m^2 of the riverbank. Minturnae exported large amounts of agricultural products from local villas, such as Falernian wine to Rome's enormous Horrea Galbae along the Tiber River in Testaccio.[82] The region overall had ample wine production and bountiful pine forests that supplied lumber for shipbuilding and other construction and resin to coat wine jars, and the town even had a group known as the *socii picarii* (association of pitch makers).[83] Not far from Minturnae were rich clay beds, which were strategically exploited by the workshops of black-gloss pottery, also known as Campana ware, in Cales, as well as workshops that produced the amphorae and dolia for the region's famed Falernian wine. Traders and vintners of Minturnae therefore had many incentives to invest in bulk shipping and could draw on ample resources and specialized potters in the region.

Of the few families involved in dolium production for these dolium tanker ships, the *gens Pirana* was most prominent in adapting dolia for this new bulk transport system.[84] Stamps bearing the family name on shipwrecked dolia testify to the family's involvement in the western Mediterranean over the course of three generations, from the late first

78. Figures from Cibecchini 2020.

79. Brun and Tchernia 2020.

80. See Heslin 2011, 157n7, for calculations based on Parker's 1992 publication, which came out to 6.2 percent. According to the most up-to-date figures and the Oxford Roman Economy Project Shipwrecks Database, dolium wrecks make up 6.7–13.4 percent.

81. Corsi-Sciallano and Liou 1985; Coarelli 1984; Carre 2020; Capelli and Cibecchini 2020. Approximately sixty dolia from shipwrecks bear manufacturer stamps; see Carre and Cibecchini 2020 for the most up-to-date list, including references to terrestrial dolia with stamps of individuals also attested on shipwrecked dolia.

82. On shipyards in Minturnae, see Johnson 1933; Ruegg 1995. On the port of Minturnae, see Ziccardi 2000. The Sulpicii Galbani had estates in the *ager Falernus*, slaves working at Minturnae, and their Falernian wine stored in the Horrea Galbae. See also Carroll 2022b, 236–237.

83. Johnson 1933, 126–127.

84. Corsi-Sciallano and Liou 1985. Carre and Cibecchini 2020; Carre 2020: most up-to-date compilation of stamped dolia from shipwrecks. Several other families are attested on dolium stamps and seem to have been from or had a connection to Minturnae: Acerratius, Cahius, Calicius, Camidius, Helvius, Licinius, Pandius, and Pomponius.

century BCE through the first century CE: first Sotericus, then Felix and Cerdo in the second generation, and Primus and Philomusus in the third.[85] Unlike the Sestius family, the Piranus family is not attested in literary sources, only in the epigraphic evidence of the dolium stamps. The stamps and the duration of the dolium production and dolium ship trade demonstrate prolific dolium production for the shipment of wine, as well as for wine storage.[86] At the very least, the Piranus family were the owners of the opus doliare workshops producing these enormous, high-quality dolia inserted into dolium ships. Yet the family might have been responsible for both the dolia and the specialized ships, because the ship's design required a close coordination among shipbuilding, dolium production, and dolium installation.[87]

These ships and their dolia were unique, forming a specialized mode of transporting wine. As mentioned before, the ships were smaller vessels, only thirty-five to forty-five tons, compared with some trade ships that were well over one hundred tons, and they were specially designed to be fitted with dolia. The dolium tanker ships usually featured ten to fourteen dolia packed together and built into the hulls. To accommodate the dolia, shipbuilders made the ship's keel wider (18–20.5 cm) than high (ca. 15 cm), with a reinforced axial frame to support the weight of the dolia.[88] Pointing to the construction of the ship and the placement of the dolia in the hull, some scholars have posited that the ships could sail only with full dolia; more recent studies have pushed back against this limitation, arguing that the massive dolia were heavy enough for the ship not to require ballast. Nonetheless, the ships' construction and furnishing restricted the cargo to wine.

The ship dolia featured greater variation in their shapes and sizes to accommodate and maximize placement in the ship's hull (Figure 4.5A–B).[89] Many of the shipwrecked dolia were massive and exhibited one of three traits: (1) having their widest point at their midpoint or belly (rather than the shoulder like their terrestrial cousins) for a more spherical shape, holding one thousand to three thousand liters; (2) featuring a more elongated and cylindrical form, with some holding up to two thousand liters (Figure 4.5B); or (3) appearing as smaller versions of dolia (*doliola*), though still large.[90] The forms and arrangement of the tanker ship dolia suggest that they were manufactured for packed and stable placement in tanker ships. With such forms, the vessels nestled next to one another, touching at their midsections, and endowed the ships with a lower center of gravity. The Diano Marino ship, for example, featured a total of ten large dolia on both sides of the hull (Figure 4.5C); between the two rows of massive dolia were four smaller, more elongated *doliola* positioned in the center, filling the tight space as much as possible.[91] Because installing the dolia was a major aspect of building these tanker ships, the two production processes, shipbuilding and dolium potting, likely occurred near each other to facilitate coordination as

85. Corsi-Sciallano and Liou 1985, 151, 173–174; Rice 2016, 179.

86. Lazzeretti 1998.

87. This was also when harbor infrastructure was intensely developed along the coasts of western Italy and southern France; see D. Robinson et al. 2020.

88. Marlier 2008, 2020; Carre and Roman 2008; Cibecchini 2020; Rice 2016; Heslin 2011.

89. Carrato 2020.

90. For the most recent discussion, see Carroll 2022b, 237; papers in Carrato and Cibecchini 2020.

91. Dell'Amico and Pallarés 2005.

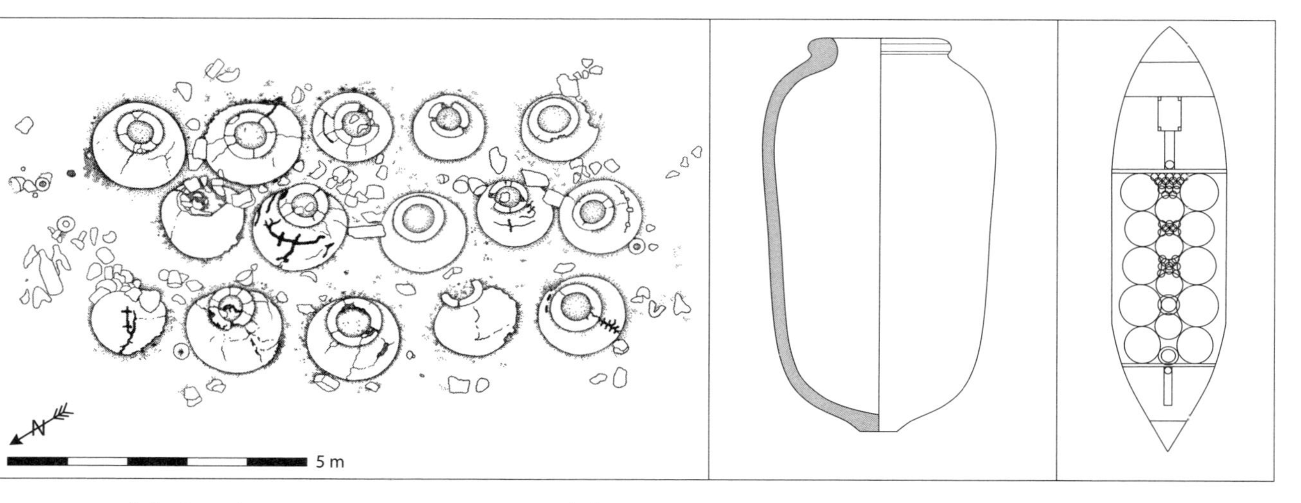

FIGURE 4.5. (*L*) Plan of the Petite Congloué shipwreck; (*C*) drawing of dolium, Diano Marina shipwreck, after Corsi-Sciallano and Liou (1985); and (*R*) reconstructed plan of the Diano Marina shipwreck, after Dell'Amico and Pallarés (2005). Illustrated by Gina Tibbott.

shipbuilders constructed hulls around the dolia and dolium potters produced dolia that would snugly fit and fill the hull.[92] Such close coordination, moreover, required dolium potters and tanker ship builders to work together with a particular result in mind.

Ship dolia did not have the typical strawberry shape characteristic of dolia found on farms and villas because they served a different purpose (Figure 4.6; Plate 12). Potters designed the (terrestrial) dolium's distinctive strawberry shape for wine fermentation and long-term storage on farms and in warehouses. The Piranus potters formed dolia, on the other hand, for overseas shipments of wine that probably had already undergone at least the primary fermentation stage, the violent, aerobic fermentation process that typically occurred in open dolia, vats, or basins;[93] some young wines today, such as Beaujolais, complete the fermentation process in as few as just four days.[94] Merchants might have been able to transport wine undergoing the second fermentation stage, a more subdued process, in the ship's dolia, a strategy some merchants attempted with amphorae. In fact, dolium lids found onboard these shipwrecks featured holes that would have enabled the wines to finish fermentation without accumulating gas.[95] Petrographic studies of the cargo have found that, although the dolia share the same ceramic paste and were made from the same clay sources, the other ceramic objects, such as the dolium lids, had mixed origins. The lids were likely replaced periodically because they were prone to breakage, perhaps because they sealed dolia containing wine that was undergoing some form of fermentation. The volatility of fermentation, whether primary or secondary, was still a risk, however, and might have ruptured the dolia, such as one dolium on the Cap Bénat B wreck, ultimately bringing down the ship.[96] Whether they held fermented or fermenting wine, tanker ship dolia functioned strictly as temporary receptacles, and not vessels for wine storage, and were therefore built with a different form.

The Piranus family's workshops produced exceptionally large, globular dolia for these ships, which moved huge quantities of wine and streamlined the transfer of contents. The large volumes of these tanker ship dolia are easily double to quadruple the amount other dolia in central Italy can hold. The dolia of the La Giraglia shipwreck were 1.5–1.7 meters in height *and* width; with their globular shape, these dolia could each hold approximately 1,400 to over 2,000 liters. The dolia from the Diano Marina shipwreck were 1.8–1.9 meters in height and width, estimated of having capacities of over 3,000 liters, while the smaller, cylindrical *doliola* were about 2 meters tall and estimated to have held 1,200 liters, much more than terrestrial *doliola*. Incisions on the dolia confirm their massive storage capabilities. For example, at least two of the dolia from the Grand Ribaud D wreck, dated to 9–1 BCE, bear inscriptions indicating their capacities: one contained "LXIII" amphorae of wine (ca. 1,660 liters) and the other "LXXV" (ca. 1,970 liters).[97] Considering that the

92. Marlier 2020; Carre and Roman 2008, 186; Gianfrotta 1998; Sciallano and Marlier 2008; Dell'Amico and Pallarés 2011, 82–101.

93. Heslin 2011, 164; Van Oyen 2020b, 51. Some scholars suggest merchants deposited must into the tanker ships, with the primary fermentation of new wines happening onboard.

94. McGovern 2003, 8: natural yeast on grape skin enables ripe grapes to ferment in a few days.

95. Cibecchini 2020, 187.

96. Joncheray 1994; Heslin 2011, 164.

97. Hesnard et al. 1988, 16; Gianfrotta and Hesnard 1987.

FIGURE 4.6. Shipwrecked dolium, Cosa, Ansedonia, Italy. Courtesy of the National Archaeological Museum and Ancient City of Cosa (Regional Directorate of Museums of Tuscany). Reproduction or duplication by any means is forbidden.

capacity of these dolia far surpassed that of even the massive second-century CE dolia made in opus doliare workshops around Rome (average capacity ca. 1,000 liters), dolium makers of the Piranus workshop had already achieved a major technological feat by the end of the first century BCE.

These dolium tanker ships had two main advantages over other forms of wine shipping. One was sheer capacity. They were built with dolia as the centerpiece, designed to increase the amount of wine they could carry. Calculations suggest that the dolia of these ships were able to hold at least 10 percent more wine than if the hull had been filled solely with Dressel 2/4 amphorae. Given that Dressel 2/4 amphorae were already more efficient than Dressel 1 amphorae, dolium tanker ships offered an even more significant capacity advantage.[98] By filling the hull with some of the largest dolia, some exceeding 1,500 liters, shipbuilders were able to construct ships capable of holding large amounts of liquids, from 10,000 to over

98. Pallarés (1985, 617–618) assumes porters stacked three layers of amphorae into the hull.

41,000 liters of wine, enough to fill roughly 400–1,300 amphorae.[99] The Diano Marina ship, for example, could easily hold over 35,000 liters of wine in the dolia alone, three and a half times the maximum amount a small farm such as Villa Regina at Boscoreale could produce in a year and enough to fill almost 1,400 amphorae; a tanker ship could thus deliver more than an entire annual batch of wine from two modest-sized vineyards.

The dolium tanker ships further amplified their wine shipments by storing wine not only in dolia but also in transport amphorae. To maximize the use of space further, porters filled the dozen or so dolia in the center of the hull with wine and packed the remaining space in the fore and aft of the ship with rows of Dressel 2/4 amphorae; it is likely that lower-quality or younger wine was loaded in bulk into the dolia, while better vintages were packaged in smaller amounts in amphorae, vessels with a smaller neck that reduced the wine's exposure to air.[100] Yet the dolia could also hold higher-quality wines that were marketed for wholesale.[101] With the additional one hundred to three hundred wine amphorae in the fore and aft, a single dolium tanker ship could carry roughly twelve thousand to almost forty thousand liters of wine, the modern equivalent of fifteen thousand to over fifty thousand bottles of wine. Calculations and estimates based on several better-preserved shipwrecks provide a sense of how much more wine the tanker ships could carry (Table 4.1). Compared with ships carrying solely amphorae (Chrétienne H and Sud-Lavezzi 3), dolium tanker ships could hold six to seven times more wine by fitting both amphorae and large dolia. The Petit Congloué shipwreck featured fifteen dolia and could fit fifty-four Dressel 2/4 amphorae, holding an impressive total of nearly forty thousand liters. Thanks to the design of these dolium ships, they could hold much more wine in their dolia while reducing the number of amphorae by four- to sixfold.

The second advantage the dolium tanker ships offered was that their smaller size and design and lower center of gravity enabled the wine-laden ships to traverse the sea and along rivers, where many of the wine markets were positioned. Merchants navigating larger ships (those holding over three thousand amphorae) would have had to offload their amphorae into smaller boats to reach fluvial ports, meaning more transshipments, vessels, labor, and time, while smaller vessels could be rowed or towed along the river.[102] Thanks to the dolium tanker ships' ability to navigate rivers, they offered new possibilities. Shippers could respond expeditiously to changing supply or demand and pick up wine from multiple vineyards and deliver wine to multiple ports, bypassing middlemen and going directly to the source or destination. Traders also did not need to transfer wine between ships and smaller riverboats, as porters could pour wine from cullei directly into the ship's dolia, saving time and containers and eliminating the need for hundreds of amphorae or other jars. Upon the ship's arrival, porters could transfer the wine from the ship's dolia into amphorae, smaller jars, barrels, or cullei for further transport, or even directly into dolia found at maritime or fluvial warehouses. Since the dolia were immobile and merchants needed to transfer the wine from the dolia, porters must have used pumps, perhaps kept

99. Pallarés 1985.

100. Gianfrotta and Hesnard 1987.

101. Djaoui 2020.

102. Dion. Hal. *Ant. Rom.* 3.44.

TABLE 4.1. Estimated volume of wine transported on various ancient ships (after Carrato 2013).

Shipwreck	Number of dolia	Capacity of storage in dolia (liters)	Number of amphorae (Dressel 2/4)	Capacity of storage in amphorae (liters)	Capacity of total storage (liters)
Petit Congloué	15 ovoid dolia	37,500	54	1,409	38,909
La Giraglia	11 ovoid dolia and 5 *doliola*	28,500	60	1,566	30,066
Diano Marina	10 ovoid dolia, 4 oblong dolia, and 2 *doliola*	29,800	50	1,305	31,105
Chrétienne H	—	—	300	7,860	7,860
Sud-Lavezzi 3	—	—	200	5,220	5,220

at harbors, or ladles and pipettes to move wine out of dolia and into their new receptacles.[103] Moving table wine in dolium ships could also save the producer, merchants, and buyer from the expense and logistics of using, accumulating, and discarding amphorae.[104] Although amphorae were generally cheap, expenses could add up from ordering and transferring wine into hundreds, if not thousands, of these single-use vessels, not to mention the time in making them. Additional opportunities to expand, streamline, and expedite this process burgeoned as various communities installed cellae vinariae along the marine and fluvial routes, not only in Italy but also in the western Mediterranean. In fact, we should keep in mind the possibility that these dolium tanker ships operated also within Italy. Although the tanker shipwrecks can help trace the journey of these specialized ships, they are the ships that did not reach their destination. Ships that completed their journey likely left behind no trace, and ships traveling along rivers, such as the Tiber River, pursued more secure routes than those in the open sea.

Profitable Packaging

Long-distance trade could reap substantial profits, but merchants needed containers that could properly protect their products. Although vintners usually had large cullei to move bulk quantities of wine, they were not suitable for overseas transport. Wine merchants turned to amphorae, a potentially lucrative industry with opportunity to consolidate and cut costs. Potters developed different types of amphorae to increase the vessel's weight-to-capacity ratio in order to produce more efficient containers. The abundance of Sestius amphora finds, for example, stemmed from a colossal pottery enterprise in the *ager Cosanus* tightly wound with viticulture and trade. The Sestius amphora workshop was wildly successful in supplying amphorae to wine traders during the mid- to late Republic. Some scholars believe that, in addition to producing amphorae, the Sestius estate also operated

103. Heslin 2011, 164–165. Fixed lead pipes aboard these ships functioned as bilge pumps that emptied sea water, but no evidence of pumps for dolia has been found. If pumps were not available, porters could use ladles, but the process would have been very slow and caused a bottleneck at the port.

104. Tchernia 2016, 141–142.

a large vineyard in the *ager Cosanus*, perhaps the massive Villa Settefinestre, producing the wine shipped in their own trademarked amphorae.[105] In the heart of a vast wine-producing region, they possessed the financial, social, and political means to bankroll not only a pottery workshop but also wine production and commercial shipping. A passage from Cicero's letters and an inscription from Delos suggest that a member of the Sestius family was involved in trade at Delos and might have owned commercial trading ships.[106] Furthermore, because sources note that the Sestius family supplied ships during the Civil War, they might have also owned commercial ships for shipping wine; they certainly could raise the funds for it. While we cannot definitively comment on the Sestius family's involvement in producing or trading wine, their ceramics enterprise could have been indirectly implicated. Even if members of the Sestius family did not directly transport and sell their wines abroad, they could have offered the amphorae or trade ships to wine merchants as an additional option to their contracts: traders buying wine from the Sestius family could therefore also purchase their amphorae or rent their ships to move and sell the wine abroad. With their resources, the Sestius family had multiple ways to exert their influence in the wine trade, and the success of their amphora enterprise shows the plentiful opportunities amphora production offered wealthy Romans.[107]

During the late first century BCE, a synergistic collaboration among dolium production, shipbuilding, and wine transportation resulted in the application of a technology, usually reserved for immovable storage, to bulk transportation overseas. Making and using dolia for ships posed unique challenges, and the Piranus workshop reached a high level of mastery. The dolia themselves already testified to expert dolium potting, but their utility skyrocketed as they were integrated into the tanker ships. Together, the dolia and ship became more advantageous than the sum of their parts, as they could offer a unique form of highly efficient bulk transport. The decision to finance such a specialized vessel was a type of investment similar to that made when commissioning the masonry shop counters so distinctive of the Vesuvian towns, as described by Ellis:[108] "Counters were sizeable, fixed points of sale that eschewed anxieties about seasonality and supply, or the uncertainties of dropping sales or depressed markets. Compared to more temporary and readily (dis)assembled wooden tables, the masonry counter and its requisite hearth demonstrated a knowing confidence that the retail sale of food and drink was a sound investment."[109] Although tanker ships differed from counters in scale, they also reflected a type of confidence in specialized investment, one for bulk wine trade across the Mediterranean Sea. Whether the tanker ship dolia temporarily carried common table wines or higher-quality wines for wholesale, traders expected to find ready consumers as well as producers in Italy and the northwestern Mediterranean. As a result, massive amounts of wine were moved from different regions of production to various markets. The dolium-based containerization system, with dolia in both terrestrial and nautical contexts, often found more lucrative markets farther afield. Entrepreneurs in west-central

105. Will 1979, 1982a; Will and Slane 2019; Manacorda 1978.

106. Cic. *Att.* 16, 4, 4 (*navigia luculenta Sesti*); *IG* XI 4.757.

107. Manacorda 1978.

108. Ellis 2018, 173–180.

109. Ellis 2018, 175.

Italy extended their reach into other pockets of the empire, with many finding opportunities in the western Mediterranean. With the high costs of acquiring and using dolia, this type of transportation was an ambitious and bold endeavor, fraught with risk, but one that forged networks linking areas of intensive viticulture to mass markets.

Dolia were traditionally used for wine and oil production, but their adoption for specialized transport was a new and peculiar way to implement this storage technology. Access to natural resources, such as wood for the ships and clay and pitch to manufacture and coat the dolia, facilitated the process but does not explain why traders would undertake moving so much wine in such an unusual and unprecedented type of ship. Furthermore, the ship design was so specialized that traders not only had to fill the dolia to make the return journey but could only move one type of product. On the other hand, the boom in wine production and export, and the potential profits, might have motivated wine traders to take risks in and experiment with specialized wine shipping. Amphora potters were already refining their products, making them thinner, lighter, and more capacious. By shipping wines in dolium tanker ships, traders were able to move even more wine, reduce the need for amphorae, and distinguish the different-quality wines in their cargo, while navigating both the sea and river to different markets.

The possible entanglement between pottery production and viticulture extends to dolia as well, as highlighted by the unique and central role of the Piranus family in dolium production for engineering specialized tanker ships, and the close relationship between building the ships and manufacturing the dolia. Karen Heslin has suggested, based on the diverse economic activities a single estate could take on, as well as Minturnae's reputation for wine, that the Piranus family might have also had a large viticultural enterprise and therefore designed and produced specialized dolium-carrying ships to transport their commodities.[110] Such an explanation would also suggest that a wealthy and influential family involved in viticulture was a driver in investing in opus doliare workshops, and perhaps shipyards too, to produce and transport bulk amounts of wine to profitable markets. Members of the Piranus family could have owned and employed these ships themselves and expanded their enterprise and network in the western Mediterranean as they targeted specific areas to sell their wines. On the other hand, the family might have constructed the ships and then sold or rented them to individual traders and merchants, who sold the Piranus family's wine or other Italian wine abroad; the Piranus family could have even had their freedmen undertake the wine transportation, an arrangement that would align with the risk-averse behavior echoed in many textual sources, while offering business ventures (and obligations) for freedmen.[111] In sum, there are three possibilities: the entrepreneurial family involved with these special dolia and tanker ships supplied dolia for these ships, supplied both the dolia and the ships, or supplied all three—wine, dolia, and ships—in a type of vertical integration to cut costs and control the supply chain. While we cannot say with certainty which arrangement the Piranus family practiced, dolia certainly connected the various ventures and offered opportunities to consolidate costs and tighten control over various steps in and people involved with the wine trade.

110. Heslin 2011.

111. Plin. *Ep.* 3.19, 7.30, 8.2, 9.36, 9.37; Kehoe 1989.

Taken together, the dolium tanker wrecks highlight a journey that not only expanded westward over time but also disseminated the dolium storage technology, stimulating large-scale viticulture in the west. As the next chapter will show, the destination for wine aboard these dolium tanker ships changed over time. While Gaul and Iberia presented desirable markets, the ships were likely bringing mass quantities of wine also to the Roman military stationed in the northwest provinces, and the regional Dressel 2/4 amphorae point to the tanker ships *bringing* wines from the northwest to Italy. By adopting and implementing the dolium storage technology, producers beyond Italy expanded the production and export of wine, oil, and fish sauce products, which became increasingly geared toward the imperial capital.

5

Dolia in Iberia and Gaul

Martial Epigram *13.107. Pitch-seasoned Wine*

Do not doubt that these pitch-seasoned wines came from
vine-bearing Vienne: Romulus himself sent them to me.

Haec de vitifera venisse picata Vienna
Ne dubites, misit Romulus ipse mihi.

Martial Epigram *13.118. Tarragonese Wine*

Tarraco, which concedes to the vineyards of Campania alone,
produced these wines, rivalling the Tuscan wine jars.

Tarraco, Campano tantum cessura Lyaeo,
Haec genuit Tuscis aemula vina cadis.

Martial Epigram *13.123. Massilian Wine*

When your sportula will discharge hundreds of citizens,
you may present the smoky wines of Massilia.

Cum tua centenos expunget sportula civis,
Fumea Massiliae ponere vina potes.

IN THE FIRST century CE, a resident of the city of Rome had many options for wine. She could purchase a jug of local wine produced just outside the capital; wine from a different part of Italy, such as Campania or Tuscany; or even vintages from farther afield such as Massalia or Tarraconensis. For connoisseurs such as Martial, the various vintages of the empire had different *terroir* (or at least a similar concept), reputations, and even histories. Although some consumers were probably content with, or could only afford, local table wines, wealthier patrons might want a more diverse or curated selection, especially if they wanted to show off their sophisticated tastes to banquet guests. Consumers in the Roman Empire appreciated food and drink from different places, including wine, and Rome was fully stocked.[1] By the time Martial was writing in the late first and early

1. Rowan 2019b. See also Moore 2011; Lawall 2011. On eastern Mediterranean wines in Rome, see Komar 2020. On imported wines from the amphora evidence, see Komar 2021.

second centuries CE, the western provinces, places where Italian traders had earlier made their fortunes, made their own wines that not only were popular and widely consumed but even rivaled Italian wines.

When dolium tanker ships entered the scene, agricultural developments in Iberia and Gaul were already in place and accelerating, and the direction of trade was shifting. Although traders initially used the new bulk dolium shipping technology to move wine to southern Gaul in the late Republic, local dolium industries and dolium use boomed in southern Gaul and parts of Iberia and ultimately catered to the imperial capital from the first century CE onward. As the previous chapter highlighted, the dolium tanker wrecks trace a trade that expanded westward, going beyond southern Gaul and into Iberia. The dolium stamp evidence from the tanker shipwrecks shows that the first phase of the dolium ship journeys followed routes similar to those of wine shipments in amphorae from earlier periods. The dolium ships manufactured by the first generation of the Piranus family enterprise, under Sotericus, were loaded in central Italy with wine, both in the installed dolia and in Italian Dressel 2/4 amphorae packed around the dolia. The dolium ships then departed from Italy to southern Gaul, where porters unloaded most of the batch of Italian wine, if not all of it. Although this was the same destination as the Sestius amphorae and other Greco-Italic and Dressel 1 wine amphorae in previous centuries, the situation at Gaul had changed. Gaul and the isolated trading system during the mid-Republic had now been incorporated into the empire and the value of wine was no longer as high. Demand was still great, especially with the Roman army stationed in various locations, but local agriculture was undergoing significant changes and growth. Yet the often mixed cargo of the dolium tanker ships adds another piece to the puzzle. While the Republican-period trips (ships containing both amphorae and dolia) brought Italian wine to western Mediterranean markets, aboard many tanker ships from the late first century BCE were sets of Dressel 2/4 wine amphorae, including regional variations associated with areas outside Italy. During the first century CE, the specialized wine ships containing dolia produced by the second generation of the Piranus family's enterprise, under Felix and Cerdo, and then the third generation, under Primus and Philomusus, continued from Gaul westward to Tarraco in Spain.[2] Consumers from Hispania purchased wine aboard these dolium tanker ships, either Italian wine that remained after initial sales in Gaul or Gallic wine that had been loaded onto the ships, and in return merchants picked up Tarraconensian wine for Italian markets.[3] Although the Iberian Peninsula seems to have been a new destination added to the tanker ships' journey, the region had long been part of the Roman Empire's trade network, and its prominent role in supplying food to the capital of the Roman Empire was already beginning to take shape well before the first century BCE.[4] The productive landscape of the western Mediterranean was expanding, and Iberian products, including wine, olive oil, garum, and metals, were exported in increasingly large numbers.[5] In conducting business in Iberia, merchants of

2. Tarraconensian Dressel 2/4 amphorae were found aboard several ships, but they could have been picked up in Gaul.

3. Only one wreck (Petit Congloué) has had Gallic amphorae identified. The absence of Gallic amphorae could be due to successful journeys and sales: sailing from southern Gaul to Hispania was not as dangerous, taking half the time (four days) as from central Italy to southern Gaul (nine days).

4. For an overview of agriculture in Roman Iberia, see Lowe 2009, 2020.

5. E.g., Bustamante Álvarez and Ruiz 2013.

dolium tanker ships tapped into preexisting opportunities and both extended and deepened trade networks in the western Mediterranean. In fact, by the late first century CE, wines from the western provinces became so popular that Spanish and Gallic amphorae became the most well-attested type of wine amphorae in and around Rome.

Thanks to recent and ongoing work by Charlotte Carrato, Javier Salido Domínguez, and Joaquim Tremoleda Trilla, among others, we now have a much better understanding of how intertwined agricultural developments were with storage and dolia in the northwest Mediterranean.[6] This chapter continues to examine the wine industry beyond Italy to see how agriculture and local dolium industries developed in the western Mediterranean. Dolium industries proliferated across parts of southern Gaul and northeast Iberia and expanded local agriculture, especially viticulture, so greatly that wine from the northwest Mediterranean began to outpace Italian wines in popularity, at least according to the amphorae evidence.[7]

Agriculture and Storage in Iberia and Gaul

Before we can understand Roman-era changes to agriculture in the northwestern Mediterranean, we need to situate it in a broader historical context. Iberia and Gaul underwent major transformations from the Iron Age through the first two centuries CE under Roman rule. Although many of these changes reflected local concerns, there generally was a transition from mixed agriculture for local needs to increasingly large-scale and export-driven production and from long-term storage with timed access to more short-term but accessible storage.[8] Iron Age Iberia and Gaul (third to first century BCE) were characterized by small, dispersed settlements and hilltop strongholds known as *oppida*. Communities practiced mixed agriculture and stored their cereals in underground silo pits, which were hermetically sealed for long-term and carefully timed storage. Silo pits relied on seals made of clay, stone, or other dried material to keep out oxygen, light, and pests for long periods of time. The lack of oxygen, and buildup of carbon dioxide, created an environment hostile to (micro)organisms, potentially leaving the contents dormant for fifty to one hundred years.[9] Once opened, however, the silos' contents would immediately begin to age and deteriorate. Silo storage was undertaken at the household and community level, and it was both a safeguard against bad harvests and a way to store for communal feasts.[10] Inhabitants

6. E.g., Carrato 2017, 2020; Carrato et al. 2019; Salido Domínguez 2017; Tremoleda Trilla 2020. Surging interest in dolia in the area is demonstrated by a conference dedicated entirely to the vessels, "Dolia in the Hispania Provinces in the Roman Period: State of the Art and New Perspectives," September 7–9, 2022. Although Roman-period pithoi/dolia have been found in other provinces, the lack of systematic work precludes a more comprehensive understanding of their development, how prevalent they were, and how they were used. Some regions seem to have low numbers of dolia, perhaps because they did not have a robust opus doliare industry or wine was not produced at a large scale for export. See Dodd 2023.

7. On pitfalls of using amphorae to gauge wine trade, see De Sena 2005; Marzano 2013a; Komar 2021.

8. For a summary of storage developments in the northwestern Mediterranean, see Van Oyen 2020b, ch. 3; Garcia 1987, 1997.

9. Varro *Rust.* 1.57, 1.63.

10. Dietler and Herbich 2011.

of Iron Age Iberia and Gaul also stored some foods in ceramic storage jars mainly in domestic contexts. Local viticulture and olive oil production was small scale and localized, having been introduced by Phoenician, Greek, and Etruscan colonists and settlers. As changes in agricultural production and culinary preferences developed, such as the market for wine, local communities began to shift their storage practices and equipment.

Wine traders from Italy (and elsewhere) made their fortunes in some parts of Iberia as early as the late third century BCE, evidenced by both Greco-Italic and Dressel 1 amphorae. At the same time, landowners in Tarraco and its hinterland constructed residences centered on a courtyard that resembled Roman-Mediterranean villas that became more elaborate and ornate over time.[11] Later on, landowners in other regions of Hispania began to build agricultural estates as well. Although the earliest farmhouses in the southern and western parts of the province were modest and followed indigenous traditions, owners of later farms and villas increasingly fashioned them after Roman villas found in Italy. Over time, estates from simple farms to lavish villas engaged in large-scale agricultural production, first geared toward supplying themselves and local markets and then toward large-scale export abroad. Numerous cellae vinariae were constructed, and local dolium and amphora production took off starting in the second half of the second century BCE. By the mid-first century BCE onward, Tarraconensis was supplying Gaul with wine in amphorae that closely followed Italian wine amphorae, and Iberian amphora producers quickly shifted to producing their own version of Dressel 2/4 amphorae, just a few decades before Italian traders brought wine to southern Gaul, where they found buyers willing to spend more on the coveted beverage.

After the conquest and integration of Gaul into the Roman Empire, local viticulture in southern Gaul grew exponentially. Starting in the late first century BCE, wealthy landowners in Gaul were building villas and agricultural estates in line with Roman concepts of *negotium* (business, or lack of leisure) and also to produce food for growing urban centers.[12] Recent studies on agriculture in Gaul show that starting in the first century BCE, farmers increasingly grew, processed, and sold cash crops to urban markets, a strategy that also included storage, and storage facilities, on a massive scale.[13] Farmers were growing surplus grain and no longer placing unprocessed cereals in the traditional underground silo pits for long-term storage and local consumption as they had in previous centuries. Instead, they stored large amounts of grain in purpose-built granaries for urban markets. This meant they deposited agricultural goods not only in aboveground structures but in structures that were also centralized and massive, bordering on monumental. Along with the emergence of these large-scale production and storage regimes, Gallic farmers also began to practice intensive viticulture using dolia, greatly expanding the limited local viticulture of prior generations that had been mostly confined to the area around Massalia. Urban centers developed across southern Gaul and Iberia, and aboveground storage in distinct structures became more common.[14]

11. Teichner 2018.

12. For an overview of agriculture in Roman Gaul, see Ferdière 2020.

13. See papers in S. Martin 2019, especially Bossard 2019. Also see Rickman 1971; Van Oyen 2020a, 2019, 2015a. For discussion of military horrea, see Salido Domínguez 2011.

14. On horrea in Spain, see Arce and Goffaux 2011.

By the time traders transported wine in dolium tanker ships, they were no longer following the trajectory of wine export in previous centuries. Both Iberia and Gaul had already been experiencing waves of local agricultural and economic growth from the second century BCE onward, if not earlier.[15] Northeast Iberia, in particular, underwent radical reorganization in the countryside as many farms and villas shifted from cereal cultivation to viticulture;[16] as a result, some regions developed their own wines that gained particular reputations among Roman authors, such as the high-quality Tarraconian wine or the abundant, yet cheap, Laietanian wine.[17] During the first century CE, estate owners and merchants in southern Gaul and Hispania developed their estates to seize on new economic opportunities, with the result that the Guadalquivir Valley became a notable region for supplying Rome large quantities of high-quality olive oil, supported by over four hundred producers and nearly one hundred amphora kiln sites.[18] By the second half of the first century CE, almost all of Gaul was producing wine.[19] While earlier shipments of wine from Italy were geared toward wine export, now merchants were also *importing* wine from the western Mediterranean to urban markets in Italy, filling ships, including dolium tanker ships, with wine from Tarraconensis and possibly also Gaul for the return journey.[20] Both survey archaeology and excavations have revealed widespread construction of massive agricultural estates in Gaul and Iberia designed for the large-scale production of wine, and occasionally olive oil too. Some of these villas, including modest ones, even had their own tile and pottery kilns capable of producing building materials and ceramic containers on-site to bolster their ability to export wine. In the late first century BCE, their exporting power materialized as potters in Tarraconensis made their own version of Dressel 2/4 amphorae, while potters in several regions in southern Gaul also developed their own local versions, such as the Lyon and the Loire basin types. Local potters in Gaul also developed another local type of transport jar, a flat-bottomed wine jar known as the Gauloise 4 that could be transported on ships, boats, and carts. Scholars have identified several Gauloise 4 amphora workshops, often connected to vineyards, wine-producing villas, and ports, in the Narbonne region.[21] The *tituli picti* (painted labels) on the amphorae testify to several different types of wine that these jars transported, including notable Italian-style wines that those vineyards imitated. While transport jars certainly enabled large-scale distribution and export of the northwestern Mediterranean's agricultural bounty, the ability to produce and store so much wine and olive oil was thanks to the widespread adoption of dolia.[22]

15. Van Oyen 2020b, 67ff.

16. Olesti Vilà 1995, 165, 192; 1997; 1998, 247.

17. Plin. *NH* 14.71; Mart. *Ep.* 7.53.6, 13.118.

18. Lowe 2020, 489; Remesal Rodríguez 1998. For sites with olive presses, see Ponsich 1991. For amphora kiln sites, see Remesal Rodríguez et al. 1997; Chic García and García Vargas 2004; Ponsich 1974. See also Keay 1984; Miró 1988.

19. Ferdière 2020; Brun and Laubenheimer 2001; Brun 2004; Poux et al. 2011.

20. Lowe 2020, 486: amphora production in Iberia began in the first century BCE, expanding tremendously during the early first century CE.

21. Laubenheimer 1985.

22. Marzano 2013b; Teichner 2018; Buffat 2018.

Dolium Development in the Northwest

In order to produce and store large quantities of wine and olive oil, landowners in Gaul and Hispania invested heavily in dolia. Although both provinces had a local ceramic storage jar tradition that extended back to the sixth century BCE during the Iron Age, greater supply and demand for wine and olive oil spurred the adoption of the supersized and specialized dolium technology in a way parallel to the development we saw in central Italy.[23] As pottery and dolium-building skills developed, local workshops began to produce large dolia that serviced local villas. Although potters in both Iberia and Gaul produced dolia for expanding agricultural production, each province, and even regions within the same province, crafted fermentation and storage jars specifically for their area and following local traditions. In her review of dolia across the western Mediterranean, Carrato has suggested that dolium size responded to local climate conditions and harvest calendars. Because wine yeast ferments at a particular optimal temperature range, dolium potters in Gaul made their products larger for the colder climate and later harvest season; in Italy and Iberia, on the other hand, potters built slightly smaller dolia for the warmer climate and earlier grape harvest.[24]

Iberian dolia mostly departed from the Iron Age pithoi tradition as dolium potters greatly enlarged the dolia, providing them with an ovoid or more strawberry shape, and often eschewed previous decorative elements and added capacity markings instead (Figure 5.1). They developed large and strictly utilitarian dolia that mostly shed their previous decorations as agricultural production grew from the second century BCE on. Gone were the decorative handles and bands. Instead, the Iberian dolium had a plain appearance and was designed to be buried in the ground. Many of the storage jars from the sixth century BCE, if not earlier, were decorated with traditional Punic motifs. Iberian pithoi were typically smaller than later dolia and had a cylindrical, spherical, or ovoid shape with nonfunctional handles that were purely aesthetic on vessel shoulders. Some pithoi might have even been painted, and the design of these jars also suggests they were placed above ground for the decorations to be seen. Centuries later, large spherical dolia appeared in Hispania from the second century BCE to third century CE, the same time wine and olive oil surplus production expanded for export.[25] The evidence from antiquity for Spanish dolium production is sparse, but six workshops have been securely identified in different areas of Iberia based on recovered misfired fragments.[26] Similar to opus doliare workshops in west-central Italy, these workshops produced not only dolia but also terracotta building materials and Dressel 2/4 amphorae for wine and Dressel 20 amphorae for olive oil export industries. Dolium kilns are difficult to identify, especially because dolia were often manufactured with amphorae and other products, which formed a greater percentage of the manufacturing debris, so there are likely more dolium workshops to be discovered. Some of the stamped dolia feature names of several Roman

23. Tchernia 1986, 58. Rogers 2003: diffusion of innovation relies on communication and influence among different sectors of society.

24. Carrato 2017.

25. Salido Domínguez 2017; Tremoleda Trilla 2020, esp. 89–91; Simon Reig et al. 2020.

26. Beltrán 1990: Marchena in Seville; El Olivar in Chopiona in Cadiz; Mortatalla in Córdoba; Torrox, Malaga, Tivissa, and Palàmos.

FIGURE 5.1. (*L*) Profile drawings of Iberian dolia, by Gina Tibbott after Carrato (2013). (*R*) Iberian dolia, Merida, Spain, by Yuntero, 2006.

individuals with known links to local large-scale viticultural activities, further suggesting an overlap between viticulture and dolium production.[27]

With their larger capacities, dolia expanded wine and olive oil production and supported local needs. The dolia in Tarraconensis and Lusitania were the largest storage vessels in Hispania (average capacity: six hundred to eight hundred liters), and several had inscriptions on their rims or shoulders that recorded their capacities.[28] Dolia from the villa of Olivet d'en Pujol in Viladamat, for example, had capacities of 1,000–1,600 liters each. The volume incisions and general shapes of the dolia show that the dolia in Iberia, on average, were becoming larger over time, especially in the northeast, due to, in part, regional variations and practices within Iberia. The smaller storage jars, which often retained the traditional decorations derived from Phoenician-Punic vessels and were probably the vessels ancient authors called *(h)orcae*, were still in use after the introduction of larger dolia. Some Spanish jars, often used for olive oil, had small, decorative, and nonfunctional handles on the shoulder. Farmers producing olive oil in Baetica and northern regions used smaller storage jars, perhaps because of temperature constraints or because olive oil yields were lower. Overall, dolium use for oil was high where, as the amphorae evidence from Testaccio testifies, large quantities of the famed high-quality olive oil was produced.

With the advent of Roman conquest, storage jars in southern Gaul likewise went from smaller, multipurpose jars in the Iron Age to incredibly massive wine fermentation and storage jars by the first century BCE. Local peoples in Gaul used ceramic storage jars as early as the fifth century BCE, a practice that was perhaps spurred in part by the influence of Iron Age Greek and Etruscan traders and colonists. As viticulture grew, local potters changed their vessels over the course of several centuries in three different stages. From the fifth to fourth century BCE, then the third to first century BCE, and finally the first to

27. Tremoleda Trilla 2020, 107–119. See Bernal Casasola et al. 2021 for a possible Italian dolium in Hispania.

28. Tremoleda Trilla 2020, 91–116.

third century CE, dolium makers progressively enlarged these ceramic storage containers as landowners in Gallia Narbonensis developed intensive viticulture and olive oil production; as a result, the capacity of locally made dolia in Gallia Narbonensis, which were often labeled with Roman numerals, became thirteen times higher during the imperial period than their Iron Age predecessors and were on par with or even larger than the largest Italian dolia.[29] Some preserved fragments show that dolia were made with rims sixty to one hundred centimeters wide, larger than most of the dolia around Rome and Ostia. This development not only allowed an increased capacity for the vessels but also created more standard and simplified forms. The local tradition of ceramic storage containers that had been in place for several centuries featured smaller containers often with decorative and idiosyncratic motifs, but by the first century BCE, potters in numerous workshops in Gallia Narbonensis produced standardized storage jars, mostly undecorated, with a morphology that closely resembled that of dolia of west-central Italy. Some of the earlier forms and decorations can still be found on the large dolia, such as decorative horizontal bands that mimicked cords around the vessel near the base, a motif not found on Italian dolia but typical of Greek storage jars, perhaps the source of influence on Gallic pithoi. Gallic dolia were now instead created for bulk storage, with some still preserving traces of local traditions (Figure 5.2).

Although identifying who exactly—local landowners, Roman entrepreneurs, or others—was involved can be difficult, the region underwent a major transformation. Some of the landowners in Gaul might have been Italians with connections to workshops in central Italy or might have shifted their dolia from Italian to Gallic estates. Carrato's comprehensive study on the dolia of Gallia Narbonensis credits some of this innovation and development to Roman entrepreneurs who established, supported, and invested in dolium workshops in the area.[30] Contact with Italian wine merchants, especially those operating the dolium tanker ships, might have applied some pressure on local producers hoping to expand their enterprise to take on the new storage technology. Early adopters of dolia, likely wealthy landowners and adventurous entrepreneurs, pioneered new practices in increasing the scale of wine and olive oil production. Dolium stamps reveal that some wealthy landowners in southern Gaul even purchased and imported dolia produced in the urban opus doliare workshops near Rome.[31] Several stamped dolia, for example, trace the dolia's production to members of the Tossius family, who are attested on dolia in Rome and Ostia. As Chapter 2 discussed, the Tossius family was particularly prominent in producing dolia, bricks, and tiles around Rome, with several generations of free, enslaved, and manumitted labor. Some stamps on dolium lids trace their production to opus doliare workshops of the *gens Domitilla,* owned by Marcus Aurelius' mother. Trade between central Italy and Narbonne was not uncommon, as other opus doliare products, such as bricks, tiles, and *mortaria,* were brought to southern Gaul. The isolated dolia found in the sea along the coast of the western Mediterranean might have been dolia sent to more distant destinations on ships that did not complete their journey. Petrographic analyses of dolia at the Vicus at Vagnari in Puglia and the Roman farm at San Giovanni on the island of Elba indicate that

29. Carrato 2017, 186–194, 653–704. See also Teyssonneyre et al. 2020.

30. Carrato 2017, 133–175.

31. Carrato 2017, 166–175, 599–638.

FIGURE 5.2. (*L*) Profile drawings of dolia from Gallia Narbonensis, by Gina Tibbott after Carrato (2017). (*R*) Gallic dolium, Vienne, by Romainbehar, 2021.

they were originally made around Rome or Minturnae and then likely shipped to the villas.[32] Maureen Carroll has suggested that the imperial villa of the Vicus at Vagnari in Puglia shuffled imperial resources, bringing dolia to a new viticultural center in southern Italy, while Emlyn Dodd, Giuliana Galli, and Riccardo Frontoni have speculated that perhaps dolia were shifted from Villa Magna to the Villa of the Quintilii as another example of circulating imperial resources.[33] Yet this kind of trade, and the central Italian dolia in southern Gaul, was anomalous. Carrato's study also identified several Roman entrepreneurs who innovated dolium production in Gallia Narbonensis by setting up workshops and bringing skilled craftspeople and plans for larger and more effective kilns.[34] The vast estate of Saint-Bézard à Aspiran included numerous facilities, such as a private bath, vineyard, and pottery workshop, and has been traced based on seals and stamps to a Roman citizen possibly from Puteoli, Quintus Iulius Primus or Priscus, and his dependents and associates (Figure 5.3A).[35] The pottery and opus doliare workshop, set up when the villa was founded in the beginning of the first century CE, produced dolia, amphorae, and building materials, presumably to build the massive villa and support its large-scale viticulture. The three massive dolium kilns (supported by ten clay-processing basins) could hold twenty to forty dolia per firing, able to manufacture many dolia that surely contributed to the villa's wine cellar of over three hundred dolia defossa, which were capable of holding 420,000–450,000 liters. The layout and dimensions of the kilns were scaled up for dolium production, the plans of which have been suggested to have been brought by Italian immigrants.[36] A stamped dolium, bearing the name of Quintus Iulius Priscus, testifies to Quintus Iulius Priscus employing potters on-site to produce dolia. The overall prevalence of Roman and Italian producers in Narbonne suggests that Italian immigrants might have

32. Montana et al. 2021; Manca et al. 2016.

33. Montana et al. 2021; Carroll 2022a, 2022b; Dodd et al. 2023, 449–450.

34. For a list of villas or farms and ports with dolia in southern Gaul, see Carrato 2017, 324–598.

35. Mauné and Carrato 2012.

36. Mauné et al. 2006; Mauné et al. 2010; Mauné and Carrato 2012; Carrato 2012. The pottery production area covered a large part of the 2.5 ha estate and featured a large circular basin to work clay and twenty kilns, three of which were dolium kilns (others were for amphorae, terra sigillata, bricks and tiles, etc.), with four heating chambers; the best-preserved dolium kiln was 10.50 × 5.90 m (Carrato 2012).

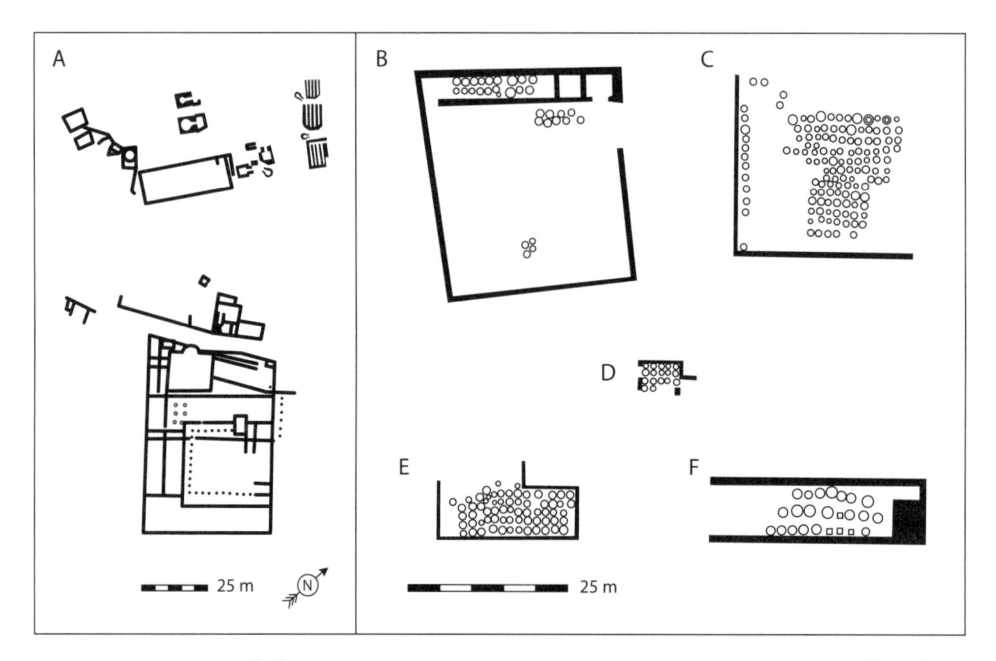

FIGURE 5.3. Plans of (A) the villa estate Saint-Bézard à Aspiran, after Mauné et al. (2010), and cellae vinariae at (B) El Morer, (C) Els Tolegassos, (D) Burriac, (E) Olivet d'en Pujol, and (F) Font del Vilar, (B)–(F) after Peña Cervantes (2010). Illustrated by Gina Tibbott.

partly driven early dolium production in Gaul. Many of these local dolium workshops produced large dolia at the same scale as those found in Ostia and Rome, if not grander, simultaneously developing the local dolium craft while also greatly increasing the productive and storage capabilities of agricultural estates in the northwestern Mediterranean.

According to petrographic analyses and stamps, most of the dolia found in Gallia Narbonensis were local products. Shipping dolia to buyers abroad entailed high costs and risks and was probably uncommon and unattainable for less affluent vintners and vintners without connections to Rome's opus doliare workshops.[37] Indeed potters in Gallia Narbonensis and Iberia developed and refined dolium potting techniques, and most of the dolia found in these provinces were installed not far from where they were manufactured. Local dolium production in Gaul began in the second half of the first century BCE and continued through the second century CE, with smaller workshops that served local customers in earlier periods and larger workshops that serviced a bigger customer base in later periods. Using petrographic and archaeometric analysis of the dolia's ceramic material, Carrato located eleven distinctive centers of dolium production in the province, all concentrated along the coast and the Rhône River.[38] Although most workshops served their local communities, some sold products to more distant destinations (100–150 km away); and several clusters

37. Additional studies on sources of dolia could yield more examples of imported dolia. Several shipwrecks found with dolia might have been transporting dolia; Heslin 2011; Rice 2016.

38. Carrato 2017, 242–274.

of large workshops supplied places as distant as 450 km away.[39] Certain workshops seem to have been especially successful, drawing clientele from farther afield, and became lucrative investments for elites.[40] Potters of a large and successful workshop in the Middle Rhône River Valley, for example, developed a ceramic paste that not only enabled building large dolia but also reduced the amount of fuel and production costs needed to fire a dolium, further increasing profit margins.[41] Wealthy agricultural estate owners might have even financed dolium production as part of their portfolio, especially if their workshops were already producing bricks and tiles, to guarantee equipment for wine production.

Perhaps motivated by opportunities and potential profits abroad, local dolium production blossomed with its own idiosyncrasies particular to the region, including size, shape, and decoration.[42] Decorative elements on Gallic dolia reflected earlier trends on local storage jars influenced by Greek and Etruscan traditions, while the form and embellishment of Iberian dolia stemmed from Punic-Phoenician storage jars. Potters in Gallia Narbonensis and Hispania had developed craft techniques for dolium production with their own aesthetic and functional preferences that reflected the distinctive local knowledge networks, traditions, influences, climate, and agricultural calendar. In mastering the dolium craft and adapting it to their local environment and needs, dolium potters of the northwestern Mediterranean supplied local estates with storage equipment that was not only supersized but also tailored to their region, enabling wine production and storage at a colossal scale.

Villas and Agricultural Production: A New Scale of Production

Over time, many agricultural estates had expansive sets of dolia, some with over one hundred on a single property. Recent publications of dolia and smaller dolium-like jars from these regions show how numerous and ubiquitous dolia were in areas practicing intensive viticulture and oil production in the western Mediterranean.[43] Found at rural agricultural facilities as well as urban storage buildings, dolia appeared in great numbers as these areas expanded their production of cash crops starting in the second century BCE in and around Catalonia and in the late first century BCE in southern Gaul. Dolia mapped onto the agricultural needs of Iberian and Gallic estates well, not only because of the region's preexisting pithos tradition but probably also because of the way dolia could be inserted into traditional local storage practices. Local viticulture during the Iron Age, albeit small scale, made use of pithoi while cereals and other dry foods were stored long term in silos. With the expansion of local agriculture and development of massive dolia, agricultural estates began to store wine in dolia, often concurrently next to silo fields or on top of the silo fields, changing both the type of commodity (from grain to wine) and the scale of

39. This is probably similar to patterns of production and scale in central Italy, where opus doliare workshops near the capital produced high-quality dolia that also supplied customers farther afield.

40. Carrato et al. 2019: the workshop studied in this paper supplied storehouses as far away as ca. 250 km; the 350 dolia at the Vareilles villa would have been from this workshop (Mauné 2003).

41. Carrato et al. 2019: potters added calcite to a fluvial clay (ca. 78 percent clay and ca. 22 percent calcite temper).

42. See Carrato 2020 for comparison of dolia from Italy, Gaul, and Spain.

43. Salido Domínguez 2017; Carrato 2017; Carrato and Cibecchini 2020.

production. Astrid Van Oyen has posited that the building of cellae alongside or on top of silo fields was a mnemonic association between Iron Age storage practices in a Roman imperial order.[44] In Catalonia, for example, several villas featured silos and dolia alongside one another. At Can Bonvilar and Torrebonica, silos and cellae vinariae equipped with dolia were constructed and in operation in the first and second centuries CE. The estate of Sentromà in Tiana shows how silos could be directly replaced by dolia, as the recesses and bases for several dolia were positioned on top of the previous generations' silos and their dimensions echoed those of the silos. The two agricultural sites of Olivet d'en Pujol and Els Tolegassos in Girona both featured cellae vinariae, with approximately seventy-five dolia installed directly over the previous generations' silos at Olivet d'en Pujol and approximately four hundred dolia in a large 690 m^2 storeroom in the expansive villa of Els Tolegassos (Figure 5.3C, E). The coupling or overwriting of silos and dolia occurred, and quite frequently. On a practical level, the reclamation and retooling of previous silo fields into cellae vinariae could have been a labor-saving way of installing dolia defossa. These spaces had already been designated for storage, and workers were already accustomed to the rhythms and mechanics of underground storage. With slight modifications too, the silos that had been dug out for storing and hermetically sealing cereals could also be retrofitted—enlarging or reducing the pits and lining the recesses—for the installation of dolia defossa widely across the northwest Mediterranean by the first century CE.

Iberia, especially Tarraco, had a long history of viticulture. Evidence from the eighth and seventh centuries BCE reveals the presence of wine production in the area, which could have been developed by local domestication or brought by the Phoenician colonists to the Iberian Peninsula. Several sites were discovered to have fermented and stored wine in silos, while others used local pithoi. Some sites in the Cadiz region even produced a local type of amphora for the transport and sale of wine as early as the eighth century BCE. Nonetheless, wine production was nowhere near as immense as in later periods. By the second century BCE, however, the scale of and evidence for both wine and olive oil production exploded across Iberia.[45] The region of Tarraconensis became especially well known for its wine production and Baetica for both wine and olive oil production, but they were not the only areas with large-scale production for export. In fact, over seven hundred sites with pressing facilities have been identified; every region in ancient Hispania used dolia, and nearly every excavated agricultural site (90 percent) featured a cellar filled with dolia.[46]

By the first century CE, dolia were installed at many sites in Iberia following local conventions. Most wine cellars were approximately 150–380 m^2, with 75–190 dolia defossa installed.[47] Landowners along the coast, especially in Tarraco and Lusitania, installed some of the most sizable storerooms in Iberia on their estates. Two kinds of cellae vinariae were found in Iberia, with regional preferences: long, narrow cellae in Tarraconensis, Lusitania, and Baetica, areas known for colossal scales of wine and olive oil production, and square cellae in northeast Iberia (Figure 5.3B-F). The largest cellae vinariae discovered in

44. Van Oyen 2019, 2020b. See also González-Vázquez 2019.

45. For the distribution of villas and cellae, see Carrato 2013, figure 2.

46. Peña Cervantes 2010, 2020.

47. Carrato 2013. For a list of cellars with dolia in Hispania, see Salido Domínguez 2017, 282–289.

Iberia thus far include Torre de Palma in Lusitania (553 m^2) with three hundred dolia defossa and Els Tolegassos in Tarraconensis (690 m^2) with four hundred dolia defossa. These were the regions not only where viticulture was thriving early on and that later developed their own, distinctive Dressel 2/4 wine amphorae but also where dolium tanker ships from Italy stopped to load Spanish wine in the first century CE.[48] Dolia were widespread, and even modest farmhouses could feature over six dozen dolia to produce wine or olive oil for urban markets. Wine-producing estates in Tarraco had ample space dedicated to cellae vinariae, where several dozen to as many as four hundred dolia were installed for the fermentation and storage of wine, rivaling the largest estates in Italy.[49] As a result, modest villas could produce 30,000 liters of wine, and some of the largest villas were capable of producing up to 320,000 liters of wine per year, enough wine for roughly 1,300–3,200 people, and Tarraco began to supply most of southern Gaul's wine.[50] Production extended beyond the countryside and bled into urban areas too. In Barcino (Barcelona), remains of dolia and other processing equipment point to a winery and garum factory in the heart of the ancient town.[51]

Dolia were not universally employed across all of Iberia. Several inland sites in Iberia did not install dolia but used smaller freestanding storage jars instead; at the sites of Las Musas in Navarra, Rasero de Luján in Cuenca, Rumansil in Murça do Douro, and Cortijo de la Marina in Sevilla, dolia defossa were notably absent, but wine fermentation and storage took place in freestanding jars scholars have identified as *(h)orcae* in Latin texts (Figure 5.4L). *Orcae* had wide shoulders, were occasionally feted with small aesthetic handles, and had little to no lip around the rim, but they were not buried in the ground and were easily removed, making the identification of wine production facilities difficult. Likewise, only a few oil cellars have been identified in Iberia, despite the region's reputation and the staggering number of olive oil amphorae that left an archaeological footprint across the Roman Empire, especially in the city of Rome. The few identified oil cellars were all elongated structures, and most rather small, the smallest being only 20 m^2 with only 10 dolia. The largest oil cellar was 240 m^2 with 120 dolia, capable of holding 1.6 million liters, but that cellar was an anomaly. Similar to the situation in central Italy, the presence or absence of dolia in Iberia is unreliable as an index of olive oil production. Carrato has suggested that the low number of dolia for olive oil production is due to producers placing olive oil directly in the transport amphorae, rather than settling the oil in dolia and then transferring the oil into amphorae, as a time- and labor-saving strategy and perhaps to export olive oil before the winter months.[52]

During the imperial period, Gallia Narbonensis had built up its own production and storage vessels and facilities and was well equipped to import and export wine and oil, often overshadowing Iberian and Italian estates. Many of these storerooms contained a multitude of large dolia, testifying to the enormous productive capacities and capabilities of Gallic vineyards from the first century BCE onward. Villas in Gallia Narbonensis commonly contained

48. For an overview of dolia found off the coast of Tarraconensis, see De Juan 2020.

49. Salido Domínguez 2017, 260; Brun 2004, 266.

50. Castanyer Masoliver and Tremoleda Trilla 2007, 275–290.

51. Beltrán de Heredia Bercero 2001b, 2001c.

52. Carrato 2013.

FIGURE 5.4. Cellae vinariae of (*L*) the Roman villa of Arellano (Navarra), by Yiorsito, 2008; and (*R*) 28 Place Vivaux, Marseille, by Robert Valette, 2008.

50–150 dolia as part of their storage infrastructure. The villa of Saint-Martin, for example, is considered an average-sized estate in southern Gaul, but with sixty-six dolia in the cella vinaria, it was easily three times the size of the Villa Regina in central Italy. Even modest farmhouses such as Prés-Bas were equipped with ninety dolia and capable of storing 150,000 liters of wine, much more than the massive Italian Villa Settefinestre and Villa della Pisanella.[53] Some villas modified their storerooms to augment the number of dolia in use. The villa of L'Estagnol in Clermont-l'Herault was outfitted with a large storeroom of seventy-one dolia in the first phase but was later expanded with new rooms with additional dolia installed to increase the villa's storage further.

The large capacities of the Gallic dolia, combined with the massive size of some of the estates, led to unprecedented levels of production and storage. Furthermore, wine-producing villas in southern Gaul were expansive and often featured multiple presses, whereas Italian estates were smaller and often had only a single press, testifying to the high levels of investment among Gallic estates. Equipped with two crushers, four presses, and an enormous cella vinaria, the site of Le Molard in Donzère in La Drôme was an immense wine-producing estate.[54] An astonishing 270 dolia defossa were packed into a long rectangular cella vinaria (67.5 × 14.9 m) that was divided by a central colonnade; the dolia were buried to their shoulders and clustered together with a space of only thirty to forty centimeters between the dolia defossa. The preserved capacity incisions in units of amphorae on several dolia indicate that the average capacity was ninety amphorae, the equivalent of 2,444 liters per dolium. The entire cella vinaria was able to hold approximately 660,000 liters of wine, enough for the annual consumption of a small village. This staggering amount was ten times more than Villa Settefinestre's storage capacity and more than seven times Villa della Pisanella's, the largest viticultural production estates in central Italy. Le Molard in Donzère was not an isolated example, but one of several enormous agricultural estates in southern Gaul. Two first-century villas in Languedoc further demonstrate incredible storage capacities. The villa of Saint-Bézard at Aspiran with the dolium workshop had a warehouse that was gradually expanded to hold about 300–310 dolia, capable of storing 420,000–450,000 liters of wine.

53. Pellecuer 2000.

54. Odiot 1996.

Approximately 150 hectares of the vast estate were planted with vines, estimated to have produced 450,000 liters of wine on an annual basis. The wine cellar at the villa of Vareilles, in operation from the second half of the first century BCE to the beginning of the third century CE, initially had at least 350 jars, then enlarged to 477 in four separate cellae vinariae in the last phase of the villa's occupation. One dolium had a capacity incision designating its ability to hold 149 amphorae (3,874 liters) of wine.[55] If that is the average-sized dolium, the whole estate could produce nearly 1.5 million liters of wine, enough to supply a small town.

The dolium-based storage technology became integral not only to large-scale agriculture but also to long-distance bulk transportation. Cellae vinariae constructed at various coastal and fluvial ports, such as Massalia, Lugdunum, Lattara, and Tarraco, boosted these trade nodes by supporting storage and transport infrastructure for wine merchants, especially those operating dolium tanker ships or moving wine in cullei.[56] Merchants could then quickly collect wines, whether local or imported, and package them for further distribution; cellae vinariae even enabled merchants to move wine without amphorae, which could be time consuming and expensive to acquire and prepare for large shipments. With the grape harvest and sailing calendar, time was of the essence. Massalia, a colony founded by the Phocaeans in the Iron Age, became a prominent port town and was outfitted with three known warehouses in the Roman period. The partly excavated warehouse at 28 Place Vivaux, for example, shows sheer storage concentrated at an ancient port (Figure 5.4R). Discovered in the middle of the twentieth century, the ancient warehouse had well over thirty large dolia and has been reconstructed as originally containing at least sixty-five dolia defossa in antiquity. Towering at nearly two meters tall and wide, they have each been estimated to contain 1,800–2,000 liters. The resinated interior walls of the dolia and the large-scale local Dressel 2/4 wine amphora production suggest that the dolia stored wine temporarily for transport in antiquity. Two other commercial warehouses have been found and partly excavated in Marseilles, one with 162 dolia defossa and the other with 115.[57] The multiple, large cellae vinariae clustered around the ancient shoreline provided flexibility and massive capacity for merchants loading and unloading wine on-site. Lattara (Lattes), another important commercial port that thrived thanks to its lagoon environment, developed as early as the second century BCE; by the last quarter of the first century BCE, the port was expanded with new infrastructure including two long warehouses (ca. 8 × 30 m; ca. 5 × 24 m), one of which contained seventy-three and the other forty-three dolia defossa closely packed in long rows. Inscriptions from the town featuring the occupation *utricularius* suggest that culleus handlers frequently worked at the port, perhaps conveying wine to and from inland sites to the port's cellae vinariae for expedited distribution; the wine in the cellae vinariae's dolia defossa had to be transferred into another container, and the cullei would have been an ideal bulk container for overland transport. Numerous fluvial and coastal Iberian settlements, such as Empúries, also constructed cellae with large sets of dolia to boost the collection point's storage capacity, bolstering the storeroom's impact as a nexus in exchange networks. Commercial ports featuring dolia defossa along the marine and fluvial routes

55. Carrato 2017, 426ff.

56. Excoffon 2020: the cella vinaria in Frejus was likely attached to a production facility.

57. On the port, see Hesnard 1994, 1995, 2004; France and Hesnard 1995; Philippon and Védrine 2009.

to markets in Gaul and Iberia trace a network of maritime and fluvial warehouses and ports to support large-scale and expedited wine transport between the coasts and their hinterland. In fact, this was a system that also extended to west-central Italy, where dolia defossa were commonly found in urban warehouses.

With the expansion of dolium use came new opportunities in the northwest Mediterranean and sources of wine for the city of Rome. As landowners in Gaul and Iberia were ramping up wine, olive oil, and garum production, they invested not only in large dolia but also in the production of export amphorae, which is well attested archaeologically in the city of Rome. The regions of both Iberia and southern Gaul had their own versions of Dressel 2/4 wine amphorae, which overthrew Italian Dressel 2/4 amphorae as the most common type of amphora in the Mediterranean from the mid-first century CE onward, leading some scholars to suggest an emergence and even dominance of provincial wines at the expense of Italian wine. New or tightened links between traders and markets of central Italy, southern Gaul, and northeast Iberia not only brought wine but also introduced technologies and opportunities into local practices. The growing consumer base of the imperial capital, moreover, shifted the flow of goods, pulling more and more from different parts of the empire. From the mid- to late first century BCE onward, villas, large farms, and warehouses spread throughout Gallia Narbonensis and Iberia as entrepreneurs capitalized on opportunities to mass produce and export wine to the ready markets of Rome. Estates in Gaul and Hispania frequently surpassed the largest villas in central Italy not only in sheer size but also in production capabilities. Gallic and Iberian villas usually had far larger sets of dolia than Italian estates, and many also featured multiple presses. While only a single press at an Italian villa was the norm (Settefinestre with its three presses was exceptional), it was not uncommon for villas in Iberia and Gaul to have several presses, with some massive villas furnished with six presses. The preference for multiple, small landholdings seen in central Italy (Chapter 3) did not apply to estates in Hispania or Gallia. Instead, estate owners in the western Mediterranean maximized landholding, equipment, and storage facilities for mass production of cash crops that they sent to the capital and other lucrative markets. In and around the capital itself, multiple cellae vinariae were set up to receive and distribute wine and other foods. A whole network of these warehouses along the coasts of central Italy, Gallia Narbonensis, and Hispania Tarraconensis extended the empire's *façade maritime*, facilitating and diversifying the wine trade.[58]

58. On Rome's *façade maritime*, see Purcell 1996.

6

Dolia in *Urbe*

EXPANDING URBAN STORAGE AND CONSUMPTION

IN THE SECOND CENTURY CE, a winery owner commissioned a large funerary relief that depicted an expansive wine cellar filled with workers busily transferring wine from dolia to amphorae (Figure 6.1).[1] Two large figures on the left shake hands, presumably over a successful agreement or transaction. A seated figure in the upper right corner takes stock by recording the sales, expenditures, and profits of the winery, underscoring the wealth and prestige a well-stocked wine cellar could generate. By this time, however, the owner might not have been a vintner, and might not have been in the countryside at all. The seated figure on the right could have been recording not how much wine the estate produced but the amount of wine an urban storeroom had collected and sold, as well as their profits. Storing and selling wine in towns and cities not only brought opportunities and profits but also cemented and elevated social distinction. By the second century CE, the depicted wine cellar could have come from a productive farm or a large villa, but it could have just as likely been a storeroom or part of a bar or shop in a town or city.

The growth of wine production and consumption led to the widespread adoption of storage containers also in urban areas, a development that we can trace most clearly at Rome and its environs. In the second century CE, Aelius Aristides delivered a panegyric to Rome, praising the empire for its command of resources that made the capital the market for the world's goods.[2] Yet in order for the imperial capital to receive and distribute so many goods, it also had to create storage facilities to store them. Scholars have estimated that, on average, one person consumed as much as 250 liters of wine each year;[3] if we multiply those amounts by the population estimate of one million, we can gauge how much wine the entire city of Rome would have needed annually. The supply of this massive amount of wine to the city required extensive facilities and organization for its storage and distribution, including, among other things, the construction of specially outfitted warehouses along the banks of the Tiber River and downstream at Ostia, some of which

1. Angelicoussis 2009; Zimmer 1982.

2. Aristid. *Or.* 26.11–13.

3. This assumes the amount of wine before it was mixed with water, probably at a 1:1 ratio. Estimates of wine consumption include Kehoe 2007, 566: 100 liters; Tchernia 1986, 26: 146–182 liters; Purcell 1985: 250 liters/year.

FIGURE 6.1. Depiction of a Roman wine cellar on a sarcophagus, World Museum Liverpool. © National Museums Liverpool / Bridgeman Images.

included dolia. The introduction of dolia to urban areas had other socioeconomic effects on local market relationships and investments as well, offering new opportunities for retailers and, in some cases, concentrating resources in the hands of the wealthy.[4]

Although most dolia were traditionally used for wine fermentation on farms and villas, this chapter explores how a technology for rural storage was introduced into urban environments and contributed to urban economic growth. A survey reveals different uses of dolia and their wider impact on urban areas in central Italy, especially on the scale of the local economy and on the relationship between towns and their hinterland. By integrating dolia into towns and cities, communities fostered not only the movement of foods but also urban food retail and service industries, opening new opportunities for entrepreneurs and investors who were not necessarily engaged with traditional food or craft production. Owning and operating a wine cellar was expensive, however, and only some had the means to invest in and operate such an enterprise. In the capital, setting up and running a cella vinaria was especially expensive and challenging, which further fueled unequal wealth and power. For those who had the means, such an investment helped secure and manage control over valuable resources and made them even wealthier. The following sections focus on three case

4. On different types of retail in the city of Rome, see Holleran 2012. For the archaeology of retail spaces, see Ellis 2004, 2018; Mac Mahon 2005, 2006.

studies—Cosa, Pompeii, and Ostia and Rome—to consider ways this new storage technology was taken up (or not) by urban communities in the heart of the empire.

Trials and Tribulations with Technology: The Case of Cosa

With its ideal position for trade and its productive countryside (Chapters 3 and 4), Cosan residents had much to gain in experimenting with the dolium-based storage technology (Figure 6.2L).[5] Many of the agricultural estates were large properties equipped with dolia that engaged in viticulture during the final two centuries BCE.[6] From the mid-second century BCE on, inhabitants brought vessels typically installed in country estates into the town. The town itself, however, was never particularly prosperous and perhaps struggled with a dwindling population over the course of its long history.[7] Although residents over the next couple of centuries sporadically used the jars, the dolium-based storage system did not align with their consumption patterns and was ultimately abandoned.

Cosa's inhabitants could buy dolia from local workshops but had low numbers of dolia in their houses—only about fifty dolia over the course of the town's occupation have been found.[8] Although delivering dolia to Cosa presented additional expense and could have been one factor among several in a customer's decision—porters likely needed a costlier carriage to ensure they could make the journey up the precipitous hill—plenty of other remote and elevated sites had dolia too.[9] Other factors affected the scarcity of dolia at Cosa—namely, whether the kind of bulk storage dolia offered aligned with the town's needs, especially in a region where amphorae were abundant.

Almost all the dolium fragments from Cosa date to the period from the middle of the second century to the first quarter of the first century BCE, the peak of the town's intermittent history. The small- and medium-sized storage vessels were often found in houses, such as the House of the Skeleton and several others on the west block, and were used to store household grain, olive oil, wine, or other foods. The large dolia, on the other hand, were used for bulk storage of local wine or oil likely brought in from the large estates that dotted Cosa's hinterland.[10] Archaeological excavations of large villas nearby, such as the Villa Settefinestre with its large dolia-laden winery, and local amphora workshops reveal the footprint of the wine industry. In addition to the Sestius family's wine and pottery empire in the *ager Cosanus*, an amphora production site near Cosa (15 km north) at the modern town of Albinia also kept

5. Carandini et al. 2002; Perkins 2012.

6. See Celuzza 1985 for an example of a small estate.

7. Livy *Ab Urbe Condita* 39.55: as early as 197 BCE, less than a century after Cosa's foundation, Rome sent another draft of colonists to the shrinking colony. More recent work on Cosa has been fine-tuning and reshaping the narrative of Cosa's occupation—e.g., Fentress 2003; De Giorgi 2018; R.T. Scott et al. 2015.

8. Gliozzo 2007, 2013; Gliozzo et al. 2014; papers in Vitali 2007; Vitali et al. 2012.

9. Examples of dolia on hilltop or remote sites include Cetamura (De Grummond 2020) and sites excavated by the Roman Peasant Project, including Case Nuove, Pievina, and Poggio del'Amore (Ghisleni et al. 2011; Vaccaro et al. 2013; Bowes et al. 2017; Bowes 2021a).

10. Bruno and Scott 1993, 1: ca. two-thirds of the settlement was for public works. Houses had gardens, but not vineyards; Fentress 2003, 17ff.: the House of Diana had a kitchen garden. The literature on viticulture in the *ager Cosanus* is vast; see Manacorda 1978, 1981, 1980; Rathbone 1981; Dyson 1978; Carandini 1985; Carandini et al. 2002.

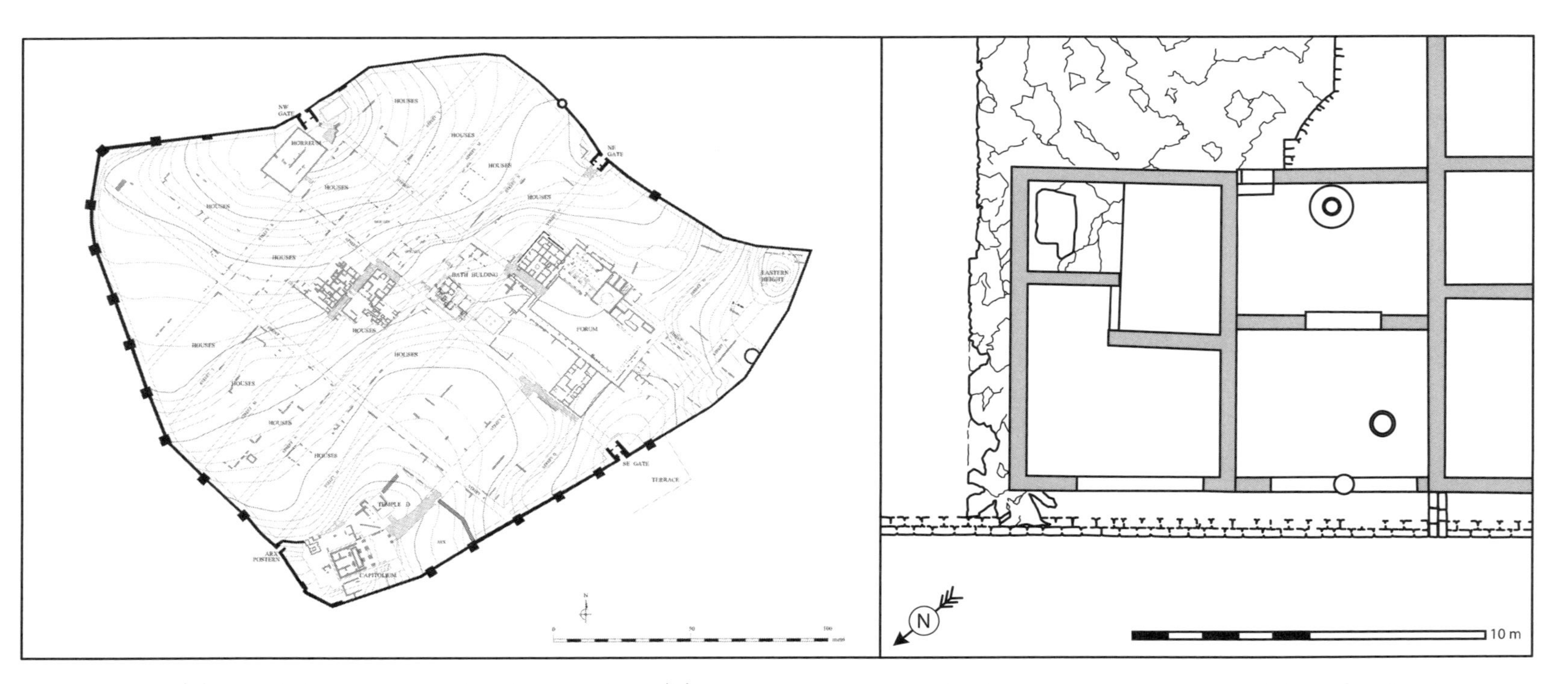

FIGURE 6.2. (*L*) Plan of Cosa, courtesy of Andrea De Giorgi. (*R*) Plan of Cosan tavern with dolium in Room 24, by Gina Tibbott after F.E. Brown et al. (1993).

vintners of the *ager Cosanus* well stocked with amphorae to ship wine from the port of Cosa to lucrative markets abroad.[11] Although the wine was exported as far afield as Gaul, some of it surely arrived at the local market of Cosa too. Merchants could transport local wine in individual containers such as amphorae or en masse in animal-hide containers on carts, decanting the wine at the destination.[12] In the latter case, workers could pour the wine carefully into individual containers one by one—for example, wine from a single ox-hide container into twenty amphorae—but a single, large dolium enabled faster decanting and reduced bottlenecking, and sellers could later package the wine into smaller vessels for distribution. Large dolia stored wine in the town's horreum, near the northwest gate, and in a shop by the Forum (Figure 6.2R).[13] A shop owner might have had arrangements with local estates or even owned both the estate and shop to sell their agricultural products in town.

It was not necessary to utilize dolium technology without (many) consumers in the town who purchased wine regularly from these shops, though; Cosa's consumption patterns, and perhaps its hinterland's declining production in later years, might explain the scarcity of dolia. As Macrobius in the *Saturnalia* notes, storing wine in dolia could be problematic: "But if you draw [wine] from it [the *dolium*] and expose space for the air to mix in, all that is left over becomes spoiled" (7.16: *ceterum si inde hauseris et locum aeris admixtioni patefeceris, reliquum quod remansit omne corrumpitur*). Today, people are advised to consume wines within three to five days of opening the bottle, and some wastage was probably common.[14] A dolium's shape, especially its large upper portion, exposed more wine to air.[15] Most ancient wines stored in and distributed from dolia (where the dolia are constantly opened for the retail of wine, rather than sparingly opened only for big transfers), known as *vinum doliare* or *de cupa*, were probably lower in quality (similar to Italian *vino sfuso*, wine on tap) and did not have a long shelf life. The impact of this shortcoming can be observed in Terence's *The Self-Tormentor* when the farmer Chremes bemoans a courtesan's extravagance as he "opened all [his] *dolia*, all [his] jars (*seriae*)" (l. 460: *relevi dolia omnia, omnis serias*), for her and her guests to try and then decline. Chremes' outrage was rooted in wine storage practices: when properly filled and sealed, these ceramic containers could store wine for extended periods of time; once opened, however, the wine was exposed to air and would begin to deteriorate. Chremes' wine cellar had several dolia and *seriae* (a smaller storage jar), likely arranged to store different vintages: more valuable vintages in smaller *seriae* and lower-quality wine in large dolia.[16] In Cosa, shops that sold food and beverages were not numerous or widespread, though.[17]

11. Vitali 2007.

12. Tchernia 1986, 285–292; Panella 1989, 162–163; De Sena 2005, 140; Brun 2004, 47–48, 284–294.

13. F.E. Brown et al. 1993, 106; F.E. Brown 1984, 495–497. The new excavations of the town's horreum, Die merkantile Zentralortfunktion der römischen Kolonie Cosa (Ansedonia), which began in September 2023, have uncovered not only wine storage, but possibly also wine production on site for the town's consumption (pers. comm. Maximilian Rönnberg). Future excavations might uncover more evidence, including additional dolia (one dolium rim was found in the horreum in the soundings in the 1970s).

14. E.g., Cato *Agr.* 148; Columella *Rust.* 12.26; Juv. 13.216, 3.292; Hor. *Sat.* 1.5.16. See James 2020, 94–125, for problems with grain wastage in the supply chain.

15. Vintners using *talhas* in Portugal add a layer of oil on the wine to prevent oxidation and drain wine from the bottom of the vessel.

16. On *seriae*, see K.D. White 1975, 185–188.

17. F.E. Brown et al. 1993.

The dolia of the town were mostly out of use by the mid-first century BCE, the same time estates in the *ager Cosanus* involved in wine production began to drop off or invest in other pursuits. By the first century CE, only a couple of large dolia serviced the town.[18]

The inhabitants of Cosa practiced not the vast, bulk storage that dolia supported but storage that was more individualized and fragmented from its countryside. While many large dolia in the *ager Cosanus* fermented wine and stored wine and oil in bulk quantities at production facilities, the town itself had different storage practices. Throughout its occupation, Cosans obtained their wine and olive oil in amphorae not only from the *ager Cosanus,* which was certainly well serviced by the Sestius family workshop—in fact, over half of the stamped amphorae found at Cosa were produced by the Sestii—but also from other regions of the Mediterranean.[19] A diachronic history of amphora use in Cosa suggests that the town (and its port) was initially an export center, but that changed during the first century BCE.[20] The once-bustling port and consignment point for the export of wine silted up at the end of the Republican period and, by the end of the first century BCE, the movement of goods was increasingly redirected to Rome. From the first century CE onward, Cosa and its *ager* were no longer producing wine for mass export to profitable markets abroad. Inhabitants of the town continued to consume wine and oil, but during the imperial period, they instead consumed Gallic and African wine and African oil and Spanish garum imported in amphorae. As the region shifted from an exporter to importer of agricultural goods, the town also pivoted from a mixed storage system that included bulk storage offered by dolia to almost exclusively smaller, individualized household storage. Amphora use suggests not only the import of different foods but also the ability and desire to institute more controlled storage and portioned access to them. Unlike the other towns we will explore in this chapter, Cosa did not have a bustling population or retail landscape and, certainly by the first century CE if not earlier, it was not a common destination. Moreover, attempts at Cosa to incorporate dolia into the urban fabric predate the widespread urban retail revolution of the first century CE that is especially salient in Pompeii.[21]

Pompeii: Dolia in Urban Retail and Service

With its extraordinary preservation, Pompeii offers the most detailed snapshot for urban dolium use in the first century CE. A town of approximately sixty-five hectares in the Bay of Naples, Pompeii was well connected by land and sea thanks to numerous roads and a port. At the time of the eruption of Vesuvius in 79 CE, the town, with its nearly two hundred storage jars, had not only embraced the storage technology the dolium offered but also adapted it widely for a variety of purposes (Figure 6.3). Because Pompeii was primarily an agro-manufacturing town in a fertile landscape, residents regularly used these jars, most of which are still in situ, for numerous agricultural and retail activities at farms, vineyards, gardens, shops, bars, and houses.[22] The town had a close relationship

18. Unfortunately, they are without provenance.

19. Will and Slane 2019.

20. Will and Slane 2019, 102ff.

21. Ellis 2018, 151–186.

22. At least fifteen dolia were removed from their contexts and were not accessible. Several dolium fragments were found in a midden by Tower 8 at Pompeii, which had no accumulated material beyond 50 CE; Romanazzi and Volonté 1986, 57; Peña and Cheung 2015, 2119–2120.

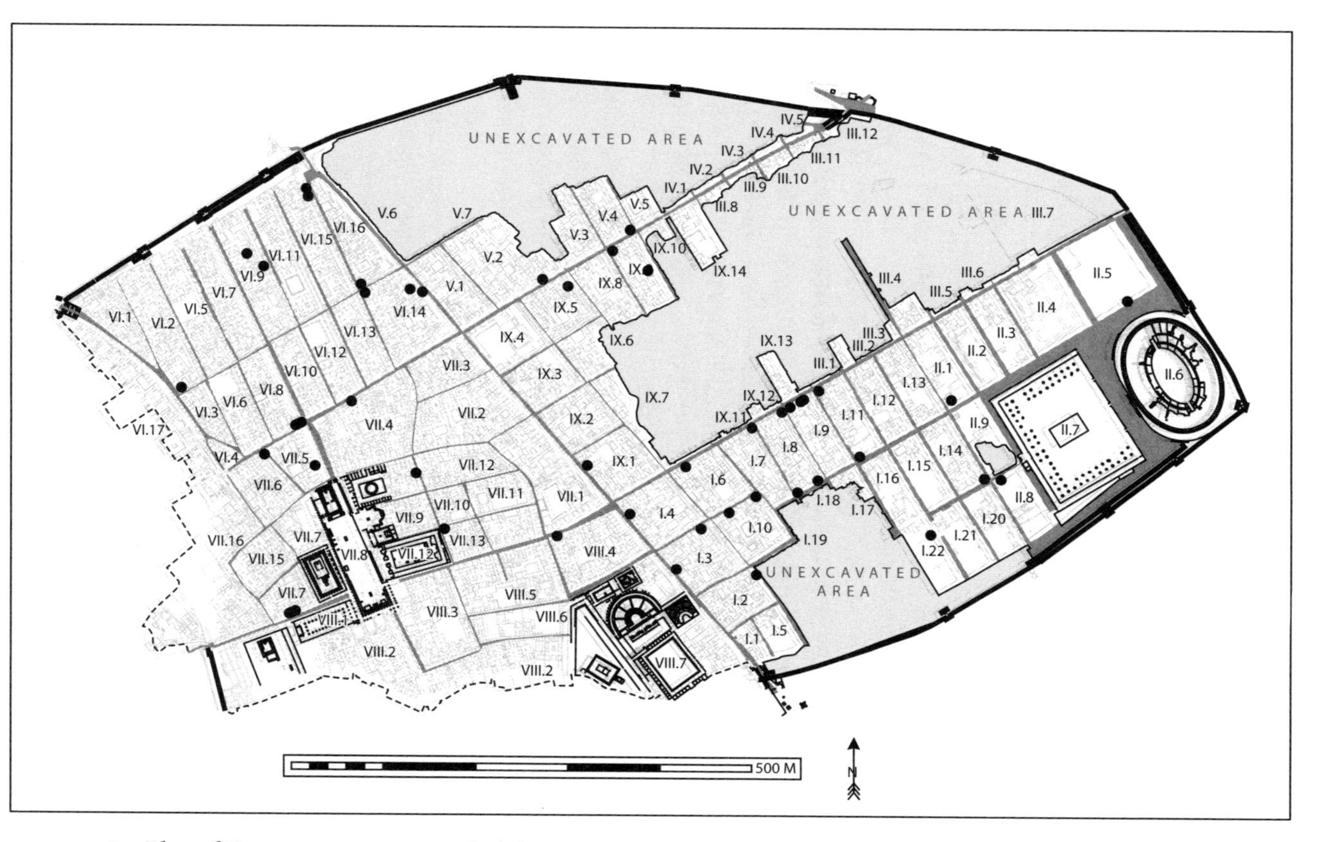

FIGURE 6.3. Plan of Pompeii, properties with dolia and storage containers marked, by Gina Tibbott.

with its hinterland, and the dolia and a second type of bulk storage jar further blurred the distinction between town and country by not only facilitating agricultural production and storage in town but also fostering service industries that centered on wine and dining.

Pompeii was in a well-known wine-producing region, and its residents drank a great deal of local wine. Many farms and villas stocked with large sets of dolia were in operation near Pompeii, such as the Villa of the Mysteries just outside the Herculaneum Gate, the Villa Regina and Villa della Pisanella in Boscoreale, and the Villa of N. Popidius Narcissus in Scafati. According to Pliny the Elder, wine production in Italy was well established by the mid-second century BCE, and the Bay of Naples was known for certain vintages, such as Falernian, Surrentine, and Massic. "For there [around Vesuvius] is the Murgentine, a very strong vine from Sicily, which some call Pompeian, fruitful only in fertile soil, just like the Horconian grown only in Campania" (Plin. *NH* 14.35: *ibi enim Murgentina e Sicilia potentissima, quam Pompeianam aliqui vocant, laeto demum feracem, sicut Horconia in Campania tantum*). The reputation of wines from the Bay of Naples and the sheer quantity of dolia and amphorae testify to the region's remarkable wine production and consumption. As mentioned in Chapter 4, a staggering number of washed and pitched amphorae stacked in the courtyard of the Villa B in Oplontis, where local wines were collected and bottled for shipment abroad, illustrates the area's massive scale of wine export.[23] Although many farms and large villas were engaged in producing wines for overseas markets, a significant portion of the wines also reached local markets, including Pompeii and Herculaneum. In this way, viticulture at Pompeii was similar to that at Cosa: large-scale agricultural production in the *ager* for both export and local consumption.

Yet the similarities between Cosa and Pompeii end there. While a few houses in Cosa had storage jars, Pompeian houses, on the other hand, typically did not have dolia for household storage.[24] The absence of these food storage vessels in residential settings suggests that dolia were not considered general household items, and residents of Pompeii instead visited shops and bars on a regular basis to buy fresh food.[25] The bars of Pompeii, with their masonry service counters, constitute one of the town's most distinctive features.[26] Called *thermopolia, tabernae, popinae, cauponae,* and even the "McDonald's of the ancient world," bars could be found throughout the town, with many clustered along the major thoroughfares.[27] Although only a few houses had storage jars on-site, the abundance of dolia in shops gave residents and visitors to the town regular access to agricultural products in various commercial spaces. As Steven Ellis rightly notes, studying these architectural fixtures as "products, or 'outcomes,' of a set of complex agencies in urban investment [can] reveal much information about the social, economic, and even political motivations behind

23. M.L. Thomas (2015) posits, based on amphora stamps, that Villa B's wine might have been sent to the eastern Mediterranean and Carthage.

24. The exception was houses engaged in agricultural production or retail.

25. Ellis 2018, ch. 7. Strickland 2016: before households in China had refrigerators, gas stoves, and ovens, people visited markets for fresh foods and local shops for prepared foods daily.

26. See Ellis 2018, 229ff.; Packer 1978, 47–48; Mac Mahon 2005, 81ff.; Monteix 2008, 2010.

27. See Ellis 2004, 2018; Monteix 2010 for an overview of food and drink establishments in Pompeii. See Ellis 2004, 371–373, for terminology. For early identifications, see Kleberg 1957; Packer 1978.

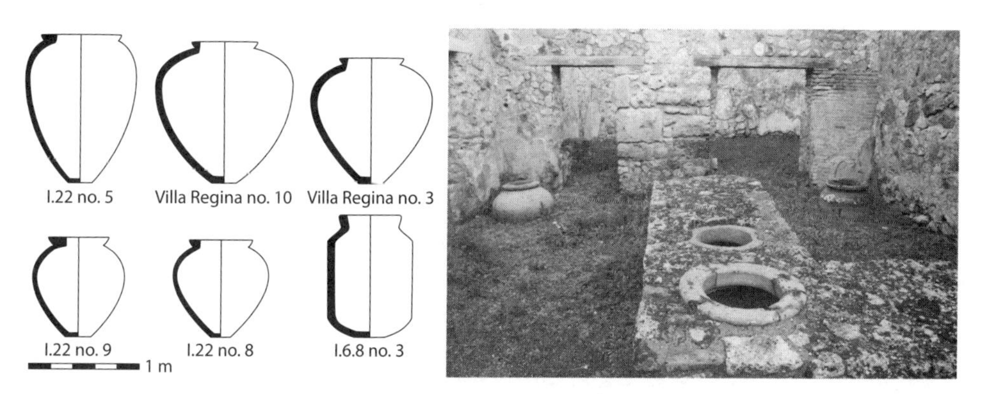

FIGURE 6.4. (*L*) Profile drawings of Pompeian dolia and cylindrical storage jars, by Gina Tibbott. (*R*) Bar with dolia and cylindrical jars installed (VI.14.36), Pompeii. Courtesy of the Ministry of Culture—Archaeological Park of Pompeii. Reproduction or duplication by any means is forbidden.

the development of economic portfolios and the creation and maintenance of the retail network."[28] The distribution of these vessels, like the bars, was strategic.

Food and drink establishments were well equipped with two distinctive types of storage jars to stock up on local and imported foods to serve the town: the dolium, which was ideal for liquid contents, and a cylindrical jar, which often stored dry foods and was built into bar counters (Figure 6.4).[29] This second type of storage jar, which represents half of all the storage jars in Pompeii, often functioned alongside but was distinct from dolia.[30] Although these cylindrical storage jars were manufactured using the same technique of coil building, there were some important differences in their production. Dolia were much more challenging to build due to gravitational forces and air-drying constraints with the vessels' strawberry shape. Cylindrical jars, on the other hand, had straighter and thinner walls (ca. 3 cm compared with the 4–6 cm thick walls of dolia) that a potter could pull up easily while using less clay. With their regular form, which allowed better stacking, they also fit more efficiently in a kiln and were hence more fuel efficient and less costly than dolia. As a cheaper option, the cylindrical jars occasionally served as alternatives to dolia, but most often dolia and these cylindrical jars complemented each other in a specialized food storage system in Pompeii.

While dolia were built for liquid goods, cylindrical jars were commonly used to store dry foods and were thus set up differently. As with dolia, the ceramic material of cylindrical jars stabilized temperatures, kept their contents dry, and prevented vermin and pests from accessing the foods. The shape of cylindrical jars, however, provided particular benefits for storing dry foods, such as enabling regular access to contents. Moreover, only a few cylindrical jars were partly buried in the ground to hold wine in shops; almost all the others were embedded in counters at food and drink establishments to store nuts,

28. Ellis 2018, 26.

29. Monteix 2010, 102–132. The cylindrical jar is similar to Carrato's (2017, 126–127) *tronconique* form.

30. Pompeii (including Villa Regina and Villa of the Mysteries) has at least 108 dolia and 101 cylindrical jars.

legumes, and other foods. With their ceramic walls encased in Roman concrete, adding extra insulation, the jars acted as a pantry by keeping foods dry and insulated.[31] Their regular, cylindrical form also facilitated building a bar counter around these vessels. Although some earlier studies have conflated the cylindrical jars with dolia, more recent scholarship has noted that their installation in counters made it difficult to maintain and clean these vessels in a way that would have made them fit to hold liquids.[32] In fact, excavations of food establishments in Herculaneum revealed that shop owners stored legumes, grains, and vegetables inside these vessels.[33]

Both the dolia and cylindrical storage jars were available in set sizes and used for specific purposes (Table 6.1). Small dolia and cylindrical jars (capacities of ca. 40–150 liters) were usually partly buried near other types of equipment, such as mills in bakeries, and stored food temporarily during processing.[34] Several shops were outfitted with one or two medium-sized dolia (ca. 150–300 liters) partly buried in the ground, likely to store wine or oil brought in animal-hide containers from local farms, in addition to or in lieu of amphorae of wine and oil.[35] A painted lararium from the Hospitium of Hermes (I.1.8), close to the Porta Stabia, suggests workers might have also poured wine from amphorae into dolia to consolidate space (Figure 6.5L):[36] in the same place, a dolium could hold two to four times more wine than amphorae. Storing lower-quality table wine in dolia was probably a common practice, while more valuable wines probably remained in their amphorae. Graffiti and shop advertisements point to shops offering a selection of wine. In a bar (VII.2.44), someone scratched a graffito, "Hedone says, 'You can drink here for an *as*; if you give two coins, you will drink better; if you give four, you will drink Falernian wine'" (*CIL* 4.1679: *(H)edone dicit: Assibus (singulis) hic bibitur; dupundium si dederis, meliora bibes; qua(rtum) (assem) si dederis, vina Falerna bibes*), noting a range in the quality of wines on offer. A painted sign next to a shop (VI.14) in Herculaneum also shows an establishment's variety of wines with four jugs, under which are different prices: two to four *asses* for a *sextarius* of wine (Figure 6.5R). Alimentary shops and bars also sold different food items, such as nuts and legumes, stored in cylindrical jars built into bar counters. These jars ranged in capacity, from under one hundred to over four hundred liters, and some were incised with numerals to label their capacities in units of *urnae* (ca. thirteen liters, half an amphora).[37] The largest jars were generally reserved for grand shops in heavily frequented areas, such as the Thermopolium of Vetutius Placidus (Figure 6.6). Overall, cylindrical jars were generally found in shops and fast-food joints throughout the town, whereas dolia were usually clustered in wine production or service

31. Cheung 2020.

32. Mac Mahon 2005.

33. Maiuri 1958, 402 (V.6), 434 (IV.15–16).

34. Estimates based on mathematical modeling (Cheung et al. 2022).

35. Some bars—e.g., Thermpolium of Asellina (IX.11.2)—kept wine in amphorae. Archaeologists have found a peculiarly low number of amphorae in the town in relation to its population. De Sena and Ikäheimo (2003) note that (re)using amphorae and other containers for local wines distorts the numbers; dolia that stored local wines further support this point. Monteix (2010, 102–127) notes that jar contents were often unknown; only one has been found with its contents: walnuts.

36. Fröhlich 1991, 249; *PPM* vol. 1, 1–9; Ellis et al. 2012.

37. De Caro 1994, 68ff.: dolia volumes were also in units of *urnae* at Villa Regina.

TABLE 6.1. Selection of properties (primarily production) in Pompeii with dolia.

Address	Property name	Number of dolia	Activity
I.20.1	Caupona of the Gladiators	4	Wine production; hospitality
I.20.5	Shophouse garden	8	Wine production
I.21.2	Garden of the Fugitives	3	Wine production; hospitality
I.22	House of Stabianus	9+	Olive grove; hospitality (but under renovation in 79 CE)
II.5	Vineyard by Amphitheater, "Foro Boario"	10	Wine production; hospitality
II.1.8–9	House of the Lararium of Hercules or Felix and Sabinus	1	Wine production; hospitality
II.8.6	Garden of Hercules	3	Perfume production
II.9.6	House of the Summer Triclinium	2+	Wine production; hospitality
III.7	Vineyard	?	Wine production; hospitality
V.4.6–7	Caupona of Spatulus	8	Caupona; wine production; storage area
VI.14.27	House of Memmius Auctus	4	Wine retail
VII.2.48	House of D. Caprasius Primus	2	Wine retail
VII.5.21	Shop	1	Shop
IX.9.6/10	House of wine seller	1 (+ many amphorae)	Wine retail and production; hospitality?
	Villa of the Mysteries	4+	Wine production

FIGURE 6.5. (*L*) Drawing of a lararium, Hospitium of Hermes (I.1.8), Pompeii, by Gina Tibbott, after drawing by Geremia Discanno, 1872. (*R*) Sign of wines and prices in shop (VI.14), Herculaneum, by Carole Raddato, 2014.

sites primarily in the southeastern part of Pompeii. Having the jars in shops and bars provided a variety of foods and drink, and a varied or well-curated selection could attract clientele. For dining establishments, where customers consumed their purchases, offering a more hospitable and social environment could lead to additional rounds of food, drink, and service.

FIGURE 6.6. Thermopolium of Vetutius Placidus (I.8.8), Pompeii, by Gary Todd, 2019.

Pompeii also featured expansive green spaces and agricultural facilities in the town, many of which were dedicated to the production and consumption of wine.[38] Almost a quarter of all Pompeian dolia, and every single one of the largest dolia (average capacity ca. 550 liters), are found chiefly at these properties, mostly clustered in the southeastern part of the town, an area distinctive for its green and open spaces.[39] Wine production facilities in town were planted with vineyards, some of which still preserved root cavities of the vines, and often had equipment such as a wine press or treading vat and large fermentation dolia.[40] Most wine production sites were small, with only two or three dolia, but other sites produced more substantial amounts, such as the vineyard by the amphitheater (II.5), which had ten dolia and was estimated to have been capable of producing up to around 10,000 liters of wine.[41] There might have even been a grape-gatherers' station (VI.16.23), just by the Vesuvian Gate in the northern part of town, where grapes from the countryside were gathered and then distributed to shops, bars, and other sites that would press them and ferment the wine.[42] Even some of the wine retail shops, such as the Caupona of Spatulus (V.4.6–7) and Shop of the *Vinarius* (IX.9.6–7), combined the sale of different varieties

38. See Jashemski 1967a, 1967b, 1973a, 1973b, 1974 , 1977, 1979a, 1979b, 1993; Della Corte 1965; Meyer 1980; Maiuri 1928, 1956, 1959; Orr 1972; Elia 1975; Bragatini et al. 1981; Rossiter and Haldenby 1989; Fiorelli 1860, 1862, 1864, 1873, 1875; Boyce 1937; Niccolini and Niccolini 1854, 1862, 1890, 1896; Pernice 1932; Sogliano 1888, 1889; Mau 1889, 1890; Brizio 1868; Breton 1970; Warscher 1935–1960; Nappo 1988; Senatore 1998.

39. Nappo 1997: row houses built in the second century BCE were replaced with gardens and vineyards.

40. On wine and oil production in the town of Pompeii, see Dodd and Van Limbergen forthcoming.

41. Jashemski 1968, 73; 1973a, 36.

42. The property's identification is based on electoral *programmata*: *CIL* 4.6672; Della Corte 1965, 87.

of wine, including imported ones, with on-site wine production.[43] Making wine was a good use of the verdant urban areas but likely not strictly an economic choice tied just to the wine itself. If the proprietor chose to operate a vineyard in the countryside, the rural vineyard would likely be less costly and more spacious. In other words, running a vineyard in a town was probably more expensive and constrained due to the denser topography and pricier property values.[44] Making wine in a town offered different advantages, though. The wine was no ordinary wine, but a rare vintage given how few wines were made in the town itself. Surely the location of production also elevated the wine's appeal and value. Producing wine on-site also offered the proprietor the opportunity to combine the charms and cultural cachet of viticulture and the countryside with urban consumption. The uniqueness of the town-made wine, combined with production and retail, opened the doors to high-end hospitality and service.

Almost every urban vineyard went beyond wine production to promote a hospitality industry in which the dolia also had symbolic prestige. Wine was a central component of Roman dining, and certain spaces and occasions elevated the beverage by creating an ambiance of leisure. Vineyards in Pompeii featured reception rooms or a masonry triclinium (though portable couches must have been common and could expand a party) in the garden where guests could sample the wine and dine al fresco. Numerous indoor dining spaces in Pompeii were commonly decorated in a lavish manner, from painted walls to tessellated floors. Outdoor dining could be extravagant too, with ornately landscaped gardens and even small pools and waterfalls, such as the outdoor dining area of the Praedia of Julia Felix. As Ellis notes, even a small bar (I.11.16) featured a large triclinium decorated with garden frescos in a small open space to provide more refined dining.[45] Urban vineyards paraded winemaking and its attendant charms, and offered a banquet space for town visitors, those who did not reside in elite homes, or homeowners who wanted another option. Some vineyards, such as the Garden of the Fugitives and the vineyard by the amphitheater, placed triclinia in the middle of the vineyard under a pergola, transporting the banqueters to a sort of *locus amoenus* (pleasant place). Other spaces, such as the Caupona of the Gladiators, also positioned the dolia in full view of the reclining banqueters. As guests enjoyed wine while reclining on triclinia in the vineyard, the dolia not only reminded viewers that the wine was produced on-site but also signaled *negotium* and brought both the fruits and ideas of the country into town.[46] Creating such an atmosphere also drew on other services. Clients enjoying the space ordered food, probably prepared on portable hearths and served by attendants, and hired various entertainers, such as singers, dancers and acrobats, musicians, and poets. Urban gardens and vineyards therefore diversified and elevated dining in town, for those who could afford it.

Implementing the jars used in such establishments involved high levels of cost, labor, and risk, but dolia and cylindrical jars fed into the investment of urban retail and dining

43. Jashemski 1967b.

44. The Praedia of Julia Felix (II.4.6; *CIL* 4.1136) is an example of renting out one's estate or property in town.

45. Ellis 2018, 240–241.

46. On *rus in urbe*, see Taylor et al. 2016, 103–113; N.G. Brown 2018; Purcell 2007, 1988.

establishments during the first century CE. Setting up and operating the town's vast system of storage jars required a steady stream of resources. Most properties featured a couple to a dozen dolia and cylindrical storage jars. In addition to installation, workers maintained, cleaned, and replenished these jars regularly to ensure the freshness and quality of their contents, keeping the town well stocked with a variety of local products, often combined with imported foods. The decision to use these jars also points to high levels of consumption that mapped well onto dolium storage. The storage jars, shops, bars, and gardens fostered Pompeii's "producer" nature and retail activities, highlighting the strong ties of an agro-manufacturing town to its hinterland, and fostering a range of urban dining and hospitality experiences.

Rome and Ostia: The Warehouse of the World

Praised as the warehouse of the world, the city of Rome was outfitted with numerous horrea, as many as three hundred by the fourth century CE according to the *Regionary Catalog*.[47] Warehouse construction in the capital began as the city became more densely occupied. After Gaius Gracchus passed legislation entitling Roman citizens to grain in 123 BCE, the Horrea Sempronia were constructed in the city, making them the earliest attested granaries in Rome.[48] Numerous other named warehouses followed suit, as prominent families, politicians, and other elites made their mark on the ancient city.[49] Storehouses became a common feature in the city, especially along the Tiber River. Many stored grain and other goods, but a significant portion was composed of specialized warehouses for specific items, including spices (Horrea Piperatria), candles (Horrea Candalaria), and writing materials (Horrea Chartaria), as well as wine (cellae vinariae) and olive oil (cellae oleariae). Since the dolia from Rome mostly lack dates and provenance, our limited understanding of the use of dolia and storage in Rome comes from a few literary sources, slivers of archaeological material that have been felicitously recovered, and comparisons drawn from Ostia. Overlaps between dolium use in Rome and Ostia are abundant and complementary enough to warrant looking at them together; moreover, Ostia's well-preserved buildings and topography inform much of our interpretation of the ancient capital, including its storage. In the first quarter of the second century CE, the settlement of Ostia underwent major renovations, not only to accommodate and house its large working population and (seasonal) residents but also to expand its storage capabilities:[50] several horrea and at least five cellae, storerooms with sets of large dolia, were constructed or renovated (Figure 6.7A);[51] three of the cellae are still accessible today. While horrea stored a miscellany of goods, the cellae and their roughly two hundred dolia stored only liquid foods such as

47. Holleran (2012, 72n37) advises against relying on the *Regionary Catalog*, noting that it lists forty-eight horrea on the Palatine Hill, which could not accommodate so many horrea due to the imperial palace.

48. Festus 370L.; Plut. *C. Gracch.* 6.3; Rickman 1971, 150.

49. E.g., the district of the Horrea Galbae and Tomb of Galba (Emmerson 2020, 75–77, 85ff.).

50. Keay 2013; Heinzelmann 2010; Rickman 1996; Lanciani 1868. By then, Portus complemented Ostia.

51. On insulae: Meiggs 1973, 235–262; Packer 1967, 1971; Hermansen 1981, 17–54; Storey 2001, 2002, 2004; Stöger 2008, 2014; Ulrich 2013; DeLaine 2018. On horrea: Meiggs 1973, 263–310; Rickman 1971, 2002; Vitelli 1980; Bukowiecki et al. 2008; Van Oyen 2020b; DeLaine 2005; Paroli 1996; Boetto et al. 2016.

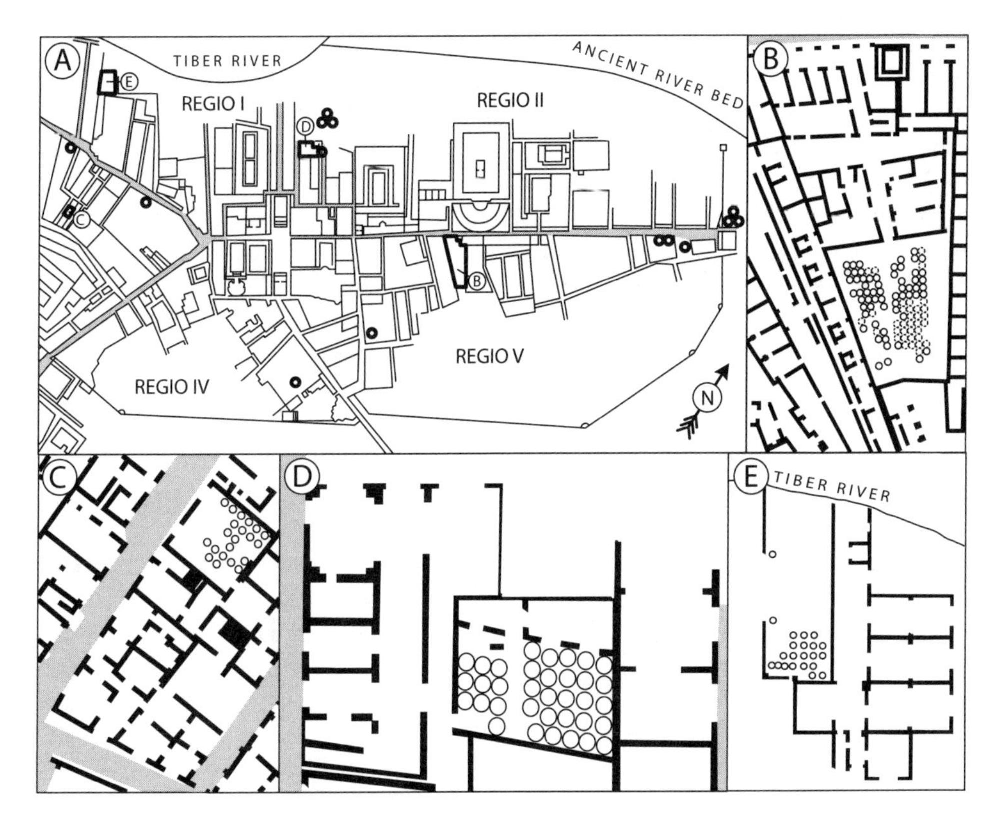

FIGURE 6.7. Plans of (A) Ostia with cellae with dolia marked, (B) Magazzino Annonario (V.11.5), (C) Magazzino dei Doli (III.14.3), (D) Caseggiato dei Doli (I.4.5), and (E) Caseggiato dei Doli (near I.19). Illustrated by Gina Tibbott.

wine, olive oil, or garum.[52] The dolia were enormous and packed close together in the storerooms to maximize space; altogether, the dolia of Ostia are estimated to have contained well over one hundred thousand liters of wine or olive oil.[53] Unlike Pompeii, Ostia and Rome were not agricultural towns. Instead, Ostia's and Rome's dolia, much larger in size, stored food produced elsewhere, whether in the hinterland or farther abroad, and brought into the city for distribution across the center of the Roman Empire.[54]

The dolia of Rome and Ostia were all standardized, very large (average capacity was over one thousand liters)—much larger than any of the dolia in Cosa or Pompeii—and used primarily in the same context: a handful of highly specialized storerooms. As

52. Portus has no dolia, except a fragment in a late antique wall (pers. comm. Évelyne Bukowiecki), perhaps because it primarily functioned as a port for large grain ships and storage facility for grain or there were dolia that were either built over or later removed.

53. Rickman 1971, 75ff.; Gatti 1903; Carcopino 1909. The plethora of oil amphorae and wine dolia in Rome suggests the dolia defossa in storerooms at Ostia primarily contained wine.

54. Heinzelmann 2010: Ostia's role changed from storage to marketplace after Portus was in use in the second century CE.

mentioned in Chapter 2, they all were similar in size and had the same shape: an extreme strawberry shape of very wide shoulders that tapered to a small base. With a capacity of over one thousand liters, the average dolium could hold two to five times more than a Pompeian dolium. The dolia of Ostia and Rome stored wine and oil to supply shops, but they were not directly featured in the urban retail spaces; only a handful of lone dolia were installed in bars at Ostia and Rome.[55] While storage jars were installed across dozens of bars, shops, and gardens in Pompeii, only a few dolia have been found in shops around the capital, and the cylindrical jars built into the service counters at bars in Pompeii are notably absent from Ostian and Roman bars. For an explanation, some scholars have pointed to the moral codes emperors put in place that prohibited hot foods in order to curb unruly crowds, the repercussions of which would have led to the dismantling of the cylindrical jars in the late first century, before Ostia's renovations.[56] Another possible factor, however, was logistics. The type of storage system we saw at Pompeii, which required frequent turnover and replenishment, might not have been effective for the capital's needs and storage and consumption patterns. Instead, grain and other dry goods were likely transported and stored in sacks around the capital, enabling quicker turnover and eschewing the transfer of contents between containers.[57] Several shops abutted the storerooms with dolia at Ostia, such as the row of shops of the Porticus of the Republican Monument connected to the Magazzino Annonario (Figure 6.8 top). Rather than putting a set of storage jars into each shop, the shops likely shared access to the storeroom, and wine was transferred from dolia into amphorae or other containers when shops needed to restock. A recently discovered *caupona* at Rome, dated to the late first century BCE, featured a single dolium encased in a marble counter and a separate storage room containing several dolia defossa.[58] The type of storage and use of storage space mixed with retail and consumption found at shops in Pompeii and Herculaneum probably worked well for smaller towns, while a large dolium in the shop and several large dolia in an adjacent room helped store the large amounts of wine needed in a dense city. With Ostia's estimated population about forty to eighty times greater than Pompeii's, it needed strategic storage; with Rome's population fifty to one hundred times greater than Pompeii's, storage had to be especially concentrated.

Almost all dolia in and around the capital were installed in specialized, purpose-built storerooms, defined by and filled with the dolia installed for the storage of wine or olive oil. While these were traditional rooms of farms and villas adjoining areas of production, cellae in towns and cities were not associated with production; instead, they were standalone rooms or a storage room adjacent to shops or other retail spaces. By grouping dolia in a systematic way in the cellae vinariae, workers could distinguish between different batches of wines, securely store incredibly large amounts of wine, and move between the vessels unhindered as they filled or emptied them. The concentration of dolia in cellae

55. For a survey of bars in Ostia, see Hermansen 1981, chs. 4–5. Bars and shops with dolia in Ostia include Cassegiato (I.16.1), shop (V.6.3), shops (II.8.1), and Cassegiato del Termopolio (I.2.5).

56. See also Cass. Dio 61(62).14. On bars and morality, see Ellis 2018, 2004; Laurence 1994; Wallace-Hadrill 1995. Ellis (2004) argues against the notion that bars were places of deviant behavior. On food at bars, see Ellis 2018, 228–238. For taverns and emperors, see Hermansen 1981, 196–203.

57. See Martelli 2013 for study of *saccarii*, grain sack carriers in Ostia.

58. Bruno 2010; 2012, 236; Carandini 2014, 266.

FIGURE 6.8. (*Top*) Magazzino Annonario (V.11.5) and (*bottom*) Magazzino dei Doli (III.14.3). Courtesy of the Photographic Archive of the Archaeological Park of Ostia Antica.

crystallizes most clearly at Ostia, where much of the ancient city has been preserved and excavated. Almost all two hundred dolia at Ostia were concentrated in just four storerooms that had restricted, or at least limited and controlled, access, no doubt for several reasons: to prevent theft, to keep out pests, and to protect investments.[59] The Magazzino Annonario, just behind a portico on the *decumanus* (road oriented east-west) by the Porta Romana, featured over one hundred dolia that likely stored wine (or oil) brought to Ostia from the

59. Rickman 1971. Paroli 1996: a fifth storeroom has been identified but not fully explored.

hinterland to supply shops and residents in the area (Figure 6.7B, Figure 6.8 top).[60] Not far from the Tiber River, the Magazzino dei Doli was connected to the House of Annius, believed to have been the lamp merchant Annius Serapiodorus, and probably stored olive oil for lamp customers in twenty dolia (Figure 6.7C, Figure 6.8 bottom).[61] The two Caseggiati dei Doli along the Tiber River, one with at least twenty and the other with thirty-five dolia, received incoming wine (or oil) from ships or boats (Figure 6.7D-E, Figure 1.2 top).[62] Although most of the dolia in Rome lack exact provenance, they were mostly also from cellae vinariae; for example, the dolia mounted on the perimeter wall of the embassy of the United States of America to the Republic of Italy originated from a wine warehouse along the Tiber River near the Church of San Francesco a Ripa.[63] Archaeological explorations under the city walls in 1789 uncovered one of the best-preserved wine cellars that showed the layout of the building: it was divided into three long narrow rooms, two of which featured rows of dolia and other large ceramic vessels buried to their shoulders.[64] Like the cellae in Ostia, Rome's many wine cellars were arranged for both concentrated bulk storage and efficient organization. In the second century CE, several warehouses were clustered along the Tiber River where wine was transported, stored, and bought and sold.[65] On the right bank of the Tiber, excavations of the Villa Farnesina's gardens unearthed rows of dolia defossa of a wine cellar, identified by an inscription as the Cellae Vinariae Nova et Arruntiana.[66] Nearby were other wine warehouses all identified and dated to the second century based on inscriptions: the Cella Civiciana, Cella Groesiana, Cellae Saeniana, and Septem Caesares.[67] These warehouses and storerooms along the Tiber received the wine and other goods that flowed from Rome's productive hinterland and from overseas.[68] Only five cellae vinariae are known from inscriptions and archaeological evidence, but surely more were included in the over three hundred horrea documented by the *Regionary Catalog*.

60. For discussion of wine and oil production in Rome's hinterland and why it is difficult to detect archaeologically, cf. Marzano 2013a; De Sena 2005. See also Rickman 2002, 358.

61. Ceci (2001, 2003) believes the Annius Serapiodorus (ANNISER stamps on lamps) is the same Annius identified on a series of terracotta plaques mounted above the house that showed a male figure on a boat, a male figure standing between dolia, and the words OMNIA FELICIA ANNI.

62. Pasqui 1906.

63. Palma 1983, 17: on May 27, 1622, and March 16, 1624, twenty-five dolia from San Francesco a Ripa were moved to the Ludovisi vineyard. By the time the embassy acquired the dolia, only fourteen remained.

64. W. Smith 1890, 390, s.v. *cella*.

65. *FUR* 28a–c, 27c; Bloch 1961, 150; Rodríguez Almeida 1984, 83–85; Holleran 2012, 79–80. Carafa and Pacchiarotti 2012; Tucci 2004; Gatti 1934. For summary, see Richardson 1992, 80, 191–192; entries in Steinby 1993–2000.

66. Fiorelli 1880; Lanciani 1880, 1890, 1897, 31; Peña 1999, 12; Purcell 1985, 12; *CIL* 6.8826. Rodríguez Almeida 1993.

67. Cella Civiciana: *AE* 1937, 16. Cella Groesiana: *CIL* 6.706. Cella Saeniana: *AE* 1971, 30. Septem Caesares: *CIL* 9.4680, 14.2886, 5.712 (uncertain). Conison 2012, 202ff.; Holleran 2012, 78–80. Chioffi 1993; Vilucchi 1993. See also Palmer 1980.

68. Marzano 2013a. Procop. *Goth.* 5.26.9 notes that the towpath on the right bank of the Tiber connecting Rome to Portus was still in use in the sixth century, though the towpath on the left bank, which connected Rome to Ostia, had disappeared; see Rickman 1971, 8; Holleran 2012, 80.

In order to keep track of dolium contents, the capacities of all the dolia at Ostia and many of the dolia in Rome were clearly and systematically marked across the various cellae (Table A1.10; Figure 2.13R). The capacities were marked on the dolium shoulder or rim, sometimes in both places, and featured Roman numerals ranging from XXIII to XLV, sometimes followed by other figures such as S or Ɔ, which were in turn occasionally followed by Roman numerals II or III; Giuseppe Gatti, the first to publish these volume incisions, determined that the first set of Roman numerals was in units of amphorae, the symbol S indicated half an amphora, and Ɔ was in units of *sextarii* (1/48 an amphora).[69] The preserved capacity incisions show that the Ostian dolia had an average capacity of roughly thirty-eight amphorae, a little over a thousand liters. The dolia of Rome were also capacious, and one especially large dolium of Rome was marked as having a volume of LVS (55.5 amphorae, ca. 1,443 liters).[70] The frequency and precision of these capacity incisions underscore a convention of designating the volume of each vessel across the various cellae of the port city and capital. Since these are most likely post-cocturam incisions (marks made after the vessel was fired in the kiln), it is difficult to ascertain whether the opus doliare workshop or the users did this, but both possibilities are interesting to consider. If the opus doliare workshops executed this, it means that they manufactured the vessels and verified their volumes to promote or validate their merchandise, perhaps for a potential customer seeking a dolium of a certain volume. Moreover, labeling the dolia with their volumes may have also simply been an expectation for the urban opus doliare workshops. These workshops produced the largest and most robust dolia and might have also provided the additional service of volume verification. On the other hand, if the dolium users incised the containers, they took the task of labeling the vessel into their own hands, checking and marking the volumes before installing the vessels. The shared format suggests a common system for labeling and keeping track of the dolia in Ostia and Rome and that the dolia could have been managed by the same personnel, possibly personnel who were working under or with officials overseeing the food supply.[71] We have already seen dolia in rural contexts labeled in such a way to help with keeping track of wine produced and sold. In effect, the capacity markings enabled quick calculations and recordkeeping, a pressing issue in the capital.[72]

The cellae vinariae amassed large amounts of wine coming into the capital, concentrating resources in certain places and in the hands of select people before distributing them later. Ostia itself was a pivotal point for importing food, and the dolia of the cellae in Ostia and Rome maximized storage space and reduced potential bottlenecks during the height of the sailing season. We have to keep in mind the possibility that, as Astrid Van Oyen has noted of granaries, cellae could and probably did hold multiple batches of goods throughout the year:

69. For volume incisions at I.4.5, see Gatti 1903; at I.19, see Carcopino 1909, 359–364; Rickman 1971, 73–76. For a compilation of dolium capacity incisions, see Carrato 2017, 709–714. Reused amphorae in the northwest provinces were incised with Roman numerals, probably in *modii* and *sextarii*, to indicate their capacity to store foods; see Peña 2007b, 130–131; Van der Werff 1989, 2003.

70. Dolia at the US embassy in Rome were also marked (pers. comm. Valeria Brunori).

71. Mattingly and Aldrete 2000, 151–153.

72. Van Oyen 2020b, 122–157. See Riggsby 2019, 83–129, esp. 110ff.; Keenan 2017, 2016.

> The bottleneck of the Tiber, combined with the seasonal peak in the arrival of goods into the ports, meant that the only response could be to increase base storage capacity at both ends. The result would have been redundant storage space, i.e. reserved for the same goods both in Ostia or Portus and in Rome. Investment in the warehouse structures and their maintenance would thus have represented significant "dead capital." Unless, of course, there was a constant turnover. Put differently, an increase in storage capacity at Ostia and Portus was as much about seasonality and speed of turnover as it was about larger volumes of supplies.[73]

Although Van Oyen is discussing horrea and the storage of cereals during the summer months, her observation holds true for wine too, as new wines were brought in the early fall and the rest of the harvest, after some aging, in the spring and summer. Workers transported wine and oil to Ostia and Rome in amphorae and other jars, animal-hide containers carted from the countryside, and perhaps dolia of tanker ships (see Chapter 4). They kept some wine and oil in jars and transferred a significant portion into the dolia in the specialized storerooms to maximize storage space in a crowded city (Table 6.2).[74] The dolium volume markings in units of amphorae suggest that porters would later transfer the contents into amphorae, containers they could move more easily, but chose to store some wine first in dolia at least temporarily. Storing wine in dolia rather than amphorae not only conserved space (storing an Ostian or Roman dolium's worth of wine in amphorae would have required about six to ten times more space) but also organized the cellae. Leaning thousands of amphorae along the walls of a cella, on the other hand, would have been chaotic, cramped, and unstable, not to mention a highly inefficient use of space. Thanks to their packed layout, Ostia's smallest cella stored at least sixty times more than a shop in Pompeii and twice the amount Villa Regina's dolia could hold; Ostia's largest cella stored at least four hundred times more than a typical Pompeian shop and ten times more than Villa Regina's storage capabilities.

A cella's astounding ability to store so much came with other advantages. Guy Métraux suggests that wealthy Romans invested in urban warehouses (rather than villa renovation projects) from the first century through the first half of the second century CE, but the investments could have begun even earlier.[75] According to Varro, already by the mid-first century BCE "there are some at Rome who have built wine cellars for the reason of profit" (*Sat. Men.* 530: *aliquot Romae sunt qui cellas vinarias fructuis causa fecereunt*). Wine cellars enabled a variety of money-making options and strengthened trade relations. With a substantial storeroom, the owner (or renter) was likely able to negotiate better prices because he or she could purchase larger volumes of wine at wholesale prices.[76] Wine merchants looking to offload wine quickly might have been tempted to sell bulk amounts to a customer, even at lower prices, rather than look for another buyer. Over time, a wine trader might come

73. Van Oyen 2020b, 156–157.

74. Leaning full amphorae against a wall could be unstable. Peña (2007b, 50) notes that the transfer of contents from amphorae into dolia would explain damage found on dolium rims and shoulders.

75. Métraux 1998.

76. The following situations would also apply to cellae renters and would show why some cellae might be more desirable than others.

TABLE 6.2. Properties with dolia in Ostia.

Storeroom	Number of dolia	Liters stored	Room size	Liters stored/area
Casseggiato dei Doli (I.19)	22	ca. 22,176	ca. 184 m^2	ca. 120.5 L/m^2
Magazzino dei Doli (III.14.3)	21	ca. 21,168	ca. 100 m^2	ca. 211.7 L/m^2
Caseggiato dei Doli (I.4.5)	35	ca. 35,280	ca. 150 m^2	ca. 235.2 L/m^2
Magazzino Annonario (V.11.4–5)	100+	ca. 100,800	ca. 450 m^2	ca. 224.0 L/m^2

to prefer certain cellae vinariae and their owners because they were reliable customers. Storerooms with only a few dolia, or dolia that frequently went out of commission, might have reneged arrangements often enough for wine traders to look elsewhere. The owner or renter of a cella vinaria could have also been a merchant, shipowner, vintner, or shopkeeper, using the cella to store the wine they already bought or produced until they could later sell it in the city at favorable prices.[77] A cella owner could also accumulate a greater volume and variety of wine, drawing more clients looking for a wider selection. Customers hoping to stock up on different kinds of wine were likely able to find more in a larger cella vinaria that could amass wine from multiple vineyards and merchants. Equally important, however, was a large storeroom's capacity to accumulate enough wines quickly and then sell when convenient or profitable. A large shop, for example, could have strategically amassed wine during the "open" sailing season to sell it at high prices during the offseason when other shops had little or depleted inventory, similar to the price gouging and hoarding of grain. Low supply coupled with high demand created a sellers' market. With a reduced supply of wine in the city, potential customers were beholden to the shops that still carried wine, until the sailing season fully reopened and vintners could ship wines en masse again. Smaller storerooms and shops, such as Pompeian bars, could only store so much, requiring regular replenishment to keep up with demand. A cella vinaria such as the one-hundred-dolium Magazzino Annonario (assuming it stored wine), on the other hand, could probably stockpile enough wine, the equivalent of ten times the amount a modest vineyard could produce in a year, to cultivate consumer confidence and a reliable customer base. Although storing wine in dolia entailed tedious and time-consuming transfers into other containers, concentrating wine in small areas bolstered a cella vinaria's storage, and hence buying and selling power.

By storing large amounts of wine at specific points, the cellae in Rome cultivated and relied on the services of commercial districts, many of which blossomed around the cellae.[78] Although the evidence is sparse, marketplaces designated for certain goods were probably quite common, such as the Forum Vinarium believed to have been along the Via della Foce in Ostia near the Caseggiato dei Doli (or in Portus).[79] Having the infrastructure to support

77. Rickman 1971, 171; Rostovtzeff M. 1941. *The Social and Economic History of the Hellenistic World.* 3 vols. Oxford: Clarendon Press. 228, 1628.

78. For wholesale retail and the commercial landscape of Rome, see Holleran 2012, 63–98.

79. On housing and apartments in Ostia and Rome, cf. Hermansen 1981, ch. 1; DeLaine 2004; Storey 2001, 2002, 2004; Packer 1967, 1971. Stevens (2005) has calculated that the apartments had four stories. On the Forum Vinarium: *CIL* 14.376, 14.409, 10.543. For support of it in Ostia, see Coarelli 1996; Pellegrino and Licordari 2018; in Portus, see Fasciato 1947.

the busy, at times frenzied, activities in these areas was paramount, likely supported by the clustering of the cellae vinariae along the Tiber.[80] Clusters, "geographic concentrations of interconnected companies and institutions in a particular field," offered advantages for suppliers, merchants, and consumers.[81] Boats carrying wine along the Tiber would have a known place to dock.[82] Dolia suppliers and repairers had set destinations. Wine customers had a market to visit and could compare prices and vintages there. Merchants, too, would enter a level playing field with customers and suppliers alike. An area along the Tiber River near the named wine storerooms, known as the Portus Vinarius, became not only an entry point for Rome's wine but also a designated wine retail district.[83] A *portus*, an unroofed yard where goods were bought and distributed, was usually associated with brickyards, but the Portus Vinarius was specifically tied to wine and located near the cellae vinariae where *negotiantes* (merchants) and *coactores* (dealers) dealt in wines.[84] The wine district fostered and depended on activities beyond just importing and selling wine, though, as illustrated by a Latin inscription outside the Porta Salaria (dated to 50–150 CE):

> To the spirits of the dead. For Gaius Comisius Successus, dealer in bottles at the wine port, Comisia Fecunda set up this monument for her well-deserving husband and fellow freedman, and for herself and their descendants.
>
> D(is) M(anibus)
> C(aio) COMISIO SVCCESSO
> NEGOTIANTI PORTO
> VINARIO LAGONARI
> COMISIA FECVNDA
> CONIVGI ET CONLIBER
> TO B(ene) M(erenti) FECIT
> ET SIBI POSTERISQVE
> SVORVM

Gaius Comisius Successus, a *lagona* dealer of Portus Vinarius, facilitated the trading that went on in this wine district by supplying *lagonae*, a type that commonly referred to small, flat-bottomed vessels but could also refer to full-sized amphorae.[85] J. Theodore Peña suggests that Gaius Comisius Successus probably acquired old, empty wine vessels from merchants and handlers and cleaned and relined them with pitch before selling them to customers who purchased wine, perhaps wine stored in dolia defossa, at wholesale quantities and prices.[86] The wine trade must have used and discarded countless vessels, which potentially led to large

80. On clustering, see Porter 1998; Goodman 2016.

81. Porter 1998, 78.

82. Casson 1965, 1971; Castagnoli 1980.

83. *CIL* 6.9189, 6.9090, 6.37807; H.L. Wilson 1910, 35.

84. Likely near Tor di Nona, Piazza di Noscia, or Ripetta. Ulp. *Dig.* 50.16.59; Coarelli 1999, 156.

85. Peña 2007b, 115–116, 115n48: "most likely classes belonging to the family of small, flat-bottomed containers produced in Sicily, central Italy, and Adriatic Italy, such as the Middle Roman 1, Spello amphora, and Forli amphora."

86. Peña 2007a, 115–116.

heaps of rubbish, as seen at Testaccio's Monte dei Cocci. Reusing containers not only managed and decreased the accumulation of waste but, in this case, also offered some, perhaps less affluent, merchants another commodity and service to deal in, one that did not require the same upfront costs as more elaborate ventures in the wine trade.[87] Activities and infrastructure for wine transfers, container (re)use, and waste management sprang up around bulk wine storage in Rome and Ostia, accommodating the transactions there and further developing the distinct character of the commercial districts and the city's topography.

The vast network of cellae vinariae supplied shops, bars, and other retail outlets in the capital. Although only a few shops in the ancient capital have been identified, Rome surely featured numerous retail establishments as early as the first century BCE.[88] Cicero's invective speech *Against Piso*, which attacks Lucius Calpurnius Piso Caesoninus, Roman senator and Julius Caesar's father-in-law, is particularly illuminating: it gives a sense of not only how common wine retail was in Rome already by the mid-first century BCE but also how wine, wine cellars, and dolia could be weaponized against the Roman elite. "He [Piso] has no baker, no wine cellar at home. His bread came from a huckster, his wine from a vat.... He drinks as long as there was anything poured from the dolium" (*Pis.* 27: *pistor domi nullus, nulla cella; panis et vinum a propola atque de cupa; ... bibitur usque eo, dum de dolio ministretur*). Cicero's underlying criticism against Piso was his lack of taste and refinement because he purchased his wine from a common tavern or shop. We have seen in Chapter 3 how agriculture and viticulture were celebrated practices of a good Roman, so Piso, a man of such elevated social status and wealth, should have his own wine cellar, rather than buy presumably cheap table wine from a common tavern. Furthermore, Cicero's charge that Piso drank until wine was poured straight from the dolium was a criticism not only that he drank excessively but also that he favored quantity over quality, since the wine was bulk, low-quality wine for the masses; as previously mentioned, *vinum doliare* or *de cupa* was lower quality than wines stored in smaller vessels in retail contexts (but a cella vinaria's wine was not necessarily low quality, since the dolia stored, and possibly aged, wines).[89] Cicero's harangue against Piso points to urban residents regularly purchasing wine from retailers, as well as a heightened discrepancy between the haves and have-nots in the imperial capital. For a man of such stature, Piso should have had his own wine cellar to collect, or even produce, his own vintages.

In a city where real estate was prime, building and operating a storehouse required substantial financial and social capital, much more than a modest shop in Pompeii. In general, the construction of a cella, whether a private one at home or a commercial one that made up part of a trade network, and transportation of dolia in and across a densely populated city drew on not only financial resources but also significant social clout to purchase or lease land, construct buildings, and broker access to ports, roads, vehicles, dolia, and the wine itself.[90] In Ostia, a single cella had as few as twenty dolia to upward of over one hundred, a huge upfront investment. Moreover, transporting the enormous

87. For informal retail—e.g., street traders and hawkers—see Holleran 2012, 194–231.

88. Livy *Ab Urbe Condita* 6.25.9. Purcell 1994, 659: "Rome was a city of shops, its people a nation of shopkeepers."

89. *Dig.* 18.6.1.4; Hor. *Ep.* 2.47; *Geoponika* 7.5, 7.6.

90. Kaiser 2011; Poehler 2011, 2017; Favro 2011.

vessels, each of which weighed several hundred kilograms, was only possible with the largest and most robust type of freight wagon. City traffic was especially difficult to navigate given the narrow, crowded streets as well as regulations regarding cart traffic. Cella owners not only set up the storerooms but also employed workers to provide regular upkeep and cleaning, such as cellar attendants, known as *cellarii*, who ensured the security of the cella and maintained the quality of the wine and kept it from spoiling.[91] To store multiple batches of wine every year, the dolia had to be cleaned and recoated with pitch several times. The dolia of Ostia were incredibly robust, but workers regularly performed upkeep, routinely inspecting the vessels and making necessary repairs to ensure they were suitable for use. A cella owner would not want any dolia to be out of order, because that would be a loss of wine, a loss of profit, and a waste of valuable space. Building and running a cella was expensive, but incredibly lucrative. It is no surprise, then, that very wealthy and powerful families, often already with trade connections, set up and operated storehouses, including granaries, general storehouses, and other specialized storerooms, and that they could often become targets of imperial confiscation.[92]

Operating an urban cella vinaria could also bring prestige and power. As Cicero remarked, not every wealthy person had their own wine cellar at home, but those who occupied the upper echelons were expected to. Yet having a wine cellar was not only a way to make profit; it also ensured access to (quality) wine for entertaining and impressing guests, displaying power and wealth, and even hosting political and triumphal banquets.[93] Dolia, with their ability to ferment and store large amounts of wine, came to symbolize wine and abundance. People chose to celebrate a small number of dolia from Ostia that supported banqueting: a funerary relief from Isola Sacra portraying Lucifer's possible occupation as a hot-water seller suggests that some dolia were used in the sale of hot water, presumably to mix with wine, while a mosaic from a house in Ostia shows a banquet scene, with a man heating water in a water heater known as an *authepsa* (self-boiler) next to a dolium (Figure 6.9; Plate 13).[94] These representations remind us that dolia were associated not only with wine but also with abundance and the service of fine banqueting. The prestige behind owning and operating cellae vinariae was also reflected in the common practice of naming these structures after the families that established them, such as the aforementioned cellae vinariae known from inscriptions and fragments of the Forma Urbis Romae. Named buildings and monuments conveyed an individual's or a family's achievements,

91. Plin. *NH* 19.188.5; Columella *Rust.* 11.1.19.3, 12.3.9.1, 12.4.2.8; Sen. *Ep.* 122; Plaut. *Mil.* l. 824, *Capt.* l. 895.

92. Rickman 1971, 164–165; *CIL* 4.4226, 6.4226a = *ILS* 1620, 6.4240, 6.3971 = *ILS* 1625, 6.31284–31285; Tac. *Ann.* 12.22; Cass. Dio 61.32.3.

93. On Julius Caesar arranging extravagant banquets, see Suet. *Iul.* 26.

94. On the funerary relief of Lucifer, the (hot) water seller, see Ellis 2018, 236ff.; Zimmer 1982, no. 176; Helttula 2007, 310–311. The inscription reads, LVCIFER/AQVATARI. Thylander (1952, *IPO* A 169) interprets AQVATARI(us) as a profession: Lucifer/the water-seller; Solin (1987, 123n10 = 1998, 284) offers another interpretation: AQVATARI as a name, "Lucifer/slave of Aquatarius." Another panel, which almost certainly accompanied the LVCIFER/AQVATARI panel on the same tomb, has two registers: the upper register features packed amphorae, and the lower shows three individuals working, two of whom are mixing the contents of two small jars on a counter, perhaps mixing water with water. For the mosaic, see Kondoleon 2000, 184–186; Dunbabin 2008, 23–25.

FIGURE 6.9. (*L*) Funerary relief of Lucifer, Isola Sacra necropolis, by Gina Tibbott. (*R*) Roman, *A Banquet in the Open Air*, early fourth century CE, marble, glass, and clay. Detroit Institute of Arts, Founders Society Purchase, Sarah Bacon Hill Fund, 54.492.

wealth, and status, as they were able to both finance and receive approval for their construction projects. Storerooms could form one component of a larger conglomeration of a monument, and perhaps even help define a particular neighborhood, putting a stamp on the city; one striking example is the Horrea Galbae, which was part of a series of buildings that clustered around Galba's tomb and became a neighborhood known for at least three centuries as the Praedia Galbana.[95]

Almost all the capital's dolia were concentrated in storerooms, but several were featured in more extravagant displays of wealth. The few dolia used for viticulture within the city walls highlight just how exclusive their ownership was: they were found only at the most opulent pleasure gardens known as *horti*, such as the *horti* of Domitia Lucilla Minor (mother of the emperor Marcus Aurelius and owner of an enormous opus doliare enterprise, mentioned in Chapter 2), and played into the performance of wealth and conspicuous production.[96] Covering expansive areas in the city, *horti* were decadent luxury estates in which the wealthy could showcase their private art collections, cultivate ornamental gardens, and escape from the hustle and bustle of busy urban living. Partaking in urban wine production further boosted the vintner's status as an estate owner, and perhaps pointed to divine favor and approval, as viticulture was associated with the religious calendar.[97] Ancient writers associated agricultural fertility and abundance with fortune,

95. Emmerson 2020, 86–88; Rickman 1971, 166–168; Platner on Horrea Galbae (also 2015, 261).

96. See Purcell 2007; Wallace-Hadrill 1998; La Rocca 1986. Ancient authors viewed *horti* as symbols of decadence and luxury, criticizing the conspicuous consumption and extravagance of such large estates in a densely occupied city. The *Horti Domitiae Lucillae* had fourteen dolia: Liverani 1996; 1998; 2004, 34–43; Ravasi et al. 2020; Haynes et al. 2018; Santa Maria Scrinari 1995; Fatucci 2012, 348. See also Innocenti and Leotta 1996 on the *Horti Sallustiani*.

97. On the urban experience of wine and oil production, see Dodd and Van Limbergen forthcoming.

highlighted by Livia's wine-producing portico and temple to Concordia, built over the spot where the cruel *eques* Vedius Pollio's townhouse and tanks of man-eating lampreys once stood.[98] As Chapter 3 discussed, villas and other production sites often placed dolia center stage in celebrations of viticulture, storage, and abundance. In the imperial capital too, the concentration of dolia and wine storage in massive storerooms expressed and reaped wealth, status, and power.

Concentrating Wine and Wealth

For some towns and cities, dolia became part of the urban landscape and facilitated bulk movements and storage of goods from the countryside and farther afield. Although Cosa, Pompeii, Ostia, and Rome all possessed dolia, their footprint differed from site to site. Over the course of its often intermittent occupation, Cosa had only a few dolia, almost all of which predate the mid-first century BCE, to store locally produced wine in elite residences, a few shops, and the town's warehouse; bulk storage of wine was common in the hinterland, but storage in the town was fragmented and individualized. After the dolia were discarded or repurposed by the mid-first century BCE, a period when the countryside's export of wine also seems to have declined, few dolia were reintroduced to the town and residents instead received and kept wine in small jars or amphorae. The type of bulk storage dolia offered did not align with Cosa's consumption needs, perhaps because of the town's low level of retail activity and changes in the hinterland's agricultural economy, and was mostly abandoned.

Other urban communities, however, embraced the dolium storage technology to increase their abilities to receive, store, and distribute foods. Some wealthy residents who could afford to buy and maintain dolia kept them for private, residential storage. The overwhelming majority of urban dolia, however, were used for some form of commercial or communal storage. Shopkeepers could expand their shops' storage capacities and retail power by installing more dolia, ceramic jars, and other storage equipment. In Pompeii, most large ceramic vessels were implemented in shops and retail spaces widely dispersed throughout the town. But the form of storage itself could also change. In more densely populated consumer cities, such as Ostia and Rome, very large amounts of wine were concentrated in dolia-laden warehouses and storerooms known as cellae vinariae. These storerooms could contain twenty to over one hundred dolia and were much more expensive and challenging to establish than a typical shop or bar. A cella vinaria's expansive storage offered merchants more flexibility, as well as opportunities to make a profit. With more storage, merchants could purchase different vintages from various distributors or even strike a better deal; a merchant who could buy (and store) five thousand liters of wine was likely able to negotiate a lower price than a merchant who could purchase only a thousand liters. Furthermore, while shops were customer-facing spaces, access to cellae was restricted, limited, and controlled, and their use and maintenance relied on regular, designated workforces. The dolia of cellae vinariae were clearly labeled with their capacities in units of amphorae, and

98. Florus *Epitome* 1.16; Plin. *NH* 19.11; Ovid *Fast.* 6.637–648; Suet. *Aug.* 29; Cass. Dio 54.23.6, 55.8.2; Flory 1984; Barrett 2002, 198–202; Bassani and Romana Berno 2019.

the layout of the cellae maximized the use of limited space to supply other outlets in town, such as shops and bars. The spatial demands of and activities around cellae could alter the surrounding topography to make way for carts, workers, and the general movement of goods, as well as open channels for new roles such as dealing in secondhand containers or brokering deals between vintners and wholesale buyers. With the expansion of storage facilities in towns and cities also came transformations of localities within the urban center, and even changes in the character of entire districts.

Investing in dolia required substantial financial resources, but could also lead to major profits, new business arrangements, and increasing power and social prestige. Although wine was mostly produced in the countryside, urban consumers could choose from a variety of options thanks to robust trade networks, markets, and storage facilities in towns and cities. For a town such as Pompeii, a vineyard with outdoor banquet space not only sold locally made wine but also provided hospitality and entertainment. Importing elements of the countryside (*rus in urbe*) was often considered a luxury, especially in more densely occupied cities, and offered another avenue to social distinction. For urban consumers, the network of cellae, shops, and bars brought a wider availability of wines, both local and imported. Operating a wine cellar in the capital also gave merchants opportunities to corner the market in a particular part of the city. Ostia's and Rome's dolia and storerooms were used exclusively for bulk, concentrated storage in a densely developed urban environment. Storerooms elsewhere with dolia solely for the storage of liquid products were not common. Most cellae were typically associated with production facilities such as farmhouses or large country villas, adjacent to the wine press or the wine press room. Ostia's storerooms, on the other hand, were completely divorced from the production process. Unlike Pompeii, Ostia had no equipment or space for wine production; there were no vats, wine presses, vineyards, or farmhouses known from the town. Within the history of Ostia, cellae with dolia defossa, as well as horrea, closely followed the fate of densely occupied *insula* apartments, which were also constructed at the beginning of the second century.[99] The situation in Rome was similar: dolia were featured primarily in cellae vinariae, storing massive amounts of wine in specialized storerooms and feeding into the vibrant retail landscape. The growing numbers of dolia and cellae in towns and cities were the pillars of large urban populations and an urban diet, and they also became expressions and instruments of status and power. Urban cellae vinariae not only belonged to the affluent; they also helped them increase their wealth and status in an intensely competitive society.

99. For a discussion of grain storage and urban growth in Ostia, see Vitelli 1980.

7
Mending Costly Investments

DIOGENES THE CYNIC fascinated people in antiquity. Known for his extreme disdain for luxuries and amenities, he lived in poverty, which was highlighted by where he lived: a pithos.[1] This was such a powerful image that, during the Roman period, artists depicted Diogenes in a dolium. A curious marble bas relief from the imperial period, centuries after Diogenes' lifetime, portrays the famed meeting between Alexander the Great and Diogenes, lying in his dolium home (Figure 7.1).[2] Juvenal's cautionary satire against avarice also mentions Diogenes' dolium abode and includes the detail that it was a mended dolium: "The dolia of the naked Cynic do not burn down. If you shatter one, another house will be built tomorrow, or else the same one will remain, patched together with lead" (*Sat.* 14.308–310: *dolia nudi/non ardent Cynici; si fregeris, altera fiet/cras domus atque eadem plumbo commissa manebit*). Juvenal's comment and the artist's portrayal of Diogenes' dolium, as not only cracked but also *repaired*—fastened together with lead—highlight both Diogenes' abject poverty and a serious concern for dolium owners and users in antiquity: broken dolia.

Dolia were both expensive and challenging to make and were expected to last several decades, but they had one major flaw: they cracked easily. This must have made a significant financial impact on the dolium buyer and user, in terms of not only the dolium itself but also the contents the vessel would have lost: according to Diocletian's *Price Edict*, a broken dolium would have been a loss of 1,000 denarii, and a broken dolium full of wine could have been an additional loss of 8,000–30,000 denarii; considering that an unskilled laborer's daily wage was 25 denarii during this period, this was an astronomical amount.[3]

Although repairing dolia was a sensitive and pressing concern in antiquity, it is rarely discussed in scholarship today. Excavation reports and catalogs might note dolium repairs, but only a few scholars have attempted to address *how* these were made.[4] Both textual and material evidence shows that dolium owners and users protected and prolonged the use of their expensive jars. They might have mended the pots themselves, but they also had options for finding repairers. Dolia were commonly repaired, and examples abound

1. Diog. Laert. 6.23. See also Desmond 2008, 21, 81–82, 112, 122.

2. Amelung 1927, 289–293.

3. According to Diocletian's *Price Edict* and archaeological evidence, an average-sized dolium held a thousand *sextarii* (545 liters). If a *sextarius* of wine could cost between 8 and 30 denarii, as the *Price Edict* listed, it was a loss of 8,000–30,000 denarii.

4. Rando 1996; Peña 2007b, 37, 213–227.

FIGURE 7.1. Relief depicting the encounter between Alexander the Great and Diogenes, with detail of dolium repair, Villa Albani. Album / Alamy Stock Photo.

across various sites. Large dolia displayed outside the Baths of Diocletian feature suture-like repairs. Ongoing excavations at Gabii have uncovered patched-up dolium fragments. Dolia across southern Gaul and Iberia were also bound in place with metal bars. A closer look at repaired dolia, especially a particular area in detail, can reveal not only repair techniques but also how the maintenance of dolia was organized.

This chapter builds on previous work and repair typologies to take stock of the various dolium repairs at Cosa, Pompeii, and Ostia and Rome before considering dolium damage and repairs more broadly.[5] The number of repaired dolia at these sites is staggering and sheds light on the circumstances, the repair methods, and even the people who made them, bringing into view the repairers' major innovations and practices surrounding these precious pots, as well as the lengths to which dolium users went to maintain and protect their costly investments.[6] Focusing on west-central Italy also gives a more holistic view of how urban storage in the heart of the empire was supported. The following sections examine the numerous dolium repairs with a focus on types of damage (why), technique (how), material of the dolium repair (what), stage in the dolium's life cycle when the repair was made (when), and the possible dolium repairers (who). Moreover, the repairs show us the level of investment and new, often specialized tasks and occupations that supported increasing the longevity and dependability of dolia. In order to understand the different dolium repairs (and the repairers) across the sites, we first need to orient ourselves with the kinds of damage dolia could suffer and the various repair types and repairers that remedied them.

5. For overviews of various dolium repair types, see Cheung 2021b; Cheung and Tibbott 2020.

6. For work on knowledge networks, see Rebay-Salisbury et al. 2015; Miller 2009, 237–245.

Damaged Dolia

All pottery was vulnerable to damage. In pottery manufacture, it was inevitable for a portion of a batch of pottery to break or crack, especially during the firing process. A certain amount of wastage was expected. Pots could crack while drying or during firing. Inclusions could burst from a pot's surface, leaving behind a large hole. Sometimes entire batches of pottery could explode in a kiln. Only rarely was pottery damaged during the production process repaired; because most pottery was so cheap, defective pottery was sold at a discounted rate or merely discarded.[7] The majority of repaired pottery we have from antiquity was therefore mended during use. Surviving ancient pottery repairs usually consists of lead elements to fill cracks or anchor and brace fragments.[8] Repairs were also made with organic materials that do not survive. Lead offered several distinct advantages. It was widely available and inexpensive in antiquity.[9] As a stable material with a low melting point (327.5°C), lead was easy to work and a material almost anyone could use. To fix cracks and reattach fragments on smaller types of pottery (non-dolium repairs, henceforth "pottery repairs"), repairers often formed clamps, though they occasionally used staples or fills.[10] Forming clamps and staples required some expertise and special tools: the repairer would drill, probably with a bow drill, one or two sets of holes on either side of the crack, introduce metal pins or legs into the holes, and then join them to a crosspiece on one side to form a staple, or on both sides of the vessel wall to form a clamp; the lead elements were likely made in a mold to control the lead and form consistent pieces.[11]

The Latin agronomists mention various circumstances under which dolia could break while in use. As discussed in Chapter 3, Varro (*Rust.* 1.13.6) noted the danger in using large ceramic jars for wine fermentation; if sealed too early during the fermentation process, the buildup of gas would make the dolium explode, which is why Columella advised closing and sealing the dolia only after the first seven days of fermentation, and Cato thirty.[12] Columella also highlighted the risks of the maintenance process for the vessels when workers applied and spread molten pitch on the dolium wall: "These tasks, however, should be done on a day free from wind, so that the vessels do not burst if the wind blows while the fire is being applied" (*Rust.* 12.18.7: *sed haec die quieto a ventis fieri debent, ne admoto igne cum afflaverit ventus vasa rumpantur*). Although dolia were designed for wine fermentation, their very use and general maintenance could damage them.

But cracks could start to form as early as during the production process: because parts of the vessel dried at different times, with clay naturally shrinking first as it air-dried and again during the firing process, it was easy for dolia to crack over the course of their lengthy

7. Peña 2007b, 235. De Caro 1994, 179n151; Peña 2014: an African cookware lid (Hayes Form 196) with a production defect from firing was used at Villa Regina at Boscoreale.

8. For overviews of pottery repairs, see Bentz and Kästner 2006; Peña 2007b; Lawall and Lund 2011.

9. Lead was a by-product of silver mining. See Hong et al. 1994; McConnell et al. 2018.

10. Peña 2007b, 232ff. There is a discrepancy between terms employed for ceramics and architecture. The present book employs the term "staple" instead of "architectural pi-clamps" (or cramps), and "clamp" instead of "double pi-clamp." I thank Lynne Lancaster for discussing this with me.

11. Rotroff 2011, 122; Peña 2007b, 239.

12. Columella *Rust.* 12.36; Cato *Agr.* 26.

production.[13] Furthermore, the dolium's strawberry shape featured areas of extreme curvature where cracks were even more likely to appear, during both the forming and firing stages. These problems during production, coupled with damage that could occur while dolia were being used, presented plenty of opportunities for a dolium to crack or break completely. As with any costly investment, when dolia became damaged, it was clearly preferable to repair them than to discard them. But repairing a dolium was inherently difficult. It could not be taken apart, tinkered with or patched up, and then put together again. In addition, dolia, like other types of pottery, could not be repaired using clay or ceramic materials; making pottery was an irreversible, chemically transformative craft, altering the materials in ways that prevented them from ever bonding properly to clay again.[14] Instead, dolium repairers used other materials to fill, bind, or anchor damaged areas.

Dolium Repairs and the Repairers

Craftspeople in west-central Italy mended dolia with a range of techniques from the second century BCE to at least the end of the second century CE. In distinguishing the dolium repairs at Cosa, Pompeii, Ostia, and Rome, however, we can see that the methods and materials for mending dolia were not static or fixed but rather varied across time and space. Here we will review the different types of dolium repairs, and the possible menders, before examining evidence from the various towns. Dolium menders formed repairs in several ways using different techniques, strengthening the material or even identifying and preemptively repairing damage to reinforce the vessel. Sometimes it is possible to determine *when* certain dolium repairs were made based on their physical characteristics, which can help narrow down where and by whom the vessel was mended. In the case of repairs made *after* the vessel had been fired (*use repairs*), drill holes are usually accompanied by cracks, and any cuts made into the vessel are apt to display irregular, generally rough or chipped edges (Figure 7.2L; Plate 14L). Almost all pottery repairs were made while the vessels were in use, and a significant portion of dolia were also mended during use. Pino Rando, in his study of the dolia from the Diano Marina shipwreck, noticed that the cuts in the vessels displayed a distinctly crisp, regular edge, and he reasoned and then confirmed through experiments that these incisions must have been made before firing, presumably after air-drying when the vessels were in the leather-hard stage.[15] Repairs that have regular edges were therefore carved or drilled *before* a dolium was fired (*production repairs*; Figure 7.2R; Plate 14R). Occasionally the stage in which a dolium repair was made is unknown, while some dolia feature repairs made both during production and then again during use.

To have a fuller picture of how dolium repairs were made, it is important to know more about the tools, the procedures, the skills of the repairers, and the repairers themselves. Yet almost nothing is known about the different craftspeople who repaired dolia and other

13. Güven 1993: as clay dries, water evaporates and clay particles are drawn closer together, resulting in shrinkage, warpage, or cracking. Contemporary potter and archaeologist Gina Tibbott estimates a shrinkage rate of 8–10 percent for heavily grogged clay used for the manufacture of dolia.

14. Miller 2009, 103–128.

15. Rando 1996.

FIGURE 7.2. (*L*) Double dovetail tenon made during *use* (I.22 no. 7) and (*R*) detailed view of one made during *production* (I.22 no. 3), Pompeii. Courtesy of the Ministry of Culture—Archaeological Park of Pompeii. Reproduction or duplication by any means is forbidden.

types of pottery. We almost never encounter these workers in textual sources and subsequently never even begin to consider who they might have been. But every repair had to be made by someone. As we review the dolium repair types, materials, and stages of execution, we should keep in mind that they often provide important clues regarding the repairers (Table A1.11). Although we cannot say who *exactly* made the dolium repairs, we can at least narrow the range of possibilities by distinguishing the requisite craft skills for the repair, when and where the repairs were made, and the frequency and intensity of the task. Although several possible occupations could be responsible for different dolium repairs, people in antiquity probably had more than one occupation or an occupation that encompassed multiple crafts, so someone who was a builder might have also been a plumber. Overall, evidence for the repair of dolia extends from the earliest period of their manufacture through to the end, demonstrating that interventions of this kind were considered a worthwhile endeavor, and crucial to the longevity and use of dolia.

To form an effective dolium repair, menders had to select the right material and technique. Similar to pottery repairs, dolium repairs were made with lead or a combination of lead and another metal. Lead was the most common material used in the repairs, affirmed by the passage by Juvenal quoted at the beginning of the chapter and Cato's discussion on mending dolia. Both Cato and Varro, however, also mention using organic material, such as rushes (*sirpata*) or dried oak wood (*materie quernea uirisicca*), to bind the dolia, probably by lacing rushes through holes drilled on either side of the crack, but these materials are not preserved in the archaeological record.[16] The use of organic material to repair dolia might have been practiced in Italy during the final two centuries BCE and simply did not survive in the archaeological record, or was a practice from a period before which menders mostly repaired dolia with lead.[17] Menders sometimes formed a stronger lead alloy.

16. Cato *Agr.* 39; Varro *Ling.* 5.137. Discussion of these texts, and Juv. *Sat.* 14.308–310, is in Peña 2007b, 214–227.

17. Peña (2007b, 227) suggests a development in dolium repairs, from farmhands mending the vessels with organic fibers to the emergence of a specialized repair industry using lead.

Although pottery repairs were typically made with lead, by itself lead often did not lend much mechanical support or tensile strength; menders modified lead repairs, often with stronger metals. Tin is most likely one of the metals that repairers used to strengthen lead. It has a workable melting point (232°C) and is a strong metal; its typical brittleness decreases when paired with lead, and it has been chemically identified on a few pithos and dolium repairs.[18] Pliny (*NH* 34.156) noted that there were two kinds of lead (*plumbum*): a black one (*nigrum*), which corresponds to lead, and a white one (*candidum*), which is actually tin.[19] Interestingly, the terms *cassiterum* in Latin and κασσίτερος in Greek mean "a mixture or alloy of lead, silver, and other metals, afterwards tin" (*TLG*); the Greek term is related to the word κασσιτερᾶς (tinker), suggesting ancient tinkers used a lead alloy mixed with tin.[20] Working with different metals, however, often required higher temperatures, which could only be achieved with a proper furnace, fluxes, and the expertise of a metal worker, requiring extra skills, materials, and cost.[21]

The dolium repair technique mattered as well. As J. Theodore Peña notes, both dolium and pottery repairs generally involved two kinds of elements: fills and bracing elements. Fills served to infill cracks or, in some cases, gaps in a vessel wall extending over a large area, ideally rendering the vessel liquid-tight (Figure 7.3A). Due to its physical properties, lead could only be applied to dolia *after* the vessels were fired in the kiln, making it difficult to establish when they were applied—that is, in the workshop soon after the vessel was fired or sometime during its use. Moreover, working lead was not a task that depended on specialist knowledge or skill. Fills could have been added during the dolium's production phase by a member of the workshop or during the vessel's use by a range of people: pottery menders, tinkers, metal workers, plumbers, or even the dolium users. Bracing elements, such as staples and clamps, limited the propagation of cracks and solidified the vessel (Figure 7.3B-C). As with pottery repairs, forming clamps required some expertise and special tools to drill holes and form the clamp or staple bars with molds, likely by pottery menders, tinkers, and other craftspeople who were not necessarily specialists.[22]

Although many dolium repairs drew on traditional pottery mending techniques (clamps and staples), they differed from pottery repairs in substantial ways, namely in the types of damage they remedied: dolia often cracked during the manufacturing process, *before* they were ready for use. Because dolia were coil-built over the course of several days or even weeks, dolia tended to develop cracks of two different kinds during production. The first of these were horizontal cracks that formed along the juncture between two coils—a natural point of weakness—known as coil fractures (Figure 2.3). The second were vertical cracks known as dunting cracks that ran downward from the vessel rim as the rim, which was the most exposed and curved portion of the vessel, shed water and

18. Slane 2011: at Corinth, a pithos was repaired with lead and small amounts of tin and iron. Giardino 2012: a dolium at Metapontum was repaired with lead and traces of tin and copper. Carpentieri and Cheung 2023: lead repairs at Ostia had small amounts of silver and tin, at Morgantina aluminum and silver.

19. Healy 1978, 1999. For Pliny's philosophical undertones on metals, see Paparazzo 2003, 2008.

20. Another possible metal for the alloy is copper, which was occasionally combined with lead to form architectural clamps; cf. Cooper 2008. One proposed etymology for the English word "tinker" also attributes it to the word "tin." For the etymology of tinker, see Ekwall 1936.

21. Rehder 2000, chs. 11–14.

22. This is one way people repair *tinajas* today; see Romero and Cabasa 1999, 116–124.

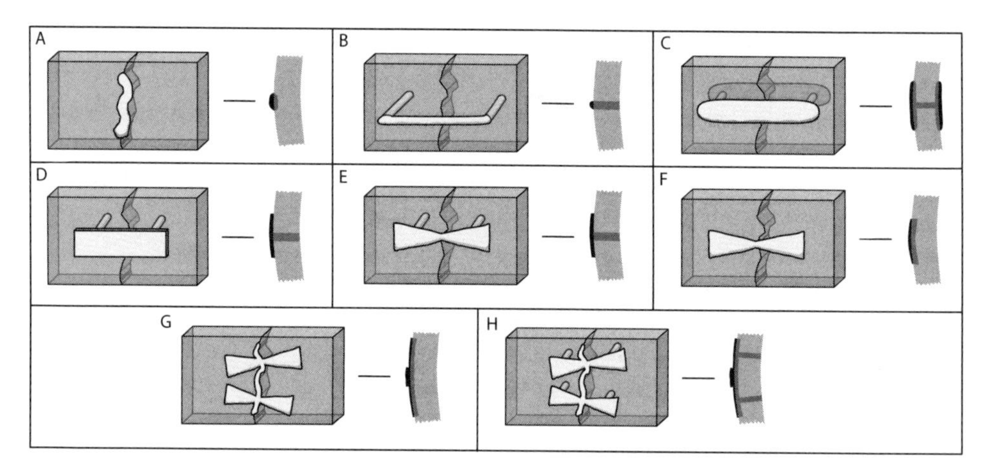

FIGURE 7.3. Illustration of various dolium repairs: (A) lead fill, (B) staple, (C) clamp, (D) mortise and tenon, (E) double dovetail staple, (F) double dovetail, (G) double dovetail tenon, and (H) hybrid double dovetail tenon and staple, by Gina Tibbott.

shrank more rapidly than the vessel's neck and shoulder during the firing phase of manufacture (Figure 7.4 top; Plate 15 top).[23] Dolium production was susceptible to certain types of production-based damage and defects not found on other types of pottery.

Dolia were also much bigger and heavier than even the largest types of pottery and often required different materials and methods to support their weight and bulk. Although clamps were commonly used to mend smaller types of pottery, dolium repairers probably noticed the shortcomings of clamps for dolia. Drilling through the thick vessel wall and coarse ceramic material with large inclusions could dislodge or displace an inclusion and cause additional cracks, damaging the big, bulky vessel even further (Figure 7.4 bottom; Plate 15 bottom). Furthermore, clamp holes could prevent a vessel from becoming liquid-tight. The lead clamp would also probably have some contact with the dolium's contents, potentially affecting their taste and quality.[24] Repair methods that were generally applied to smaller pottery could thus result in more damage to a dolium, and some repairers instead sought to improve or adapt their techniques, modify their repairs, and experiment with various materials and methods, drawing on techniques from the architectural realm.[25]

One way to strengthen a dolium repair was to combine two different methods, the traditional pottery mending clamp and a different technique, to form a hybrid repair that was used in architecture, the mortise and tenon (Figure 7.3D). The mortise-and-tenon technique is a traditional type of joinery widely found in carpentry and architecture: pieces of wood are joined by fitting together a mortise, which is a slot, with a tenon cut

23. Hamer and Hamer 2004, 119–122: dunting occurs due to silica inversions at 573°C and 226°C; this happens most frequently during firing's cooling phase. P. White 2016, 122: a Georgian *qvevri* maker said, "If it cracks horizontally, then the error is in my building; if vertically, then the fault is in the firing."

24. Romans occasionally heated wine in lead containers, so the taste might not have been distinctive.

25. Davis 2000, 12, 284–287.

FIGURE 7.4. (*Top*) Dunting on rim repaired with lead double dovetail (III.14.3 no. 1), Ostia. Courtesy of the Photographic Archive of the Archaeological Park of Ostia Antica. (*Bottom*) Crack that formed at clamp pin hole on dolium (I.22 no. 18), Pompeii. Courtesy of the Ministry of Culture—Archaeological Park of Pompeii. Reproduction or duplication by any means is forbidden.

specifically to fit into the mortise.[26] To form it on stone or ceramic, workers carved or chiseled a rectangular slot into the surface, drilled two pin holes near each end to form the mortise, and applied metal into the slot and pin holes to form the tenon. Dolium repairers

26. Bilde and Handberg 2012, 464. Ulrich 2007, 61–64: mortise-and-tenon joints were found in the last phase of the second millennium BCE at Stonehenge. Adam (2005, 96) and Ulrich (2007, 61ff.) discuss different mortise-and-tenon joinery techniques found in Roman carpentry and other construction.

familiar with architectural techniques, such as pottery menders and builders, adapted hybrid mortise-and-tenon staples or clamps to strengthen their repairs and to fill voids.

Dolium repairers sometimes modified and improved hybrid repairs with a more effective type of mortise and tenon known as the double dovetail (Figure 7.3E). Carpenters often used dovetail mortise-and-tenon joints in case-piece construction for items such as boxes, dressers, and other furniture, as well as in shipbuilding and temple construction.[27] Double dovetail mortises on stone and ceramic surfaces were cut or chiseled on the surface and then filled with metal to form the tenon.[28] The double dovetail shape prevented the wider end of the mortise (the two ends of the double dovetail) from withdrawing, acting as both an anchor and a fill. The rare instances in which double dovetails were used to repair small, fineware pottery involved scratching away only a small amount of ceramic material from the surface to form the mortise into which the lead tenon would be added.[29] Many dolia repaired with this method, however, featured a double dovetail *without* a staple or clamp, suggesting that this method was an effective way for the mender to avoid drilling into the vessel (Figure 7.3F).

Various dolium repairs show that, in developing these techniques, dolium *makers* could treat emerging damage during the production phase, and sometimes with stronger materials. In building a dolium, telltale signs of manufacturing defects could emerge during the drying process, or even as early as during forming. When the dolium maker saw potential damage, he or she could prepare a *production phase* clamp or hybrid double dovetail by drilling holes along both sides of the crack before firing the dolium and then adding the lead pegs and crossbars (Figure 7.5L; Plate 16L); drilling holes before the vessel was fired lessened the risk of further cracking the vessel since the bow drill would no longer come against the same resistance that fired ceramic posed. Most production phase interventions, however, involved double dovetails (Figure 7.5R; Plate 16R). To make double dovetails with the characteristic clean, crisp borders and consistency seen on many dolium repairs, craftspeople cut double dovetails when the dolium had air-dried, was leather-hard, and already exhibited structural defects, but *before* it was fired. They positioned double dovetails so that they would straddle both sides of the crack, which was usually minor. To mend more extensive cracks adequately during the production phase, dolium makers developed a more elaborate technique called the *double dovetail tenon* that featured multiple double dovetails connected by a channel, a method that seems to have been employed exclusively for the repair of dolia (Figure 7.3G; Figure 7.6; Plate 17). To execute this kind of repair, the

27. On double dovetails, see Korn 2003, 106ff. Evidence of carpentry only survives in arid or waterlogged contexts. Builders used wooden double dovetails to hold stone blocks in constructing the Forum of Augustus (Ganzert 1996, 124).

28. Ural and Uslu (2014) discuss the effectiveness of different metal connectors under shear stress. Greek and Roman craftspeople often used a faster, less precise method by hammering prefabricated mold-made tenons into mortises. Lugli 1957, 239: iron pi-clamps, and occasionally T-clamps, encased in lead appeared in Roman architecture around the end of the Republican period. Adam 2005, 96–100: pi-clamps were popular; fashioning T-clamps required time, effort, and precision.

29. Of the 35,000 vessels Rotroff (2011) examined, 160 were repaired, 11 with the mortise-and-tenon technique. Peña (2007b, 246ff.) notes that double dovetails appear on only 1.1 percent of pottery from the Museum of London collection. See also Dooijes and Nieuwenhuyse 2006; Bilde and Handberg 2012. See also Koob 1998.

FIGURE 7.5. Production phase repairs: (*L*) drill hole for clamp pins (VI.14.27 frag. 1) and (*R*) double dovetails (I.21.2 no. 1), Pompeii. Courtesy of the Ministry of Culture—Archaeological Park of Pompeii. Reproduction or duplication by any means is forbidden.

FIGURE 7.6. Tool marks on surface of double dovetail tenon (I.22 no. 3), Pompeii. Courtesy of the Ministry of Culture—Archaeological Park of Pompeii. Reproduction or duplication by any means is forbidden.

craftsperson produced a shallow (ca. 0.5–1.0 cm deep) cut into the surface of the vessel. Lead elements of both double dovetails and double dovetail tenons commonly exhibit densely spaced marks on the surface from the chisel, spatula, or similar tool the repairer used to push and spread the lead into the cuttings. This repair type was made typically during the vessel's production, by members of the workshop—whether the dolium maker, someone involved in the production of other opus doliare products, or a specialist repairer.

Most of the insights on the dolium repairers and their industries come directly from the archaeological evidence, but several come from other sources of information that can deepen and improve our interpretations. Cato's agricultural treatise felicitously offers detailed information and instruction on mending dolia as he enumerates tasks for rainy days:

> When the weather is bad and work cannot be done, clear out dung for the compost heap; clean up the oxen stalls, sheepfolds, courtyard, and farmstead; and fasten dolia with lead, or bind them with thoroughly dried oak wood. If you repair it carefully, or bind it well, closing the cracks with a sealant and applying pitch to it thoroughly, you can make any dolium a wine dolium. Make sealant for a dolium as follows: Take one pound of wax, one pound of resin, and two thirds of a pound of sulphur. Put these all into a new vessel. To this add pulverized gypsum, so that it becomes thick enough like a plaster, and mend dolia with it. To make it the same color after mending, mix together two parts of raw chalk and a third part of lime, then form into small bricks, bake in the kiln, grind, and apply to the dolium.
>
> ubi tempestates malae erunt, cum opus fieri non poterit, stercus in stercilinum egerito: bubile, ouile, cohortem, uillam bene purgato; dolia plumbo uincito uel materie quernea uirisicca alligato. si bene sarseris aut bene alligaueris et in rimas medicamentum indideris beneque picaueris, quoduis dolium uinarium facere poteris. medicamentum in dolium hoc modo facito: cerae p. I, resinae p. I, sulpuris p. c' c'. haec omnia in calicem nouum indito: eo addito gypsum contritum, uti crassitudo fiat quasi emplastrum: eo dolia sarcito. ubi sarseris, qui colorem eundem facias, cretae crudae partes duas, calcis tertiam commisceto: inde laterculos facito, coquito in fornacem, eum conterito idque inducito. (*Agr.* 39)

This passage presents a rare picture of dolium repair activity and highlights that, during the second century BCE, farmhands were expected to mend damaged dolia.[30] According to Cato's instructions, they could create simpler repairs to use dolia for dry foods. To make a damaged dolium suitable for wine storage again meant not only bracing the vessel wall with lead or organic matter but also forming a putty or cement to fill cracks before coating the interior surface of the vessel with pitch; the instruction to use lead or dried oak wood also indicates that menders could repair dolia in ways or with organic materials that are no longer preserved today. Cato also included directions for masking the repairs, which must have been more of an aesthetic preference than functional necessity but was important enough to include, a topic we will return to at the end of the chapter. His detailed instructions suggest that farmhands could learn to repair dolia as a sort of do-it-yourself project, perhaps because dolia menders might be hard to come by in the countryside or this was not yet a defined profession.

Examples of traditional pottery repair from more recent, well-documented periods provide useful insights into the industries for the repair of these vessels. In several preindustrialized societies, specialized pottery menders set up repair stations in cities and towns, such as Paris and London, whereas rural itinerant repairers made rounds through the countryside to offer their services to potential clients.[31] In Luigi Pirandello's early twentieth-century work "La giara," Don Lollò Zirafa, a padrone (wealthy landowner) in rural Sicily hires an itinerant repairman to fix his broken olive oil storage jar.[32] The

30. Peña 2007b, 214.

31. See Thornton 1998; Albert 2012; Garachon 2010 on mending porcelain.

32. Pirandello 1927.

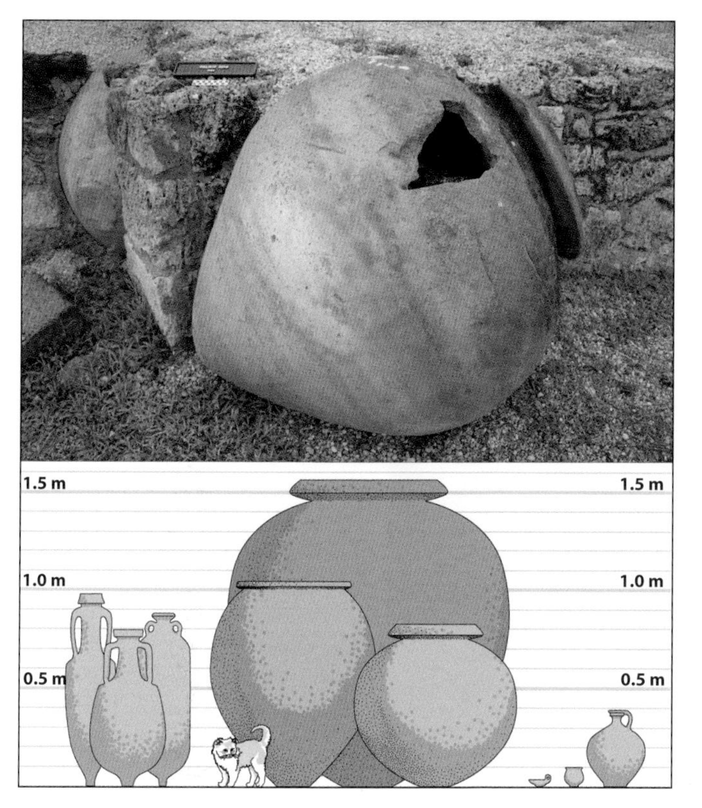

PLATE 1. (*Top*) Dolium lying on its side (I.22 no. 5), Pompeii. Courtesy of the Ministry of Culture—Archaeological Park of Pompeii. Reproduction or duplication by any means is forbidden. (*Bottom*) Dolia compared with amphorae (large bulk transport jars, *left-hand side*) and other pottery (*right-hand side*). Illustration by Gina Tibbott.

PLATE 2. Dolia defossa (*top*) in Caseggiato dei Doli (I.4.5), Ostia, by Jamie Heath, 2009; and (*bottom*) in Villa Regina, Boscoreale. Courtesy of the Ministry of Culture—Archaeological Park of Pompeii. Reproduction or duplication by any means is forbidden.

PLATE 3. (*Top*) Crack between dolium base and first body coil (I.22 no. 7); and (*bottom*) horizontal cracks between coils (I.22 no. 5), Pompeii. Courtesy of the Ministry of Culture—Archaeological Park of Pompeii. Reproduction or duplication by any means is forbidden.

PLATE 4. (*Top*) Seam between coils on dolium interior wall (I.22 no. 5); and (*bottom*) scored or paddled coil edge, Pompeii. Courtesy of the Ministry of Culture—Archaeological Park of Pompeii. Reproduction or duplication by any means is forbidden.

PLATE 5. (*Top*) Seam between rim coil and lip (I.22 no. 9); and (*bottom*) paddling marks on rim coil (VII.6.15 no. 1), Pompeii. Courtesy of the Ministry of Culture—Archaeological Park of Pompeii. Reproduction or duplication by any means is forbidden.

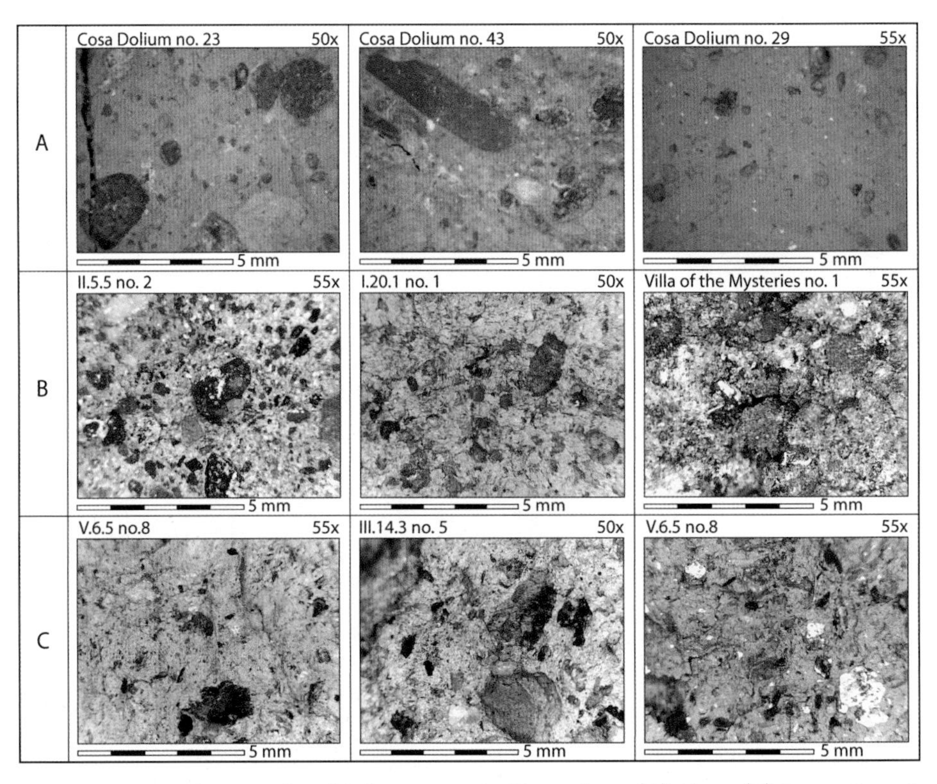

PLATE 6. Microphotographs of dolium ceramic fabrics from (A) Cosa, (B) Pompeii, and (C) Ostia.

PLATE 7. (*Top*) Dolium rim, stamped with "H" in triangular border; and (*bottom*) dolium rim, stamped with L·REMIO·C·F, Cosa. Courtesy of the National Archaeological Museum and Ancient City of Cosa (Regional Directorate of Museums of Tuscany). Reproduction or duplication by any means is forbidden.

PLATE 8. (A) Large dolium fragment; (B) dolium stamp on rim: "C·TVRI"; and (C) incision of an anchor or stylized phallus on shoulder, Cosa no. 19. Courtesy of the National Archaeological Museum and Ancient City of Cosa (Regional Directorate of Museums of Tuscany). Reproduction or duplication by any means is forbidden.

PLATE 9. Dolium rim with two stamps (VI.14.36 no. 4), C NAEVI/VITALIS, Pompeii. Courtesy of the Ministry of Culture—Archaeological Park of Pompeii. Reproduction or duplication by any means is forbidden.

PLATE 10. (*L*) A dolium from Magazzino Annonario (V.11.5 no. 67) and (*R*) dolium with capacity incision XLIIƆII (42 amphorae + 2 sextarii, 1,101.5 liters) (I.4.5 dolium no. 12), Ostia. Courtesy of the Photographic Archive of the Archaeological Park of Ostia Antica.

PLATE 11. (*L*) Stamp on dolium rim (III.14.3 no. 10): C VIBI FORTVNATI/C VIBI CRESCENTIS; and (*R*) stamps on dolium rim (III.14.3 no. 1): PYRAMI ENCOLPI/AVG DISP·ARCARI (*left-hand side*) and AMPLIATVS·VIC·F (*right-hand side*), Ostia. Courtesy of the Photographic Archive of the Archaeological Park of Ostia Antica.

PLATE 12. Shipwrecked dolium, Cosa, Ansedonia, Italy. Courtesy of the National Archaeological Museum and Ancient City of Cosa (Regional Directorate of Museums of Tuscany). Reproduction or duplication by any means is forbidden.

PLATE 13. (*L*) Funerary relief of Lucifer, Isola Sacra necropolis, by Gina Tibbott. (*R*) Roman, *A Banquet in the Open Air*, early fourth century CE, marble, glass, and clay. Detroit Institute of Arts, Founders Society Purchase, Sarah Bacon Hill Fund, 54.492.

PLATE 14. (*L*) Double dovetail tenon made during use (I.22 no. 7) and (*R*) detailed view of one made during production (I.22 no. 3), Pompeii. Courtesy of the Ministry of Culture—Archaeological Park of Pompeii. Reproduction or duplication by any means is forbidden.

PLATE 15. (*Top*) Dunting on rim repaired with lead double dovetail (III.14.3 no. 1), Ostia. Courtesy of the Photographic Archive of the Archaeological Park of Ostia Antica. (*Bottom*) Crack that formed at clamp pin hole on dolium (I.22 no. 18), Pompeii. Courtesy of the Ministry of Culture—Archaeological Park of Pompeii. Reproduction or duplication by any means is forbidden.

PLATE 16. Production phase repairs: (*L*) drill hole for clamp (VI.14.27 frag. 1) and (*R*) double dovetails (I.21.2 no. 1), Pompeii. Courtesy of the Ministry of Culture—Archaeological Park of Pompeii. Reproduction or duplication by any means is forbidden.

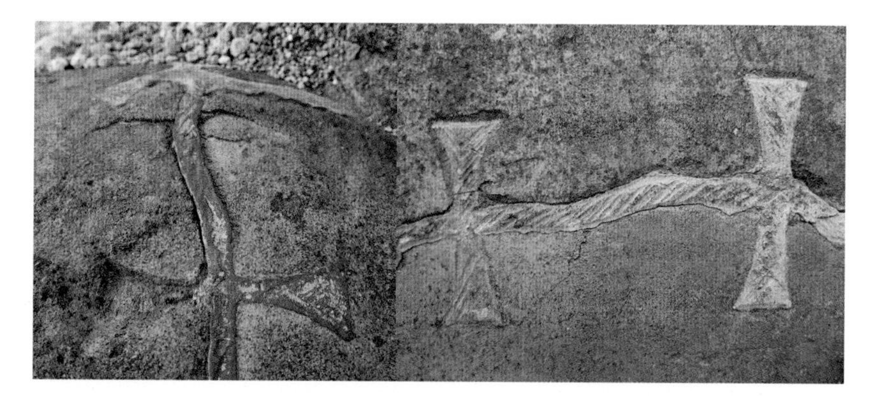

PLATE 17. Tool marks on surface of double dovetail tenon (I.22 no. 3), Pompeii. Courtesy of the Ministry of Culture—Archaeological Park of Pompeii. Reproduction or duplication by any means is forbidden.

PLATE 18. Clamps on (*L*) a dolium fragment (Cosa no. TC) and (*R*) mortarium, Cosa. Courtesy of the National Archaeological Museum and Ancient City of Cosa (Regional Directorate of Museums of Tuscany). Reproduction or duplication by any means is forbidden.

PLATE 19. Dolium rim repaired with screws (Cosa no. 1), Cosa. Courtesy of the National Archaeological Museum and Ancient City of Cosa (Regional Directorate of Museums of Tuscany). Reproduction or duplication by any means is forbidden.

PLATE 20. (*L*) Interior lower wall and base with lead clamp and (*R*) exterior underside of base with half of lead double dovetail preserved (Cosa no. 29). Courtesy of the National Archaeological Museum and Ancient City of Cosa (Regional Directorate of Museums of Tuscany). Reproduction or duplication by any means is forbidden.

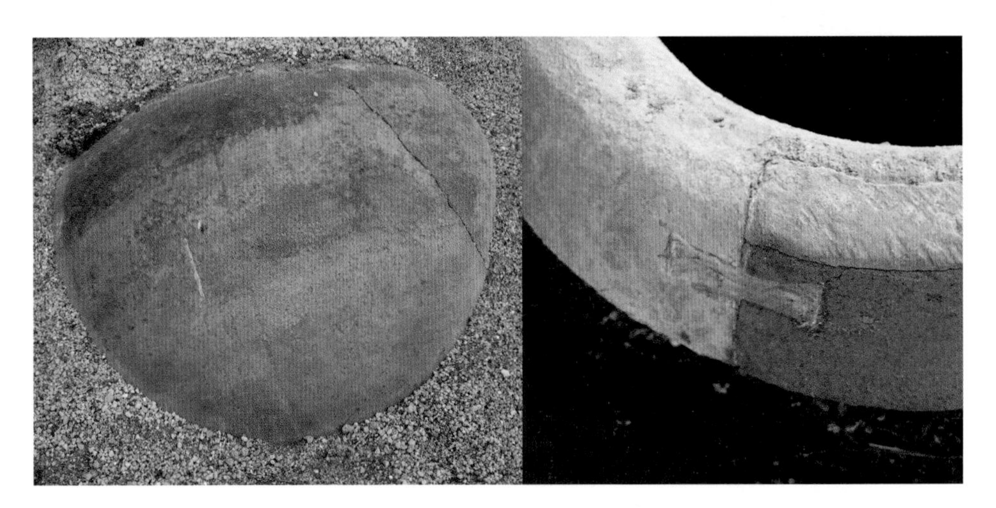

PLATE 21. (*L*) Lead fill on dolium shoulder (I.22 no. 1) and (*R*) large lead fill (and double dovetail) on rim (II.5.5 no. 10), Pompeii. Courtesy of the Ministry of Culture—Archaeological Park of Pompeii. Reproduction or duplication by any means is forbidden.

PLATE 22. (*L*) Partially preserved lead triangular clamp and (*R*) hybrid mortise-and-tenon staple on dolium rim (I.21.2 no. 2), Pompeii. Courtesy of the Ministry of Culture—Archaeological Park of Pompeii. Reproduction or duplication by any means is forbidden.

PLATE 23. (*L*) Double dovetail tenons; (*R*) double dovetails and clay smeared over horizontal crack (I.22 no. 3), Pompeii. Courtesy of the Ministry of Culture—Archaeological Park of Pompeii. Reproduction or duplication by any means is forbidden.

PLATE 24. (*L*) Horizontal and vertical lead double dovetail tenons around shoulder (I.4.5 no. 28) and (*R*) repaired dunting on rim and shoulder (I.4.5 no. 17), Ostia. Courtesy of the Photographic Archive of the Archaeological Park of Ostia Antica.

PLATE 25. Lead double dovetail tenons on dolia (*L*) no. 8 and (*R*) no. 4, Museo Nazionale Romano. Courtesy of the Ministry of Culture—The National Roman Museum, Baths of Diocletian. Any other or further use of the photograph must be authorized by this institute.

PLATE 26. Interior walls repaired with (*L*) lead fill (V.11.5 no. 61) and (*R*) lead alloy fills (V.11.5 no. 8), Ostia. Courtesy of the Photographic Archive of the Archaeological Park of Ostia Antica.

PLATE 27. Lead double dovetail tenon on interior lower wall of dolium (V.11.5 no. 52), Ostia. Courtesy of the Photographic Archive of the Archaeological Park of Ostia Antica.

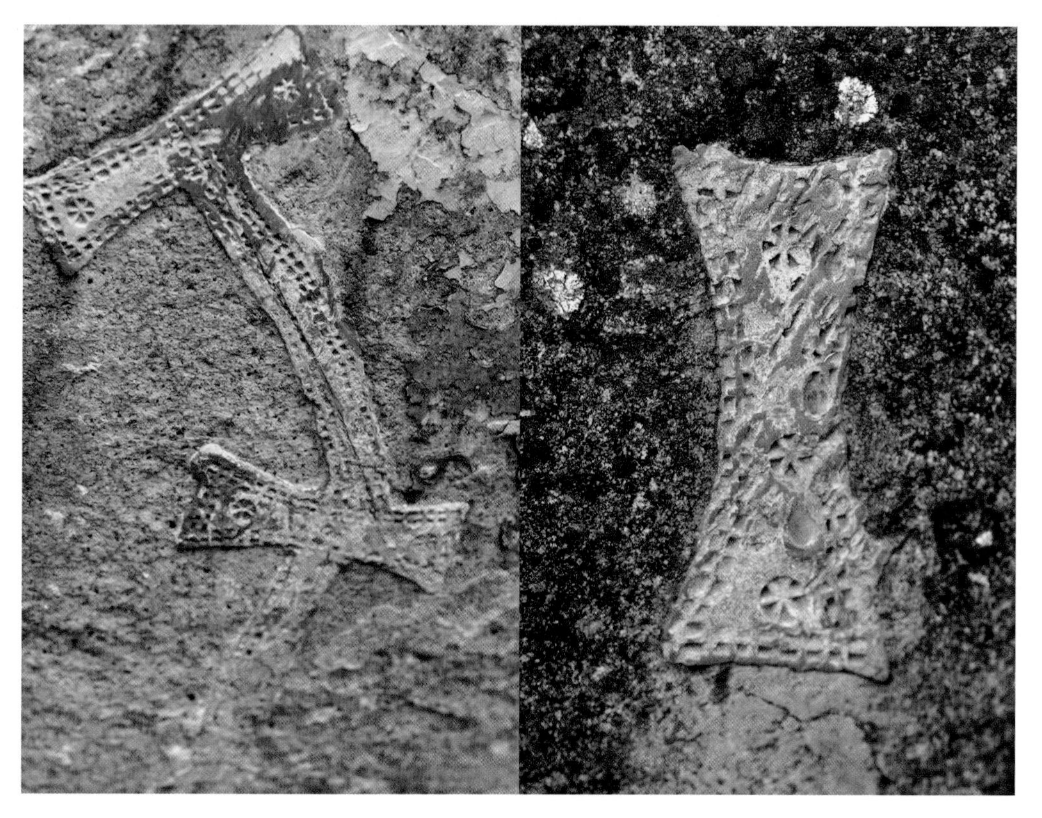

PLATE 28. Embellished lead repair, Museo Nazionale Romano in Rome (no. 10). Courtesy of the Ministry of Culture—The National Roman Museum, Baths of Diocletian. Any other or further use of the photograph must be authorized by this institute.

PLATE 29. (*Top*) Bar with five jars installed, three of which were reused dolia; (*bottom*) view inside a repurposed dolium, fragments held together by bar counter (VII.9.54), Pompeii. Courtesy of the Ministry of Culture—Archaeological Park of Pompeii. Reproduction or duplication by any means is forbidden.

PLATE 30. (*Top*) Wine production, Mausoleum of Santa Costanza. (*Bottom left*) Sarcophagus representing a Dionysiac vintage scene, 290–300 CE. J. Paul Getty Museum, 2008.14. (*Bottom right*) Miracle of Cana, Basilica of Santa Sabina in Rome, by Sailko, 2017.

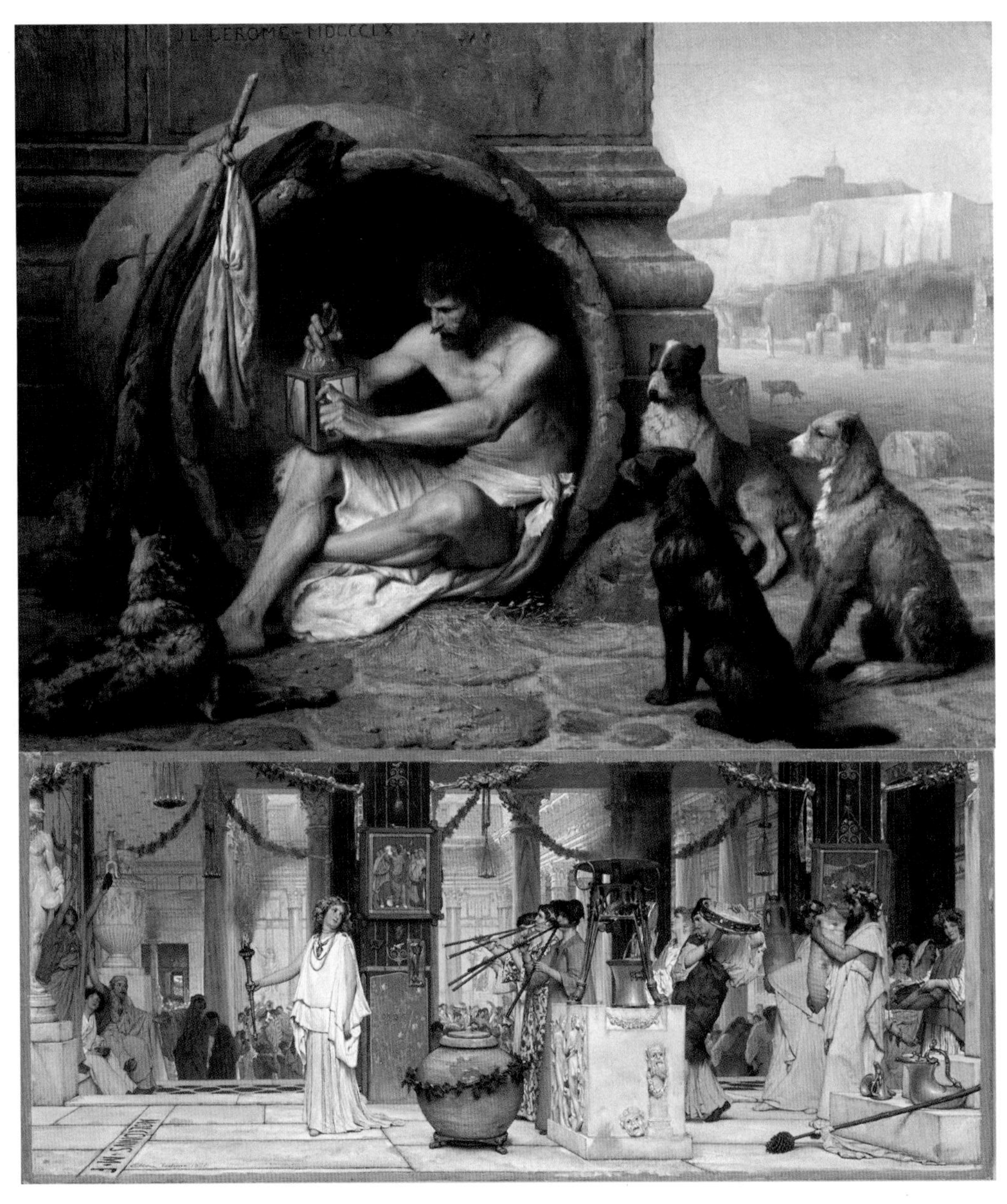

PLATE 31. (*Top*) *Diogenes*, by Jean-Léon Gérôme, 1860. Walters Art Museum. Acquired by William T. Walters, 1872. (*Bottom*) *The Vintage Festival*, by Lawrence Alma-Tadema, 1871, National Gallery of Victoria, Melbourne.

TABLE 7.1. Proportion of repaired dolia at Cosa, Pompeii, Ostia, and Rome.

Site	Number of repaired dolia	Number of dolia not repaired
Cosa	4	38
Pompeii	30	78
Ostia	48	76
Rome	9	11

repairman carried his kit of tools and materials with him as he traveled through the country. Turning to sixteenth- to eighteenth-century Europe, porcelain was a high-value, exotic commodity. Because it was considered a precious and collectible commodity that was expensive and difficult to acquire, its repair became an important occupation and emerged as a separate industry, with repair specialists who worked entirely independently from the porcelain production industries. In cities, specialized workshops or repairmen who set up workstations in heavily frequented areas mended porcelain, while itinerant repairers served the countryside.

Ethnographic examples cannot tell us what happened with dolia in antiquity, but they can help us interpret the ancient evidence, demonstrating three things. First, it is possible that different types of repairers worked in different areas, with specialists who set up stations in cities and nonspecialists, such as tinkers, who served the countryside. Densely populated urban areas probably had more demand for different services, fostering "thick" markets where specialists were needed and demand was high, whereas customers of the sprawling countryside probably had intermittent needs. Second, a dolium owner could influence the repair type a mender used. Recruiting a mender probably entailed some negotiation, during which the customer would have some say in what he or she was paying for. Lastly, the repair of an object can be entirely separate from its production, and two crafts for pottery, one for its production and one for its repair, could emerge. Not only was it expensive to transport dolia, but the workshop did not necessarily repair its wares because the techniques were so different from production. Pottery and dolia were craft goods that, if they were damaged during normal use, would likely fall under the aegis of craftspeople who had nothing to do with the manufacture and instead used their own sets of equipment, skills, and techniques. The many dolium repairs and repair types archaeologically visible in individual towns and cities, moreover, reveal varying levels of investment, specialization, and success (Table 7.1).

Cosa: A Motley of Dolium Repairs

The dolia of Cosa are mostly dated from the late second century BCE to the first century CE; almost all the dolia were broken in antiquity, but only five were repaired and with wildly different methods, resulting in varying degrees of success (Table A1.12). Almost every repair used only lead and appears to have been made during the vessels' use, reflecting a variety of repair techniques and general lack of expertise among the repairers.

FIGURE 7.7. Clamps on (*L*) a dolium fragment (Cosa no. TC) and (*R*) *mortarium*, Cosa. Courtesy of the National Archaeological Museum and Ancient City of Cosa (Regional Directorate of Museums of Tuscany). Reproduction or duplication by any means is forbidden.

Except for one, all the repaired dolia from Cosa were found in discard, rather than use-related, contexts, so we are not in the position to know more about the background of their repair and (re)use, but we can say that the repairs failed at some point and the dolia were thrown away.

One of the repairs is closely aligned with the clamp method used on smaller types of pottery (Figure 7.7L; Plate 18L; A2 C TC). In one example, a craftsperson mended a large dolium, which had broken along a horizontal crack stemming from a production flaw during the coil-building process, with a clamp made of a dark-red metal with a whitish-gray surface, likely lead combined with another metal such as iron or tin to strengthen the repair.[33] Clamps were used not only on smaller types of pottery but also on other types of heavy terracotta objects at Cosa. A large *mortarium*, another common product of opus doliare workshops used for grinding food, cracked while in use (Figure 7.7R; Plate 18R). To make it usable again, a repairer mended it with several lead clamps that straddled deep cracks that had formed on the base. Indeed, mending heavy terracotta objects with clamps seems to have been common in the Mediterranean; a severely cracked terracotta bathing tub, now in Siracusa, was also extensively repaired with lead clamps. The techniques,

33. Rotroff 2011; Giardino 2012. Scientific studies show the use of primarily lead with some iron or tin.

FIGURE 7.8. Dolium rim repaired with screws (Cosa no. 1), Cosa. Courtesy of the National Archaeological Museum and Ancient City of Cosa (Regional Directorate of Museums of Tuscany). Reproduction or duplication by any means is forbidden.

materials, and tools (bow drill and molds) used to form lead clamps while the dolium was in use point to the handiwork of pottery menders, tinkers, and perhaps even builders.

Drilling was a frequent part of the repair process at Cosa. Two dolium rims were modified and repaired with nails or pegs drilled into the middle of the rim; one rim still had the iron pegs, which were connected by a thin strip of lead (Figure 7.8; Plate 19; A2 C no. 1). The screws or pegs drilled into the rim were intended to reattach or secure the rim onto the neck of the dolium, a technique parallel to architectural dowels, where marble and stone workers often encased iron rods or bars in lead to join segments in sculpture or column drums to provide vertical support and bind pieces.[34] The technique, which did not fall in the realm of pottery repairs, suggests that people with experience in architectural or stone-working industries, such as builders and masons, repaired this vessel.

We see one example of a hybrid repair on the shoulder of the town's most sizable dolium (stamped with C TVRI) dated to the first century CE (Figure 7.9; A2 C no. 19). To mend the vertical crack that had formed on this large vessel's shoulder, the craftsperson formed a hybrid mortise-and-tenon clamp following a process similar to that used by repairers to add clamps. The craftsperson drilled holes into the vessel wall and then added crossbars on both sides of the vessel to form clamps. In this case, however, the mender also excised some of the dolium's ceramic material from both the exterior and interior surfaces. As a result, the repairer was able to apply and shape the metal in the grooves (rather than use a mold to form the crossbars) and keep the metal pieces flush against the vessel wall, which both allowed the mender to customize the repair and lessened the chance of damaging the repair later.

At least one dolium mender at Cosa harnessed the double dovetail technique to augment clamps on a dolium base for an especially tricky repair (Figure 7.10; Plate 20; A2 C no. 29). The mender carefully excised a double dovetail groove on the outer surface of the base and then added lead into the double dovetail groove and mold for the clamp crossbar on the interior of the base. Adding the metal must have been challenging, judging from the excess lead

34. On this technique to join statuary, cf. Claridge 1990; Wootton et al. 2013. On similar techniques in architecture, see Dinsmoor 1922, 1933; DeLaine 2010.

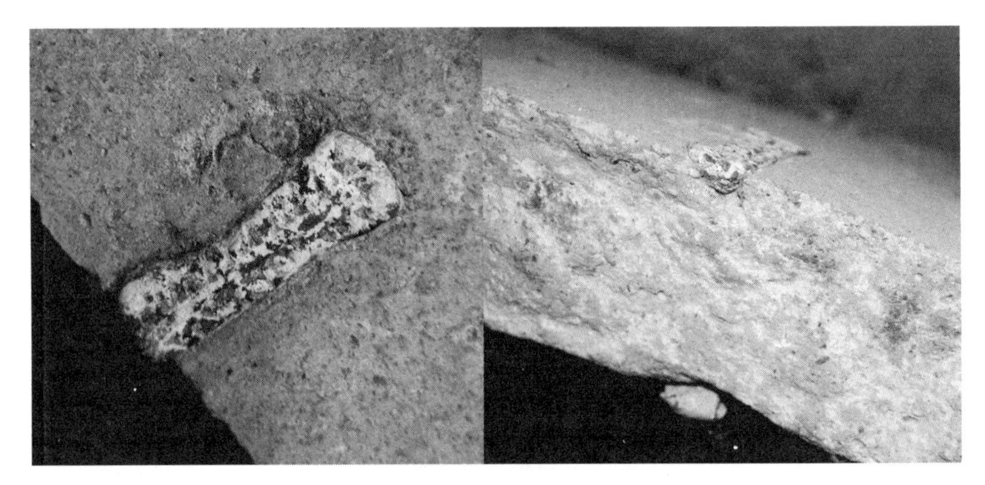

FIGURE 7.9. Lead clamp on dolium shoulder (Cosa no. 19), Cosa. Courtesy of the National Archaeological Museum and Ancient City of Cosa (Regional Directorate of Museums of Tuscany). Reproduction or duplication by any means is forbidden.

FIGURE 7.10. (*L*) Interior lower wall and base with lead clamp and (*R*) exterior underside of base with half of lead double dovetail preserved (Cosa no. 29). Courtesy of the National Archaeological Museum and Ancient City of Cosa (Regional Directorate of Museums of Tuscany). Reproduction or duplication by any means is forbidden.

that had flowed around the clamp crossbar. The double dovetail shape, on the outer part of the base, kept the repair flush against the vessel surface. Controlling the flow of molten lead would have been difficult as the mender poured it along a curved surface, and those who carried out these repairs developed methods to overcome this problem. Repairers could add linseed oil to form a putty or use a lead-tin alloy, and probably moved the vessel to direct the

FIGURE 7.11. Dolium rim with drill hole and half double dovetail (Cosa no. 11), Cosa. Courtesy of the National Archaeological Museum and Ancient City of Cosa (Regional Directorate of Museums of Tuscany). Reproduction or duplication by any means is forbidden.

flowing lead.[35] For this dolium, the repairer had to manipulate the molten metal not only between the outside and inside of the vessel but also in the innermost portion of the dolium, to control and apply the lead. The flow of the excess lead on the inside of the base suggests it was a two-person job: one applied the lead on the outer part of the base (the double dovetail cutting and drill hole) into a crossbar mold that the other held. With the large size of the dolium (at least half a meter tall), the other repairer, probably a petite man, woman, or even child, went inside the dolium in order to hold the crossbar mold at the base.

On a dolium rim, one of the two rims mentioned earlier that also preserved a drill hole, a mender excised ceramic material in the shape of a double dovetail from the rim (Figure 7.11; A2 C no. 11). Although it is unclear which repair was added first, the drilling or the double dovetail cutting, two very different repair techniques on the same dolium rim indicate one repair was attempted and, when that failed, another was applied. The double dovetail repair could have been a second attempt that avoided the clamp's shortcomings that eventually emerged, or the double dovetail could have been a production repair and the clamp added later to remedy use-life damage. Regardless of the order of repairs, the dolium owner attempted to mend the vessel twice, but eventually discarded it.

Every preserved dolium repair at Cosa was different and featured methods and materials also employed in pottery repairs. The repairs documented at Cosa, each differing from

35. On putty in repairs, see Rosenfeld 1965, 139–141; Thornton 1998, 11; Peña 2007b, 220ff. I thank Lynne Lancaster for sharing video documentation of her experiment with pouring lead, which demonstrated how challenging it can be to work with molten lead.

one another in significant ways, suggest that there was no accepted set of methods in the town. Instead, individuals with widely varying levels of skill and areas of expertise, rather than an established set of specialist craftspeople, repaired the dolia adventitiously. Although we cannot determine who exactly these repairers were, we can at least tease out some of the salient skillsets involved to better understand their possible occupations. Some of the dolium repairers drew from and combined the existing repertoire of repairs found on smaller pottery to form clamps. These could have been the handiwork of pottery menders or tinkers, who were probably itinerant, nonspecialist repairers.[36] On the other hand, construction workers, builders, or masons who worked on architectural projects in Cosa likely made the double dovetail repairs. This type of knowledge exchange can be seen in bath construction, where technical knowledge in vaulting techniques spread for building baths and interactions heightened among builders, terracotta workers, and potters, including the use of particular tools; the spread of knowledge went both ways, as tile workers likely adapted the use of lead clamps on terracotta from pottery menders.[37] Overall, various kinds of craftspeople, for whom work of this kind probably represented only a minor sideline, visited the town to make a repair, each using his or her own methods and materials. Judging by the repairs, Cosa was a place that represents a low level of sophistication in dolium repair technology. A few dolium owners tried to extend the life of their dolia, while most broken dolia were simply jettisoned, but all the repairs preserved at Cosa were found on dolia that eventually broke where the repairs were applied and had been discarded in antiquity, suggesting they were ultimately unsuccessful.

Pompeii: Experimentation in the Field and within the Workshop

Of the approximately one hundred dolia at Pompeii, nearly a third featured the telltale types of dolium damage and were repaired with a wide range of methods, some of which were intricate and labor intensive, while others were simpler and more ad hoc (Tables A1.13–14). There were only so many options repairers had to treat damage that appeared during a dolium's use. But the entire range of dolium repairs in Pompeii brings out differences not only between how repairers at Cosa and Pompeii mended dolia but also concerning *when* during the vessel's life this occurred and *who* the repairers were.

The most basic repair type was lead fills. Pompeian dolium repairers replicated this repair method found on smaller types of pottery to mend dolia for a variety of superficial damage, from minor cracks to missing surface areas (Figure 7.12; Plate 21). Cracks commonly formed on dolia, especially on areas of extreme curvature such as the rim, shoulder, and base (A2 P I.22 no. 1). Lead fills plugged cracks or voids and could even prevent some from deepening. Menders could fill in large areas, where the surface was abraded from wear and tear, to make the surface level again, such as one dolium rim where filling in the worn area

36. There is no Latin word for "pottery mender" or "tinker." There was a term for "bronze vessel mender," *refector ahenorum*, and a Greek word for "tinker," κασσιτερᾶς (*BGU* 9 IV 22 1087). See Peña 2007b, 249, 381n41. On a modern tinker, see Harper 1987. Historically, tinkers were itinerant.

37. Lancaster 2012. Joshel 1992: Latin occupational titles for builders included *faber*, *abietarius*, and *faber intestinarius*.

FIGURE 7.12. (*L*) Lead fill on dolium shoulder (I.22 no. 1) and (*R*) large lead fill (and double dovetail) on rim (II.5.5 no. 10), Pompeii. Courtesy of the Ministry of Culture—Archaeological Park of Pompeii. Reproduction or duplication by any means is forbidden.

enabled a tight fit, and hence proper sealing, with a lid (A2 P II.5.5 no. 10). Because lead was applied to dolia after firing, it is difficult to determine whether they were production phase or use-life repairs. Some examples, however, are clearly "touch-ups" that tinkers, pottery menders, or other craftspeople added to vessels that became worn with use.

Lead fills, though convenient, offered no structural support. To mend larger cracks and damage on more vulnerable areas, dolium repairers often used clamps. Many of the clamps at Pompeii are similar to those found at Cosa: they were made of lead and added during the vessel's use, usually around the rim and shoulder, an area prone to damage. For more severe and irregular cracks that had begun to expand and proliferate, dolium repairers at Pompeii fashioned and arranged clamps with three crosspieces in a triangular form and also attempted to remedy the damage with staples that combined mortise and tenons or double dovetails. To have a fuller understanding of these repairs, let us look at one particular dolium (A2 P I.21.2 no. 2) of three in the house (I.21.2) connected to the Garden of the Fugitives (I.21.6).[38] The dolium set in the center suffered two major vertical cracks from the rim through the shoulder. The repairer applied several lead clamps to brace the vessel, even to areas that were difficult to reach, such as the join between dolium neck and rim. One series of clamps terminated with a triangular lead clamp on the belly, mitigating damage on a precipitous area (Figure 7.13L; Plate 22L). The repairer might have been someone who repaired other types of pottery or large utilitarian objects, such as bakery millstones in town.[39] Clamps, however, posed problems with the use and routine maintenance of wine dolia. During the removal of residual pitch lining, which involved using a torch to melt the pitch, the lead clamp could accidentally be melted off. Workers could also simply knock off a clamp's crossbar in cleaning or using the dolium. Broken or missing crossbars were

38. Jashemski 1993, 69–70.

39. The Bakery of Popidius Priscus (VII.2.22) has a millstone mended with clamps.

FIGURE 7.13. (*L*) Partially preserved lead triangular clamp and (*R*) hybrid mortise-and-tenon staple on dolium rim (I.21.2 no. 2), Pompeii. Courtesy of the Ministry of Culture—Archaeological Park of Pompeii. Reproduction or duplication by any means is forbidden.

common in Pompeii. Their vulnerability was perhaps why this dolium was positioned so that the clamps faced the other two dolia, rather than the path where foot traffic could brush against the repairs. To repair the rim surface on which the dolium lid was set, the repairer did something different. He or she not only drilled into the rim to form staples but also excised ceramic material for the crossbar, forming a hybrid mortise-and-tenon staple, so the metal would not protrude above the surface and the lid and rim could still form a tight seal (Figure 7.13R; Plate 22R). Chiseling and drilling into fired ceramic material, however, often not only resulted in uneven edges but could also propagate additional cracks. In general, use-life repairs on dolia were problematic, and recognizably so.

Production-based damage and defects seem to have been problematic and perhaps predictable enough that members of the workshop began to treat damage during manufacture. Dolium makers occasionally formed production phase clamps; by drilling through the dolium *before* it was fired in the kiln, the mender could avoid dislodging inclusions and damaging the dolium further. Production phase clamps might have worked, but almost every dolium repaired during production at Pompeii was mended with double dovetails or a new repair technique. We have seen double dovetails made during use at Cosa and at Pompeii, but production ones were better executed and more effective. Double dovetail tenons were commonly applied to areas with more extensive damage. Some combined the double dovetail tenon with staples (with the pins on the ends of the double dovetails; Figure 7.3H). Most of the double dovetail tenons, however, only featured shallow cuttings filled with lead or lead alloy, without any drilling into the vessel walls (Figure 7.14L; Plate 23L; A2 P I.22 no. 3). Instead, repairers excised ceramic material along and from the cracks, preventing them from deepening during the firing process, and cut grooves for double dovetails that followed and straddled the crack. Menders occasionally smeared clay over emerging cracks to "patch" the damage; on one dolium, a worker spread clay over a horizonal crack that began to form and applied double dovetails to brace the vessel

FIGURE 7.14. (*L*) Double dovetail tenons; (*R*) double dovetails and clay smeared over horizontal crack (I.22 no. 3), Pompeii. Courtesy of the Ministry of Culture—Archaeological Park of Pompeii. Reproduction or duplication by any means is forbidden.

further (Figure 7.14R; Plate 23R). Several of these production repairs were placed on areas where damage did not form, underscoring the artisans' attempts not only to treat but also to prevent damage and reinforce dolia.

Overall, dolia at Pompeii were repaired with a range of techniques during both their production and use. Workshops supplying dolia to and dolium owners of Pompeii invested heavily in repairing and reinforcing their dolia, and the repairs seem to have been largely successful. Dolium makers employed preventative measures, prepared lead alloys, and formed double dovetails or double dovetail tenons to repair defects that appeared in production and prevented damage that could form later. Yet substantial damage could still occur during the use of the vessel. Depending on the severity of the damage, various workers, such as pottery menders, tinkers, architectural workers, and craftspeople who worked with lead (*plumbarii* or *artifices plumbarii*), repaired the vessel using materials and techniques that they knew, from lead fills to staples and clamps to hybrid repairs.[40] These were the same techniques, and probably the same general workforces, that repaired other large terracotta objects and vessels in Pompeii, and there have been at least three lead workshops and four iron workshops identified in the ancient town.[41] Excavations at Herculaneum have also uncovered a workshop that likely manufactured lead repairs as well as lead and lead-tin objects, testifying to the importance of lead for repairs in antiquity.[42] These workshops and urban marketplaces were places where dolium owners, from both the town and the countryside, could find a mender for dolia that broke while in use.[43] Although menders, whether craftspeople in the workshop or working in the town, used different types of repairs, they were systematic in their approach, reflecting substantial developments in mending dolia.

40. Joshel 1992. Although there were specialist occupations associated with metals, lead workers were associated with plumbing, perhaps because lead working was not considered a specialized craft.

41. On iron work in Pompeii, see Gralfs 1988; Gaitzch 1980; Amarger and Brun 2007; Amarger 2009. On Pompeian lead workshops—e.g., VII.5.28—see Monteix and Rosso 2008; Monteix 2017, figure 7.2.

42. Brun et al. 2005, 329–337: various lead and lead-tin objects and a crucible were found.

43. On how people found jobs in the city of Rome, see Holleran 2016.

Ostia and Rome: A Trend toward Production Repairs

For any discussion on dolium repairs, the capital represents the pinnacle of their technical accomplishment. Perhaps the most important port for Rome and the center of Roman food storage, Ostia had almost two hundred buried dolia, about a hundred of which are still visible today, and at least nearly half of these were repaired. In Rome, we see a similar situation, with over half the dolia repaired in antiquity. Although so many of the dolia at Ostia and Rome were repaired, there are only three types of repairs: fills, double dovetails, and double dovetail tenons (Tables A1.15–17). These three types, each of which was very consistent, shed light on specialization not only within the urban opus doliare workshops producing and preemptively reinforcing dolia but also among the workforces responsible for maintaining these storage vessels in the heart of the empire.

In urban opus doliare workshops, dolium makers took preemptive measures to anticipate and treat potential damage early on. More than half of repaired Ostian dolia were mended with production phase double dovetails or double dovetail tenons. In Rome, most of the repairs were formed during the dolium's production, with half featuring extensive double dovetail tenons, some of which covered almost the entire span of the vessel. Repairers placed double dovetail tenons over extensive cracks emerging on the vessel, carefully and strategically shaping the repairs and filling them with stronger lead alloys (Figure 7.15; Plate 24; A2 O I.4.5 nos. 28, 17). Some areas were especially challenging to reach and reflect the mender's skill and precision in forming these repairs during production. To form double dovetail tenons on the base of one dolium, for example, the mender had to lie on the ground and elevate the dolium or turn it on its side or upside down to cut the grooves (Figure 7.16L; Plate 25L; A2 R NMR no. 8); a mender repaired another dolium extensively with double dovetail tenons radiating from the base on large swaths of the vessel wall all the way up to the rim (Figure 7.16R; Plate 25R; A2 R NMR no. 4). Unlike the production repairs found at Pompeii, many of the repairs at Ostia and Rome appeared on the rim's upper surface to mediate dunting cracks that might form during the firing process but did *not* appear, testifying to the dolium maker's ability to anticipate damage.

Fills appear on almost every other repaired vessel in Ostia and on almost a third of the vessels in Rome to repair minor cracks, even on the interior wall of the vessel, meaning repairers would often climb into a dolium to identify and seal cracks that had formed. Fills could, in principle, have been production or use repairs, but the choice of material might have been more limited depending on whether they were added during production or in use. Lead was relatively easy to manipulate and could have been applied anywhere; an open flame would have been enough to bring the metal to a liquid and malleable state (Figure 7.17L; Plate 26L; A2 O V.11.5 no. 61). Lead alloys, on the other hand, required some metallurgical expertise and special equipment, suggesting that they were prepared in the workshop but probably not used in the field for use repairs (Figure 7.17R; Plate 26R; A2 O V.11.5 no. 8). This is supported by repairs in Pompeii, where no lead alloy has been positively associated with a use repair. Scientific analyses indicate that the lead alloys at Ostia primarily featured lead with significant amounts of silver and tin, traces of which might have originated from lead extraction processes, but some amount was probably intentionally mixed with lead to increase its hardness and strength.[44] Alloys of lead and tin and lead, tin, and silver were common

44. Carpentieri and Cheung 2023.

FIGURE 7.15. (*L*) Horizontal and vertical lead double dovetail tenons around shoulder (I.4.5 no. 28) and (*R*) repaired dunting on rim and shoulder (I.4.5 no. 17), Ostia. Courtesy of the Photographic Archive of the Archaeological Park of Ostia Antica.

FIGURE 7.16. Lead double dovetail tenons on dolia (*L*) no. 8 and (*R*) no. 4, Museo Nazionale Romano. Courtesy of the Ministry of Culture—The National Roman Museum, Baths of Diocletian. Any other or further use of the photograph must be authorized by this institute.

according to Pliny, and are used today as solder since they have a low melting point yet high malleability. Since all double dovetails and double dovetail tenons were made during production, half of the dolium repairs at Ostia are securely identified as having been made during production of the vessel. If we consider all lead alloy fills as production repairs (in addition to double dovetails and double dovetail tenons), then an astonishing majority of

FIGURE 7.17. Interior walls repaired with (*L*) lead fill (V.11.5 no. 61) and (*R*) lead alloy fills (V.11.5 no. 8), Ostia. Courtesy of the Photographic Archive of the Archaeological Park of Ostia Antica.

dolium repairs at Ostia were production repairs.[45] If this is the case, it means that repairers, who were most likely the dolium makers themselves or at least a member of the workshop, preemptively reinforced a third of the dolia they made using sophisticated techniques.

The great skill developed to form repairs is further attested by another example: one dolium featured a double dovetail tenon on the *interior* wall, which suggests that dolium makers were able to climb into the dolium and mend the interior surface (Figure 7.18; Plate 27; A2 O V.11.5 no. 52). This was a challenging and very unusual feat and might explain one benefit of the wider rims (average diameter ca. 62 cm) typical in Rome and Ostia. It is unclear when this double dovetail tenon was made. If it was made during the production process, this suggests that members of the opus doliare workshops (or perhaps [enslaved] children who could fit into the jar) commonly inspected the interiors of dolia, probably by lamplight, before firing.[46] To do this, one needs both an extremely high-quality and well-prepared ceramic material for the vessel and for the potter to know exactly when the unfired vessel would be strong enough to support a person's weight. Groggy clay, such as clay used for dolia, could have incredibly high levels of "green strength" (the strength of a clay body in dried form, but not yet fired, which aids in handling during production), perhaps enough to support the weight of a person when it dried between a leather-hard and a bone-dry state.

This level of intense investment at such an early stage is unparalleled. At the same time, there were no staples or clamps. None of the dolium repairs in Ostia or Rome involved

45. Some dolia were repaired with both lead and lead alloy materials. Menders likely filled a crack during production with lead alloy, but another applied lead during its use-life.

46. Apul. *Met.* 9.6.

FIGURE 7.18. Lead double dovetail tenon on interior lower wall of dolium (V.11.5 no. 52), Ostia. Courtesy of the Photographic Archive of the Archaeological Park of Ostia Antica.

any drilling that would have made the vessels more vulnerable and susceptible to further breakage. The quality and consistency of the dolium repairs at Ostia and Rome indicate that specialized dolium makers in opus doliare workshops repaired dolia with double dovetails, double dovetail tenons, and fills during production. These reinforcements were so successful that the repairers who serviced Ostia and Rome only used fills to mend minor damage on dolia during their use. By the second century CE, craftspeople in opus doliare workshops repairing dolia used well-established techniques for damage emerging during a dolium's production, while repairers at Ostia and Rome only mended dolia during their use with lead fills. What is most notable about these repairs is how many were made during the production phase. The large number of double dovetails and double dovetail tenons, as well as the absence of clamps and staples, could mean two, not necessarily exclusive, things. When a dolium cracked or broke while in use, it was simply discarded. Repairing serious cracks that formed after a dolium had already been fired with staples or clamps was a risky endeavor that could easily lead to the vessel breaking entirely. Rather than find (and pay) a repairer to drill through the vessel to make staples or clamps that might have broken the vessel, the owner could have directed resources to replacing the dolium entirely. The other possibility was that the rate of breakage among these dolia was low and that the dolia were more robust and less likely to crack during use, probably because dolium makers manufacturing these large storage vessels became better at not only making dolia but also both anticipating and preventing potential damage. The manufacturing defects that necessitated the repair of these vessels during the production phase probably stemmed from the vessels' exceptionally large size, with unusually thick walls that were highly susceptible to crack formation during drying and firing. To construct, and then be able to sell, such large

FIGURE 7.19. Embellished lead repair, Museo Nazionale Romano in Rome (no. 10). Courtesy of the Ministry of Culture—The National Roman Museum, Baths of Diocletian. Any other or further use of the photograph must be authorized by this institute.

storage vessels, then, manufacturers also had to develop the technology for reinforcing dolia during the critical stage when the repair could be successful without risking further damage to the vessel. In fact, a third of the repaired dolia at Ostia also featured a workshop stamp; what we see by the beginning of the second century CE, if not earlier, is a departure from Cato's advice on dolium repairs. Rather than hide the dolium repairs, opus doliare workshops stamped their repaired dolia. Some even seem to have decorated the repairs: while repairers often left telltale tool marks on the lead when they worked the metal into the cut grooves, some craftspeople embellished the surfaces with their tools, artfully adding patterns to the borders and center of dovetails (Figure 7.19; Plate 28; A2 R NMR no. 10). With over two hundred dolia left in situ in Ostia and all dolia in Rome complete and in good shape, the preventive repairs were largely successful, and we can see why dolium makers who constructed and reinforced the vessels celebrated their craftwork.

Reinforcing and Repairing Dolia

Mending dolia was vital not only for the vessels but also for the general success of the dolium production industries. Once they were out of the workshop, dolia were no longer attached to or dependent on dolium makers to function. To support a growing wine industry, dolium makers took on the enormous task of creating a bigger and better dolium. But its construction was so complex and time consuming that, in the end, the vessel was very expensive yet cracked easily. Use repairs were more makeshift and often of uneven and sometimes inferior quality. For at least half the history of the dolium industry, the cracks that so frequently appeared on dolia were fixed with materials and methods that were not suitable or effective. It was only well into the second half of the industry's history that the development of dolium repairs reached its pinnacle, with repairs that prevented future damage, reinforced the vessel, and even made the final product much more robust. (This is not to say that use-life repairs—the staples, clamps, hybrid techniques, and repairs with organic materials—were no longer in use; dolium menders likely practiced a diverse range of repair techniques, and what we see across the different sites reflects not only preservation bias but also asymmetrical access to resources and labor markets.) By repairing production flaws well, then, menders ensured that dolia were not as likely to suffer any more than minor cracking during use, as seen in the capital's large storerooms. The workshop's tremendous input at the outset to treat, detect, and anticipate damage during production made dolia more robust and cultivated an expectation for quality that freed the dolium users and owners from complicated repairs down the line.

These new, innovative dolium repairs, however, were not developed by the craftspeople traditionally tasked with *repairing* dolia—that is, the pottery menders and tinkers, among others. Instead, they were developed and refined by those *making* the vessels in the workshops. Dolium makers worked together on distinct materials in workshops where they shared resources and knowledge. Moreover, dolium workshops, especially opus doliare workshops, were situated near good clay sources and might have been in the vicinity of other major workshops, not only for the sharing and exchange of space, materials, and equipment but also for social and economic reasons.[47] Workshop location alone might have been how the right kinds of people got together. The proximity and potential interactions at least could have inspired dolium makers to experiment with different methods and alloys and exposed them to new tools. Moreover, the production of dolia in brick and tile workshops helped bridge the domains of large-scale pottery production and construction. As dolium makers achieved a more sophisticated understanding of the properties of the materials they were working with, they were able to develop techniques for the remediation of defects that occurred in the course of the production process.[48] Members of the workshop might have noticed double dovetails used in building projects to join ashlar masonry, fix breaks in marble statues, or seal stone funerary urns and sarcophagi; they could have adapted the technique of double dovetail joints that were used in manufacturing wooden brick and tile molds.[49]

47. Goodman 2016: workshops usually cluster in urban areas around social and professional networks. On cross-craft interactions (the interaction between different types of crafts), see Miller 2009.

48. On craftsmanship, experimentation and innovation, and workshop dynamics, see Sennett 2008; Ingold 2013; Korn 2013; Harper 1987; Csikszentmihalyi 1996.

49. Few brick molds with mortise-and-tenon joints are preserved from pharaonic Egypt; e.g., from Kahun at the Manchester Museum, acc. no. 51, and from Hatshepsut's temple at the Metropolitan Museum of Art, acc. nos. 22.3.252 (fifteenth century BCE), 30.8.7 (fifteenth century BCE), and 25.3.108 (sixteenth to thirteenth century BCE). Ulrich 2007, 66: Roman carpenters used dovetail joints to connect the sides of boxes.

But there also had to be a serious catalyst that spurred the curiosity and motivated the attempt to test these different and often risky methods on dolia, especially during the production process. Since these were production repairs, their use and success meant that not only did dolium makers decide they needed or wanted to make repairs, a task that was generally done when the vessel was in use and by outside craftspeople, but they also had to devise methods to form anchors on *prefired* vessels, a feat that traditional potters did not typically do. This was risky. If they made a mistake forming the repair, something would go wrong when they fired the dolium. Cutting too deeply into the vessel wall to form double dovetail tenons could undercut the structural stability of the vessel; the dolium could collapse during firing. Not cutting enough could do nothing or, worse, deepen the nascent crack. The workshop therefore had to provide a financial cushion and incentives for taking on the risk. For opus doliare workshops that successfully reinforced their wares during production, they could manufacture larger and more robust dolia that potentially needed little to no mending later. By investing in dolium repairs and their specialization, workshops could offer a superior product that customers would likely pay more for or at least prefer over other options.

Both the nature of the organization of dolium workshops and the advantages that were to be gained from developing a reputation for the manufacture of superior, long-lasting products induced dolium makers to work collaboratively with a view to developing methods that would allow them to manufacture increasingly larger, more durable vessels. As Chapter 2 discussed, dolium workshops, especially the larger opus doliare ones that produced multiple products, could have had many workers with different roles and expertise; some of these workers worked seasonally and might have had jobs in other industries, such as agriculture or construction, where they saw different joining or clamping techniques or even interacted with dolium users or repairers who reported their impressions (and complaints) of the vessels and repair techniques.[50] Large opus doliare workshops were perhaps better positioned to perfect these methods because of their more extensive set of manufacturing tools and facilities, more expertise, a stable and specialized workforce, and relatively large output. Perhaps mastering dolium production also entailed repairing and reinforcing dolia, and one who achieved this level of production could advance in the workshop, maybe becoming an *officinator*. We can imagine, for example, that the workers of these large manufactories practiced and mastered the repair techniques on bricks and tiles before adopting these methods for the substantially more challenging task of repairing dolia.[51] The creative interaction between the workers, dolium workshops, and other kinds of craft production rendered the dolium industry a vibrant and creative one capable of adapting innovative production and repair techniques that were probably vital to manufactories producing dolia.[52] Dolium repair became a specialized activity within

50. Kang 2015: for example, a traditional Korean *onggi* potter faced many challenges when trying to revive the craft of *jeolla*-style *onggi* pots (they have a prominent bulge at the belly); because these *onggi* pots were particularly difficult to make, they were also expensive, and customers demanded replacements if there was damage.

51. Lancaster 2015: terracotta bars at Fregellae and Massa had clamps like those used to mend pottery and dolia, suggesting overlaps of techniques between the industries.

52. On learning in pottery workshops and through generations at Sagalassos, see Murphy 2017. On workspaces, see Murphy 2016.

opus doliare workshops and promoted more robust products and perhaps growing consumer confidence.

Environment mattered just as much for production phase repairs as for use repairs (Table A1.18). At both Pompeii and Cosa, many dolia became damaged during use and were repaired by various nonspecialists who applied an assortment of techniques. The stability and change in repair techniques were to some extent driven by local considerations, including the number and density of workshops and the attitudes of and interactions between the specific craftspeople involved; another factor was market thickness, the number and flow of buyers and sellers—that is, with enough demand, specialists could exist. The low numbers of dolia at Cosa probably meant repairers did not have much experience with mending dolia. The "thick" markets at Pompeii, on the other hand, promoted systematic approaches to remedying use-life damage and fostered a community of craftspeople. Moreover, dolium owners could probably find menders at workshops and markets in the town. At Ostia and Rome, dolia damaged while in use were patched with lead fills made in a consistent way. The wide range of repair techniques at Pompeii and Cosa was likely related to their varied ownership and dispersed installation in the towns. Owners of dolia in shops, gardens, vineyards, and houses found different craftspeople to mend dolia when it was needed, whereas the few massive storehouses in the capital were regularly maintained by a dedicated staff that followed certain protocols. Investors setting up and operating cellae vinariae in the capital not only acquired dolia but also arranged for their regular upkeep and maintenance. In purchasing well-made and reinforced dolia from opus doliare workshops, cella vinaria owners probably expected only the systematic, minor "touch-ups" their staff provided in maintaining the storeroom. An incredible amount of material, labor, skill, and money went into dolia and their reinforcements and repairs for vintners and traders to implement them in a new, supersized wine industry—from production, transport, and storage to the sale of wine. The stakes were high.

Traders installed well-made and reinforced, as well as repaired, dolia in huge numbers, concentrating these massive vessels in cellae vinariae in Rome and Ostia, as well as in the northwest Mediterranean, and even for transport in the tanker dolium ships.[53] Using dolia aboard these specialized ships could be dangerous; if a dolium cracked, the ship's center of balance could be thrown off, potentially sinking the ship. It is difficult to imagine how someone could mend dolia cemented in the hulls during a seaborne journey, let alone find someone with the requisite skills to do so. Manufacturing dolia was full of risks and potential production defects, and these dolium producers found ways to reinforce the vessels. To prevent spilled wine from tipping the ship's balance, dolium potters in the Piranus workshops preemptively repaired the dolia using the same type of production phase repairs seen on dolia in and near the capital. The Le Petit Congloué and Diano Marina shipwrecks, for example, contained numerous dolia that had been repaired during their production.[54] Every dolium on the Diano Marina shipwreck off the Ligurian coast was repaired with either a double dovetail tenon or lead filling, demonstrating both the skill of the dolium makers and the integration of repair with dolium production; the workshop

53. On harbor infrastructure increasing confidence in shipping, see D. Robinson et al. 2020.

54. Rando 1996; Corsi-Scialliano and Liou 1985.

invested heavily in the reinforcements, increasing the longevity of the vessels and safeguarding the ship and its crew from potential disasters. The Piranus family workshops not only built the same types of large vessels but also identified and preemptively treated potential damage using the same techniques as other opus doliare workshops, suggesting that the Piranus workshop participated in, and perhaps even pioneered, the same knowledge networks as the opus doliare workshops supplying dolia to Rome and Ostia[55] (in fact, some dolia found in Rome were made in Minturnae).[56]

The advancements in repairing and reinforcing dolia supported the longevity of these fragile giants as well as the specialized rooms, buildings, and ships in which they were installed. Archaeological evidence of discarded dolia and the wrecked tanker ships also shows, however, that the repairs were not always successful and could not guarantee the vessels would not eventually break. In fact, the discard and even widespread abandonment of dolia point to the vulnerability and disadvantages not only of the dolium vessels but also of the dolium-based container system.

55. Dolium makers might have learned dovetails from shipbuilders in cooperating to build these tanker ships. Ulrich 2007, 64–66: double dovetails at the Forum of Augustus and Nemi.

56. Lazzeretti 1998.

8

From Valued to Trash

THE DISAPPEARANCE OF DOLIA

> I have provided for our little supper today. Just look at how the dolium, which is always empty, uselessly takes up so much space and really does nothing except to block our living-space. Well, I sold it to someone for six denarii and he is coming to pay and take away his property. So while we are waiting why don't you gird yourself and give me a hand, so that we can dig it out and hand it right over to the buyer.
>
> hodiernae cenulae nostrae prospexi. vide sis ut dolium, quod semper vacuum, frustra locum detinet tantum et re vera praeter impedimentum conversationis nostrae nihil praestat amplius. istud ego sex denariis cuidam venditavi, et adest ut dato pretio secum rem suam ferat. quin itaque praecingeris mihique manum tantisper accommodas, ut exobrutum protinus tradatur emptori.
>
> APULEIUS, *METAMORPHOSES* 9.6

IN APULEIUS' *Metamorphoses*, Lucius tells a short, but scandalous, story of a couple living in poverty.[1] The husband, a humble builder, feels proud that he has found a buyer for their old, empty dolium, which he bemoans is taking up valuable living space in their home; the story does not tell why the dolium was there in the first place—perhaps it had been installed before they even moved into their home—but its lack of use spelled out the couple's extreme impoverishment. The rest of Lucius' tale recounts the salacious rendezvous between the wife and her illicit lover. The oblivious husband believes the man is in their home to buy the dolium and goes inside the dolium to clean off the turf that had grown on the inside of the pot. The story brings up a few intriguing points relevant to this study. An unused dolium, without maintenance, could become not only unusable but also a nuisance. Dolia required constant cleaning and maintenance, while burying them in the ground probably exposed them to an accumulation of vegetation and pests. In this case, the years of neglect led to a thick layer of moss growing on the inner surface of the pot. Dolia, like other unglazed terracotta materials, were porous environments hospitable to moss and microorganisms. The husband also complained about how the old dolium took

1. For the symbolic importance of the dolium in this tale, see Murgatroyd 2005.

up a large space in their living quarters, becoming a hassle as it obstructed their ability to move freely in the courtyard. Even though the dolium was buried, it altered the room, forcing people to walk around it to avoid tripping over, stepping into, or smashing their shins against it. With some dolia reaching almost two meters in width, it is understandable how they could make a living space feel small, and why someone would want to dispose of a dolium he or she no longer needed. The episode also shows that one could offload a previously used dolium by selling it. The husband and wife were no longer using the vessel and were happy to make a bit of money from its sale. Although a new dolium was prohibitively expensive for most people, many dolia could, and did, end up on the secondhand market for a much lower price.

Dolia were expensive specialized equipment for storing wine and occasionally other foods. Despite efforts to keep them in good condition and repair them, dolia inevitably cracked, broke, or fell out of use. How did dolium owners, who financed these large pots, recover their investments? When and how did they dispose of such large hardware? This chapter surveys the types of reuse (using the vessel again for the same purpose—e.g., as a food container), repurposing (using, and sometimes modifying, the vessel for a different purpose—e.g., as a doghouse), and discard of dolia to understand how people might attempt to recuperate their costly investments or when they might cut their losses.[2] Dolium users tried to make the most of their assets in different ways, even when broken or damaged, but they could also decide it was no longer worthwhile to hold on to them. Individual cases of dolium discard could stem from unexpected breakage, but more organized and systematic approaches to both salvaging and disposing of the large, heavy materials also began to proliferate, in some cases when the technology did not align with their needs.

Starting in the third century, though, widespread abandonment of dolia began to grip parts of central Italy, as well as the northwest Mediterranean, as merchants and other dolium users reconsidered their choice of storage container. At the same time, circumstances began to change. Opus doliare activity, including dolium production, began to fall. Packaging preferences changed. Although dolia had been valuable storage containers, merchants and vintners increasingly turned to other containers, including a new container technology from the north: the barrel.[3] The shift to barrels was not simply a change in container, but sheds light on a move from a "specialized" container system that revolved around dolia to more "general" and versatile equipment. Moreover, it reveals changing priorities in the wine and container industries, as well as the pitfalls of a dolium-based storage system.

Dolium Reuse and Discard

Plenty of scenarios emerged in which people no longer wanted their dolia or found them unusable. Regardless of how successful dolium maintenance or repairs were, the vessels almost always fell out of use or broke completely. Dolia could be damaged beyond repair,

2. See Peña 2007b, 194–196; Cheung 2021b. There is overlap between reuse and repurposing, but distinguishing them can highlight (economic) choices.

3. The difference between barrels and casks today is their size; all barrels are casks, but barrels contain 180–200 liters. I use the more widely used term "barrel" to refer to both.

perhaps cracking during the violent fermentation process, and no longer suitable for their original intended purpose. They were also often contaminated by their contents to the point that the vessels were no longer deemed usable.[4] Old wine dolia might have absorbed previous batches of wine that would accumulate and then adulterate and ruin their contents. Dolia also fell out of use when the facility in which they were installed was abandoned or renovated for a different activity. At the Villa of N. Popidius Maior in Scafati, for example, most of the dolia had been removed from the winery, leaving behind large recesses.[5] The reason behind this abandonment is unknown. Whether it was related to damage stemming from the earthquake of 62 CE or changes in agricultural practice, the dolia were nonetheless removed and perhaps reused or sold as secondhand vessels. Removing the dolia seems to have been common, as excavations of villas such as Villa Magna and Villa Vagnari uncovered recesses in the cellae vinariae with either few or fragmentary dolia.[6] People were able to find different ways to retain the vessels' value, from reusing them as containers to modifying them for entirely different purposes.[7] Putting a dolium out of use could require substantial resources, though. Because dolia were such enormous architectural fixtures, even deciding *not* to use them called for their proper removal or altering the room entirely.

People often tried to prolong their investments by reusing (damaged) dolia for functions similar to their initial use. With its massive volume and strawberry shape, a dolium unfit for wine or oil storage was still useful for holding other contents. Dolium owners might have pivoted from winemaking to other industries. Residue analysis shows that, in Pompeii's Bottega del Garum (I.12.8), dolia previously used for wine were reused to store fish sauce, another lucrative and popular food product.[8] The dolia might have been considered too old or soiled for wine, but the strong flavors of the garum overpowered any residual wine that the dolia had absorbed. In other cases, the damaged dolia were perhaps no longer reliable for holding wine or high-value foods, but dolium owners instead could transition them to hold inexpensive liquids. In the Garden of Hercules (II.8.6), for example, the owner of the property found a new use for a few wine dolia that had been extensively repaired but were no longer liquid-tight (A2 P II.8.6 nos. 1–2).[9] To support the intensive flower-growing in the garden for the perfume industry, the dolia were given new jobs as they were set up in different areas to collect and channel rainwater to help irrigate the garden, which water channels divided into garden beds. One dolium, which had been extensively repaired at least twice with a variety of techniques, was set on a small masonry pedestal to collect rainwater. Another dolium, mended around the rim and shoulders, was directly embedded in the garden and held rainwater that people poured through a sawn piece of amphora; once the dolium was full, the water would have overflowed and fed into the channels along the garden walls.

The dolium's ability to hold liquids was one of its characteristic appeals, but dolia often cracked to the point where they could not be used for any liquids, though they could hold other goods. Excavations of the peristyle garden of the House of Meleager (VI.9.2) in

4. *Geoponika* 6.3; Hor. *Ep.* 1.2 ll. 69–70 mentions vessels seasoned with wines they contained.

5. De Spagnolis 2002.

6. Fentress et al. 2017; Carroll 2022a, 2022b; Dodd et al. 2023.

7. *Dig.* 50.16.206. For how dolia were reused, see Peña 2007b, 194ff.; Cheung 2021b.

8. Pecci 2020.

9. Jashemski 1979b, 279–288.

Pompeii uncovered a dolium builders reused to hold lime for ongoing restoration work of the house at the time of the eruption.[10] A leaky or porous dolium might not have been useful for liquids but could still function as a container for stiffer substances, such as plaster, and the dolium's large size and shape facilitated mixing, similar to modern concrete mixers. Owners of the House of Stabianus (I.22), which featured a market garden or orchard, also repurposed their damaged dolia for renovations on the house at the time of the eruption.[11] Construction workers had placed the dolia near the house among heaps of stone and *cocciopesto* to prepare to sort or mix materials in the dolia.[12] According to Latin agronomists, dolia could also hold dry foods, such as grain. Farmers likely repurposed leaky dolia as general food storage jars for their cereals, legumes, and vegetables, even using dolia to pickle vegetables such as turnips. At the Villa della Pisanella, for example, workers stored grains, nuts, and legumes in dolia alongside wine and olive oil dolia.

In Pompeii, where dolia and cylindrical jars for dry foods frequently complemented each other in shops, bars, and taverns, it was not uncommon for bar owners to insert damaged dolia into their masonry counters, often installing them alongside cylindrical ceramic jars to expand their establishment (Figure 8.1 top; Plate 29 top). Coupled with the bar's masonry architecture and insulation, dolia, even damaged ones, kept foods cool and dry and protected them from temperature fluctuations and pests.[13] Some reused dolia were severely damaged and cracked more than others, but, as the reused dolia at one food shop (VII.9.54) show, the external concrete framework of the masonry bar held the fragments together, even piecing together otherwise broken dolia, some of which were no longer complete (Figure 8.1 bottom; Plate 29 bottom). Installing dolia in bar counters was probably convenient for several reasons. Many bars already used dolia for wine and the cylindrical jars for dry foods, so workers would have been familiar with the necessary routine upkeep. In fact, some of the damaged dolia might have even been from the same property and inserting them into the bars became a way for the barkeeper to recuperate the loss and expand another component of the business, all without having to move the dolia. But some of these dolia could just as well have come from elsewhere. Estate owners with the means and who prioritized reusing dolia might have decided to reallocate dolia from their wine-producing country estates to wine-serving taverns in town. As various scholars, most recently Steven Ellis, have noted, many who operated bars were subelites who might have rented bars from or operated bars on behalf of wealthier patrons;[14] we therefore might see cases of landowners shuffling dolia between properties, as the imperial family did with their dolia that landed in Puglia. Salvaged dolia could have also been sold on the second-hand market, which seems to have been common according to a Republican-period painted sign in Pompeii (III.7) advertising the sale of various types of roof tiles salvaged from demolition or renovation projects.[15] As J. Clayton Fant, Ben Russell, and Simon Barker have shown, the salvage, resale, and reuse of marble from structures such as those

10. Jashemski 1993, 138.

11. Cheung and Tibbott 2020; Jashemski 1979b, 251–265; 1993, 73.

12. Pers. comm. Salvatore Ciro Nappo.

13. Cheung 2020.

14. E.g., Ellis 2018, 85–126.

15. *CIL* 4.7124 = *ILLRP* 1121; Della Corte 1936; Frank 1938.

FIGURE 8.1. (*Top*) Bar with five jars installed, three of which were reused dolia; (*bottom*) view inside a repurposed dolium, fragments held together by bar counter (VII.9.54), Pompeii. Courtesy of the Ministry of Culture—Archaeological Park of Pompeii. Reproduction or duplication by any means is forbidden.

damaged during the earthquake of 62 often directed precious marble to eating and drinking establishments.[16] Another way bars could have benefited from renovation projects was through the acquisition of dolia.

16. Fant et al. 2013. Ellis (2018, 167–173) cautions that these marble-clad bars were not necessarily built after the earthquake but likely earlier and given a facelift at a later date.

Dolia could crack or break too severely for any kind of storage, and some dolium owners found creative ways to repurpose the vessel after cutting or modifying it in some other way. The horizontal cracks that could emerge during a dolium's production, and vertical dunting cracks during its firing, were points of weakness where someone could saw a dolium along an almost natural line. We have seen in Chapter 1 how someone could saw a large jar to convert it into a doghouse in the Garden of Hercules. People might have also sawn or cut dolia to repurpose them as basins, wellheads, architectural elements, or even elements of a kiln. Although these dolia might not have been able to store anything as complete vessels, the bottom portion was at least able to serve as a small container. At the Villa Regina in Boscoreale, excavators found a dolium base, sawn and removed from a large dolium, near the main entrance of the farm and possibly used as a water basin for chickens.[17] This kind of reuse was common at the farm, as a large terracotta basin had also been modified into a makeshift trough for the farm's pig.[18] People could also retool dolia into conduits. In the House of Medusa (I.12.15) in Pompeii, the upper half of a dolium, preserving the rim and shoulder, was set above a well to serve as a wellhead;[19] a dolium was similarly reconfigured and cemented over the well in Herculaneum at the House of the Bicentenary (V.15–16), pointing to the utility of this type of repurposing.[20] Thanks to their size, shape, and material, dolia could also serve as quasi-architectural elements. The wine cellar in the Caupona of Gladiators (I.20.1) featured two makeshift skylights, consisting of dolia with the bases removed, which provided light and much-needed ventilation for the cellar's excess carbon dioxide produced in fermentation.[21] Workers also refashioned dolia into makeshift bricks. A second-century BCE furnace of a bath complex in Cabrera de Mar in Spain, for example, was built with a line of dolium rims, cut and removed from the vessel, that were arranged to accommodate the structure's shape, while the dolia's refractory material (previously fired ceramic material, which can withstand much higher temperatures) endowed the furnace with even more heat resistance and thermal insulation.[22] By cutting and sawing broken dolia, dolium users could fashion the broken vessels into objects that could serve new purposes.

The heavy ceramic material of opus doliare objects such as dolia also provided them with properties useful for construction projects. As a result, communities established and followed organized and systematic approaches to repurpose dolia for various construction projects. In Pompeii, a massive tile pile, consisting of broken dolia, tiles, and bricks, in the garden of the House of Stabianus (I.22) sat next to a construction site for the construction workers to access and repurpose the heavy materials in their renovation project.[23] Roman construction projects often raised the floor level or filled voids with large and bulky items of refuse, such as discarded pottery, amphorae, and dolia. At Cosa, broken dolia often found their way into buildings and fills, such as the later phases of the town's bath complex.

17. De Caro 1994, 123.

18. At the time of the eruption, the pig, now at the Antiquarium of Boscoreale, retreated into the basin, where it was discovered, leading excavators to posit that the terrified pig sought refuge in its basin.

19. Jashemski 1993, 55.

20. Jashemski 1993, 269.

21. Jashemski 1979b, 227–228.

22. I thank Lynne Lancaster for alerting me to this kiln.

23. Cheung and Tibbott 2020; pers. comm. Salvatore Ciro Nappo; Peña 2007b, 317–318; Dicus 2014.

Large ceramic objects were ideal fill materials given their mass, bulk, sturdiness, and abundance and could build sturdy foundations.[24] The dolium owner might have been able to offload or even resell the unwanted dolium or dolium fragments as building materials, sometimes for more targeted building projects. At the end of the first quarter of the first century BCE, residents of Cosa moved several dolia from their place of use to the Arx, where they amassed the dolia to build an artificial defensive terrace between the Capitolium and town walls.[25] The deposit has been interpreted as "'clean-up' after a sack by pirates ca. 70–60 B.C." and as evidence the site was fortified against raids of Sextus Pompey in the 40s BCE, but could have been part of the Augustan reorganization of the site.[26] Although their value as containers was not retained in this type of repurposing, using dolia as fill and building materials quickly set up a strong foundation for construction projects and perhaps saved the dolium user from expending time, energy, and money to remove them farther away to areas of trash accumulation.

Not all broken or unusable dolia were worth reusing or repurposing though. To discard dolia, dolium owners had to arrange for their removal and transportation to dump sites. In both Pompeii and Cosa, some dolia ended up in refuse mounds outside the town walls, areas close enough to be convenient for dumping, but set apart from the living quarters to minimize obstructing day-to-day affairs and affecting hygiene.[27] The discarded examples at Pompeii and Cosa were combined with a wide range of other types of refuse, such as bone, glass, and pottery in the case of the midden just outside Tower 8 of Pompeii, suggesting that they were mixed with other trash and collected, transported, and dumped all together.[28] Because dolia were such large and heavy vessels, disposing of them also depended on significant resources, but the removal of unusable dolia freed up considerable amounts of space, opening new opportunities and uses for the room. Perhaps the most striking type of dolium abandonment, especially after we have seen the lengths to which people would go to try to salvage their dolia, are the examples in which usable, intact dolia were purposefully abandoned, sometimes systematically and in a widespread manner. In fact, these changes, and decline, in dolium use spanned across Italy and the northwest Mediterranean, reflecting shifting priorities in wine storage and transport.

Moving Away from a Specialized Container System

Despite the investments in and widespread use of dolia, abandonment of dolium use began as early as the mid- to late first century CE, with the most glaring and sudden decline in the dolium ships. Tanker dolium ships had been traversing the western Mediterranean starting in the late first century BCE, but by the middle of the first century CE,

24. Peña 2007b; Dicus 2014.

25. F.E. Brown 1980, 73–74.

26. Will and Slane 2019, appendix 1; Marabini Moevs 2006, 7ff.; A.R. Scott 2008, 1–6, 177–179. On cleanup after pirate attacks, see Bruno and Scott 1993, 161.

27. Poggesi 2001: according to Frank Brown's 1951 excavation notebook, explorations outside the city walls revealed an unpublished refuse dump and the discovery of an entryway in the wall.

28. Chiaramonte Treré 1986.

about a half century after their inception, the highly efficient tanker ships declined.[29] The absence of tanker shipwrecks, after just three generations of the Piranus workshop's activity, signaled an abrupt abandonment of the storage and transport technology. Without more evidence we cannot know with certainty why that was, but, besides the possibility of less wine being transported, perhaps the risks and heavy losses were too costly for merchants financing and operating the dolium ships. Each ship could sell potentially 3,400 denarii worth of wine (when a denarius could buy four loaves of bread), so a disaster that brought down a tanker ship was a major loss, not only for the wine but also for the crew, the ship, and the dolia.[30] The wreck of a tanker ship was also likely a larger forfeiture than the wreck of an ordinary ship packed with amphorae, but if it was simply dangerous and unreliable technology, merchants probably would have abandoned the ships earlier. Given the nature of their design and homogeneous cargo, and the great costs for the dolia too, tanker ships were specialized vessels that were more expensive to produce and operate, and some merchants might have tried to find more efficient alternatives.[31] Due to their special design, tanker ships were made by a small number of shipbuilders who had both the expertise and connections with workshops that could produce robust and seaworthy dolia of the right size and shape. The variations in tanker ship size and their number and type of dolia indicate that the ships were likely custom made. Tanker ships were also challenging and perhaps expensive to operate. Their smaller and shorter design probably restricted the type of sailing routes they could follow, as they likely sailed more secure routes along the coast, rather than in open water. Their design also restricted the kind of cargo they could carry and the way the cargo was arranged. Other commercial ships had greater flexibility in the cargo and ballast they could carry, probably prioritizing items that were readily available or profitable to convey. Traders of tanker ships, on the other hand, could only place wines in the dolia aboard and would then face bottlenecks in transferring the dolia's content into other containers and awaiting other batches of wine as return cargo. Although we cannot quantify the number of tanker ships in operation or their percentage among other types of transport or containers in the wine trade—and we cannot for amphorae or ox-hide containers either, as the ancient evidence is just too fragmentary—tanker ships and the new kind of expedited bulk shipping enabled and reflected desires for new strategies. This method of transporting wine in bulk amounts was eventually abandoned, but the increased contact between different regions of the Mediterranean and the development of port facilities and wine transportation vessels left a mark that lasted through the course of the Roman Empire. In fact, wine trade continued, though with different methods of delivery.

29. There is some evidence of dolium tanker ships in the second century: Carrato 2022a, 2022b; Tchernia 2022a, 2022b.

30. A single serving (in *sextarius*, ca. 0.545 liters) of wine could cost 1–4 *asses* according to a painted sign at a shop in Herculaneum (IV.14); see Cooley and Cooley 2014, 236–237. The financial estimate uses the capacities of dolia (41,200 liters) from the Diano Marina shipwreck and amphorae from the Grand Ribaud shipwreck (over 200 amphorae, or 5,220 liters). At bar prices, 34,000 denarii would be the highest amount, but a conservative estimate would be ca. 8,500 to nearly 26,000 denarii. On food costs and consumption at Pompeii, see Bowes 2021b.

31. Ellis 2018: bar counters and investments in retail infrastructure in the early imperial period, a moment of innovation in different forms of retail, later fade in favor of less specific, multipurpose infrastructure.

The way of receiving and temporarily storing large shipments of wine began to change at some of the ports, as the cellae vinariae with dolia defossa that had been strategically employed at or near ports were undergoing major transformations. In the last quarter of the first century CE, the port of Lattara's dolia were either removed or filled and put out of use and the ground level raised to make way for a new phase of use, one without dolia defossa. This was not an isolated incident, but one of the earliest known examples of purposeful renovations to cellae vinariae near ports and in urban areas for the collection and distribution of wine. Renovations were made at Ostia too, where the dolia were mostly intact and functional at the time of abandonment, when they were intentionally and systematically put out of use. Building styles, various artifacts, and the citywide renovations point to the construction of the various cellae and installation of dolia in the first quarter of the second century, but pinpointing their abandonment is trickier. Due to generally undocumented early explorations of the city and rushed excavations of most of the site under Benito Mussolini's orders in the twentieth century, we lack the end dates for most buildings. The best-documented cella is the Caseggiato dei Doli (I.4.5), which was excavated later in the twentieth century. Archaeologists excavated the dolia themselves and found that less than a century after renovations in the second century, occupants of the Caseggiato dei Doli filled the dolia with discarded terracotta molds (for bread, pastries, or wax souvenirs) datable to the late second and first half of the third century, before covering the dolia with a rough pavement, putting the vessels out of use and changing the purpose of the storeroom.[32] By the mid-third century, dolium users disregarded the value of these vessels. Rather than repurposing the dolia or even removing and transporting them to a trash heap, as villa owners did in earlier periods, the occupants of the Caseggiato dei Doli raised the floor level and transformed the entire room by loading the dolia with discarded materials, putting the once expensive pots out of use, and out of sight.

Although doing so was a cost-effective way to alter the function of a previously specialized room, it effectively dismantled what had been a resource-intensive investment. Building such a specialized storeroom was an expensive endeavor, with high initial costs for acquiring and installing all the dolia, and regular costs for operation and maintaining the dolia and storeroom. Yet the highly specialized nature of the building also precluded it from other uses until the dolia were removed or covered up properly. By filling and then covering the dolia, though, a major loss was rooted in the decision to abandon the dolia completely, rather than try to sell them, use them elsewhere, or repurpose them for other tasks. Attempts to recuperate at least some of the costs by selling or reusing dolia, even large sets, were common at other sites and in other periods. Workers of the Villa Magna, for example, transferred dolia from the second-century winery into storage about a century after their installation, dismantling the cella vinaria while retaining the dolia for potential future use.[33] Even at Cosa, where inhabitants mostly abandoned the dolium storage technology in the first century BCE, they still gave new meaning to the dolia by building them

32. Pasqui 1906; Floriani Squarciapino 1954; Salomonson 1972; Bakker 1999; Paroli 1996. Pavolini 1983, 86: the building was built during the late Hadrianic or early Antonine period (based on *opus mixtum* technique) and turned into a cella during the Severan period.

33. Fentress et al. 2017.

into a defensive terrace. The decision *not* to salvage the dolia of Ostia suggests, then, that the storeroom owner did not find it worthwhile to (re)use or sell these vessels. The situation at the Caseggiato dei Doli might have been the best-documented and best-excavated of the cellae in Ostia, but it was not an isolated case. Every cella in Ostia and at least three cellae in Rome preserve the original dolia set in situ. In southern Gaul, too, this was a time when ports abandoned their dolia defossa. The port of Lattara had already dismantled the dolia in one warehouse by the end of the first century, but by the mid-third century the other warehouse as well as Massalia's cellae vinariae were also abandoned. As dolium ships dropped off and port cellae vinariae gradually put their dolia out of commission, other changes to the storage and packaging system were taking shape, revealing new strategies in delivering wine.

One clear development in how warehouses received different forms of wine deliveries that is archaeologically visible was a new type of amphora packaging that departed from the previously specialized forms for seafaring. When areas of southern Gaul and Iberia produced their own wines, local potters initially followed Italian amphora potters and manufactured their own versions of the Dressel 2/4 amphorae, designed with their pointed and sturdy toe and hardened salt-skin coating for long-distance and seafaring trade. A significant development in amphora design took hold in central Italy and the northwest Mediterranean in the first century CE, around the time that the dolium ships had been making their way to southern Gaul and southeast Iberia. More versatile flat-bottomed amphorae, which workers could pack onto both carts and ships, began to proliferate. In southern Gaul, potters made the Gauloise 4 amphorae that spread across the Mediterranean from the mid-first through the third centuries (Figure 8.2L).[34] In central Italy too, potters produced various flat-bottomed amphorae, some of which were mostly regional but others that reached Rome. A central Italian jar, known as the Spello amphora, became widely used in central Italy starting in the first quarter of the first century CE through the second century (Figure 8.2C). Originating from the town of Spello in Umbria, the Spello amphorae packaged table wines for riverine shipments to towns along the Tiber River and Rome and Ostia. Other flat-bottomed amphorae, such as the Empoli amphorae, originating from the Arno Valley and used to package table wines from Etruria, were regional vessels that also supplied Rome and Ostia. Flat-bottomed amphorae departed from previous designs and touted more "generalized" forms that enabled their use across different modes of transport between coastal, inland, and riverine sites.

The disappearance of dolium tanker ships and maritime and fluvial cellae vinariae with dolia defossa, coupled with the growing popularity of flat-bottomed transport jars, spelled just the beginning of a major shift in the use of the dolium-based container system. Although nearly archaeologically invisible, two other containers were likely used by merchants for bulk storage and transport at these sites in lieu of or in addition to dolia. The first is the familiar culleus, useful for temporarily holding large amounts of wine for overland and perhaps fluvial transport (Chapter 4). Numerous inscriptions commemorating

34. Gauloise 4 amphorae: Laubenheimer 1985, 2001, 2003, 2004; Laubenheimer and Gisbert Santonja 2001; Widemann et al. 1979. Spello amphorae: Cherubini and Del Rio 1997; A. Martin 1999; Panella 1989; Rizzo 2003; H. Patterson et al. 2005. Lyon amphorae: Desbat 1987, 2003; Dangreaux et al. 1992; Becker 1986; Martin-Kilcher 1994; Schmitt 1993; Liou 1987; Hesnard et al. 1988. Italian flat-bottomed amphorae: Ceccarelli 2017.

FIGURE 8.2. (*L*) Gauloise 4 amphora and (*C*) Spello amphora, from the Archaeology Data Service. (*R*) Wooden barrel, by Arnoldius, 2014.

utriclarii, handlers or transporters of cullei, at port sites in Gaul including Lattara, as well as other sites in the Roman Empire, attest to the widespread use of cullei for bulk trade around Gallic ports.[35] The second type of container that merchants and vintners turned to in order to supply and deliver mass amounts of wine was a different and new container known as the *cupa*, or "barrel" (Figure 8.2R). Barrels, cylindrical wooden containers made of individually shaped pieces known as staves bound by hoops, constituted a radically different container technology.[36] Sometime in the first millennium BCE, craftspeople in more temperate areas of northern Gaul, Germany, and Britain developed barrels for the fermentation and perhaps short-term storage of beer.[37] Barrels were not containers easily adaptable to the Mediterranean, though. They were made of wood from trees—fir, spruce, and larch—found not in the Mediterranean but in temperate forests of central and northern Europe.[38] The need for quality straight-grained wood such as silver fir was high, especially for larger barrels, and Rob Sands and Elise Marlière have even suggested that fir consumption contributed to the decline of fir in pollen and charcoal records during the first few centuries CE.[39] Remains of Roman-period barrels are mostly from northern provinces, where barrels were by chance preserved in anaerobic conditions, and are nowhere near as well or abundantly preserved as pottery.[40] Their material and structure made barrels leaky

35. Kneissl (1981) and Deman (2002) look at Gallic and Dacian evidence.

36. Twede 2005b, 254: hoops were made of organic material in antiquity.

37. Twede 2005b, 254.

38. See Sands and Marlière 2020, figure 2.

39. Sands and Marlière 2020; Küster 1994; Nakagawa et al. 2000. One large two-hundred-year-old oak tree could potentially provide enough wood for two large barrels; Edlin 1973, 98; Marlière 2002, 29.

40. Only a few ancient barrels have survived in central and northern Europe (Marlière 2002), only one in Italy (Gianfrotta 2012). Barrels were frequently dismantled for reuse; see Sands and Marlière 2020; Tomlin 2016. See also Baratta 1994a, 1994b, 1997, 2001, 2004, 2005, 2006, 2017, 2020.

and porous, which was beneficial for beer but degraded the quality and taste of wine.[41] Nonetheless, their gradual adoption was not merely a change in the type of container being used but presented several distinct benefits to counter the downsides of the dolium-based container system. Overall, barrels became poised to scaffold a growing shift from a more specialized storage and packaging system to a more general one.

Although they were not ideal storage containers, especially not for wine, barrels could hold enormous quantities while being portable and versatile.[42] Waterlogged barrels, which were often preserved because they were reused to line wells, reveal how enormous these vessels were: they had capacities ranging from seventy-five to over one thousand liters, comparable to dolia and cullei and as much as seven times the amount large oil amphorae could hold and more than twenty times the amount standard wine amphorae could.[43] Barrel makers made barrels more capacious than the largest amphorae, creating a larger yet lightweight and versatile bulk transport container the size of dolia, with portability that rivaled that of the amphora and cullei (Table 8.1). Various authors mention the use of barrels in different contexts, and several iconographic representations from Italy and the northern provinces show traders traveling with barrels in boats or carts. Two first-century CE marble reliefs, one from Campania depicting a barrel on a wagon and another showing a boat with a barrel passing through a jetty arcade, highlight the utility of these wooden vessels for both overseas and overland trade.[44] Archaeological remains, iconographic representations, and ethnographic parallels suggest that barrels offered a significant advantage, especially compared with ceramic containers, in being lightweight, capacious, rollable, and stackable. Workers could also handle the bulk containers without cranes, other devices, or additional labor, as depicted in a marble relief from Ostia (Figure 8.3L). Porters could roll them onto carts and ramps, then stack them in ships, without transferring the contents. Their ability to store bulk amounts of wine and be delivered expeditiously with boats and carts made barrels particularly effective for supplying the army, as evidenced by finds at military sites as well as iconographic representations such as panels on Trajan's Column showing Roman soldiers around warehouses loading supplies, including barrels, onto boats along the Danube River in Dacia (Figure 8.3R).[45] Thanks to its porous nature, a barrel might have even been able to transport wine as it was fermenting, without the risk of bursting the container, so the most revolutionary change to the wine trade that barrels offered was, in principle, that they could serve as the sole container for the fermentation, storage, transport, and distribution of wine. Even if merchants needed to transfer wine from barrels into another container, they could simply decant the barrel's contents through a spout. With barrels, the labor and coordination for acquiring transport jars (e.g., amphorae) and the tedious transfers between containers (dolia, cullei, amphorae), which must have taken many hours if not days, were no longer needed, and the elimination of the transfer process also diminished wastage, as

41. On the uses of barrels for beer or wine, see Tchernia 1997.

42. Tchernia 1986, 285–292; McCormick 2012, 60–77; Bevan 2014, 392–397.

43. Marlière 2002; Baratta 2001.

44. Schreiber 1896; Rostovtzeff 1911, 105; Helbig 1966, 187n1381; Tchernia 2016, 288, figure 2; Marlière 2002, figure 181. For other visual evidence, see Baratta 1994b, 2005.

45. Baratta 2004.

TABLE 8.1. Comparison of barrels, ceramic containers, and cullei.

	Barrels	Dolia and amphorae	Cullei
Material	Straight-grain wood (not readily abundant in Mediterranean)	Clay (many good sources in Mediterranean)	Ox hides (many in Mediterranean)
Tools and equipment	Metal (iron) tools	Vats; turntable, wheel; large kiln	Cutting and scraping tools, stretching frame, tanning agents
Labor for production	Highly specialized labor, can only be done by craftspeople who have trained for years	Mix of somewhat specialized to highly specialized labor; some steps require less skill	Ranges from household production to specialized workshops
Labor for repair	Highly specialized, done by barrel makers	Production phase repairs were highly specialized; use-life repairs could be made by nonspecialists	Ranges from household level to specialized leatherworker
Use	Leaky vessels, probably used mostly for bulk amounts of lower-quality wine, transport, and shorter-term storage	Hermetic seal with plaster (amphora) or pitch (dolia); long-term storage possible	Short-term storage for bulk transport, especially overland, perhaps fluvial

FIGURE 8.3. (*L*) Dockworker carrying a barrel, from the House of the senator Rosa, Ostia, third century CE. National Museum of Rome, by Carole Raddato, 2014. (*R*) Soldiers loading supplies, Trajan's Column, by Carole Raddato, 2021.

transferred liquid could be spilled, absorbed into vessel walls, or lost due to breakage or inadequate sealing.[46]

The barrel's portability presented an alluring trait, and when not in use, they could be stowed to free up even more space. As early as the first century CE, Lucan recounted in an episode of the Civil War that at Brundisium's port, "empty barrels upheld a raft on all sides, a series of which, bound together by long chains in double rows, supported the alder planks laid sideways" (*BCiv.* 4.420–422: *namque ratem uacuae sustentant undique cupae/ quarum porrectis series constricta catenis/ordinibus geminis obliquas excipit alnos*).[47] The portability of barrels could be increased even further, as barrel makers could label individual staves so barrels could be disassembled for transport and reassembled for use. Coopers from the sixteenth to twentieth century often produced barrels where oak was plentiful and cheap, labeling individual staves so that they could send disassembled barrels to their destination, where a different set of coopers arranged the marked staves to fit them together again.[48] Barrels found at the town of Silchester in Britain composed of staves labeled with Roman numerals suggested such a procedure was also practiced in antiquity, allowing barrels to take up minimal space when they were not in use, a practical feature in storing and shuttling barrels between places of use.[49]

Despite their perishable nature and the low numbers of preserved examples, a mounting pile of material and textual evidence points to growing barrel use starting from as early as the first century BCE, with the possibility that traders used barrels in addition to or as substitutes for dolia at various port sites. In the first century CE, Strabo noted that tribes in the region of Aquileia in northern Italy used "wooden pithoi" (οἱ πίθοι ξύλινοι), often larger than houses, to store and move their wines and olive oil, and loaded them onto

46. As Augustine (*Conf.* 9.8.17–18) tells us, pilfering from barrels occurred. Gilding 1971: in later periods, coopers regularly skimmed some of the product from barrels, which they called "waxers."

47. See also *SHA Max* 22.4; Baratta 2005.

48. Ross 1985.

49. Manning and Wright 1961; Kilby 1971, 99–100. For other markings on barrels, see Baratta 1994a.

wagons to transport them to different markets farther inland (*Geographica* 5.2). A funerary relief from Neumagen shows a boat with oarsmen and one or two levels of barrels lying on their side and snugly against one another (Figure 8.4L); Lothar Schwinden has estimated, based on the size of the barrels and the way they were arranged, that this shipment could hold over seven thousand liters of wine, not an insignificant amount and about the same amount as the Chrétienne H shipment that carried three hundred amphorae.[50] Given the prevalence of barrels at port sites and the growing evidence for earlier barrel use, such as pipettes and other instruments that served to provide samples of the wine within barrels, scholars such as David Djaoui have suggested that barrels were likely used starting in the mid-first century CE to ship bulk quantities of wine between central Italy and southern Gaul, which would explain why merchants abandoned specialized dolium ships.[51] While dolium ships could only transport one product and drew on a unique design and specially made dolia, barrels offered flexibility and a streamlined process. Djaoui also suggests that perhaps barrels were in use even earlier, while tanker ships were still in operation. Merchants transitioned from shipping table wines in the dolia of tanker ships to packaging the wines directly in barrels, allowing traders to then transfer the wines into flat-bottom amphorae or other containers or transport the barrels themselves to the destination.[52] In fact, a second-century epitaph set up by Sentius Victor for his wife Sentia Amarantis in Emerita Augusta, Spain, shows how barrels could function in retail settings (Figure 8.4R).[53] The epitaph depicts the woman as a tavern keeper standing at a bar ready to serve customers, holding a small jug in her left hand while grasping with her right a spout or spigot of a barrel set up on a rack in a way similar to how wine amphorae were stored in the Bottega of Neptune and Amphitrite in Herculaneum (V.6).

Around the same time barrels were becoming more widely used, dolium use at farms and villas also began to fall. This seems to have started across the province of Gallia Narbonensis in the second century CE. Although most cellae vinariae were installed in the first century CE, many dolia defossa were abandoned or built over beginning in the second century, with a rapid drop-off by the following century. Only a few holdouts persisted in their use of dolia defossa in the third and fourth centuries. In central Italy, the decline in dolium use at villas and farms seems to have started slightly later and at a more gradual pace (Figure 3.4). Most of the central Italian dolia were installed by the first century CE as part of a concerted effort to produce wine and build country homes. Only a few sites installed dolia in the second century, in line with trends of villa building in central Italy, where most villas were built by the first century CE and only undertook limited renovations in the second century CE.[54] Dolium use at these rural sites progressively declined after the first century, plunging to low levels by the fourth and fifth centuries CE.

Despite the removal or abandonment of dolia defossa at various farms and villas, many estates continued to produce wine. The second-century jurist Scaevola's reference to "wine

50. Schwinden 2019.

51. Djaoui 2020; Liou and Marichal 1979, 147; Liou 1987; Djaoui and Tran 2014.

52. Djaoui 2020.

53. *HAE* 1639; Edmonson et al. 2011, 65; Hemelrijk 2020, 167 n.98; Berg 2019; Varga 2020, 2; Brun 1997, 150; Peña Cervantes 2005/2006, 109; 2010, 181. Dating of the epitaph ranges from the first to the third century CE.

54. Métraux 1998.

FIGURE 8.4. (*L*) Neumagener Weinschiffs, wine merchants with barrels, ca. 220 CE, Rheinisches Landesmuseum Trier, by Stefan Kühn, 2004. (*R*) Funerary stele of Sentia Amarantis, second century CE, Museo Nacional de Arte Romano de Mérida, by Ana Belén Cantero Paz, 2012.

vessels, that is barrels and dolia, which are fixed in a room" (*Dig.* 32.93.4: *vasa vinaria, id est cuppae et dolia, quae in cella defixa sunt*), suggests that some cellae vinariae were using barrels for wine storage, and perhaps production too. Even before Scaevola's work, Pliny the Elder, in his discussion on viticulture, noted that vintners in colder regions in the north used wooden vases:

> There is a great difference to already vintaged wine in climate. Around the Alps they put it in wooden vessels and surround them with tiles and in a freezing winter also hold off the severity of the cold with fires. It is rarely mentioned, but has been seen occasionally, that with the vessels having burst, frozen blocks of wine stand—almost a miracle, since wine's nature does not freeze: usually it is only numbed by cold. More mild districts store [wine] in dolia and bury them in the ground entirely, or else up to a part of their position.
>
> magna et collecto iam vino differentia in caelo. circa Alpes ligneis vasis condunt tegulisque cingunt, atque etiam hieme gelida ignibus rigorem arcent. rarum dictu, sed aliquando visum, ruptis vasis stetere glaciatae moles, prodigii modo, quoniam vini natura non gelascit: alias ad frigus stupet tantum. mitiores plagae doliis condunt infodiuntque terrae tota aut ad portionem situs. (Plin. *NH* 14.132)

In this first instance in Latin texts in which barrels are associated specifically with viticulture, Pliny contrasts them with dolia, identifying barrels as the cold-climate version of dolia, which were used only in warmer climates.[55] Although dolia were sturdy vessels, terracotta pots, such as Ovid's wine jars in the Black Sea, could crack and burst in cold weather when their porous material absorbed moisture and then expanded under freezing conditions.[56]

55. See Baratta 1994b.
56. Ov. *Tr.* 3.10.22–23.

Barrels might not offer the stable insulation terracotta vessels did, but they did not freeze or crack as easily (and did not require some of the fine-tuning that dolia did to adapt them to particular climates, as potters in southern Gaul provided), and vintners could shelter or move them in extreme weather to provide the proper protection and conditions.

Around the same time dolium use plummeted across central Italian villas, barrels gained a foothold in Italian viticulture via imperial planning. According to the fourth century *Historia Augusta*, the emperor Aurelian earmarked funds for barrels in his plan to revitalize viticulture in Italy during the third century, indicating that barrels became integral for not only holding wine for transport but also viticulture itself:

> In Etruria along the Aurelian Way as far as the Maritime Alps, there are vast fields, fertile and wooded. He [Aurelian] therefore planned to pay their price to the owners of these uncultivated lands, as long as they were willing, and to settle there families captured in war, and then to plant the hills with vines, and by this work to produce wine, in order that the treasury receive none of the profit, but yield it entirely to the Roman people. He had also made provision for the vats, barrels, ships, and labor.
>
> Etruriae per Aureliam usque ad Alpes maritimas ingentes agri sunt iique fertiles ac silvosi. statuerat igitur dominis locorum incultorum, qui tamen vellent, pretia dare atque illic familias captivas constituere, vitibus montes conserere atque ex eo opere vinum dare, ut nihil redituum fiscus acciperet, sed totum populo Romano concederet. facta erat ratio dogae, cuparum, navium et operum. (SHA *Aurel.* 48.2)

Although the *Historia Augusta* was composed during the century following Aurelian's reign, it points to the prominence of barrels in the fourth, and probably already in the third, century. Aurelian was killed before this plan was brought to fruition, but his plan suggests that at least some Italian vintners were already employing barrels for wine storage, at least enough for him to provision barrels as the default container. Indeed, the wine cellar of Villa of Russi in Emilia Romagna, which produced wine from the second century through at least the third century, featured interior aisles with racks for barrels mounted on the walls, similar to how amphorae were stored in Vesuvian towns.[57] Jean-Pierre Brun suggests there were likely more villas with barrels in cellae vinariae in the region, such as the Villa Venezia Nuova of Villa Bartolomea and the Villa of Ambrosan in San Pietro in Cariano, both in the area of Verona.[58] At sites in Britain and Gaul too, storehouses with recesses in the ground seem to have had barrels installed.[59] In fact, the earlier phenomenon of retrofitting storage spaces for new storage technologies—overlaying dolia in silo fields, especially in the northwestern Mediterranean—continued, this time with barrels. While silo fields such as those at the villas of Tiana, Oliver d'en Pujols, and Els Tolegassos were built over with dolia-laden cellae vinariae, in later periods cellae vinariae owners replaced dolia with

57. E.g., the Bottega di Nettuno e Anfitrite (V.6) in Herculaneum. See Baratta 2005 for discussion of *SHA Max.* 22.4's bridge made of barrels. See Munteanu 2013 for discussion of bridges on inflated animal hides.

58. Brun 2004, 48–49.

59. Bakels and Jacomet 2003, 553: barrels seem to have been reused as storage containers at Vindonissa, where they were partly buried in a storeroom. Buffat 2018, 228; Pellecuer 1996.

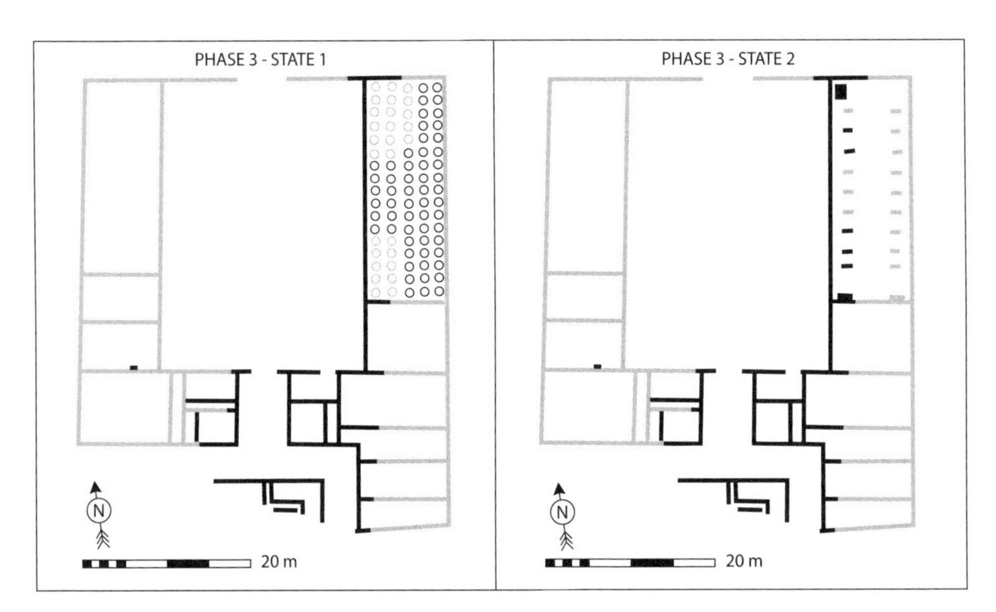

FIGURE 8.5. Plans of the villa at La Maladrerie in Saillans during the third phase of occupation. Illustrated by Gina Tibbott, after Carrato (2017).

barrels. At the Villa of Russi in northern Italy and the villa at La Maladrerie in Saillans, the cellae vinariae were renovated, whereby the dolia defossa were removed and metal fixtures were installed to hold and stabilize barrels (Figure 8.5). New container technologies could be and often were slotted into previous storage and packaging spaces, and this practice continued with barrels as they became the new defining feature of cellae vinariae.

Dolium production began to drop as workshops and their activities decreased during this period as well. Supply and demand must have gone hand in hand. In Gallia Narbonensis, dolium production and use were changing across the province as early as the second century. While dolium production expanded across the region starting in the mid-first century BCE, production seems to have ground to a sudden halt by the end of the second century CE, judging by both the abandonment of dolium production sites and the absence of newly installed dolia during this period. Charlotte Carrato has attributed this to a dip in demand.[60] Thanks to their durability, dolia could be in use for decades, perhaps even over a century. After supplying agricultural estates and warehouses, and meeting demands, dolium workshops in southern Gaul were no longer needed and the specialized potting skills stagnated and even disappeared. In the following century, building activity slowed down and the vitality of the opus doliare workshops, where dolia and amphorae were usually produced, precipitated across central Italy. Storerooms were not the only construction projects that ceased after the second century. The intensity of building projects, especially imperial ones, had reached its height in the first and second centuries CE, but nearly came to a complete stop halfway through the third century. Wolf Liebeschuetz notes that during the years 240–250 CE in particular, the construction of monumental buildings and the setting up of commemorative

60. Carrato 2017.

inscriptions came to a standstill empire-wide, marking a turning point in the disappearance of evidence for monumental commemorations of civic patriotism and religion.[61] Although there was probably some building activity and signs of recovery in parts of the empire, construction during this period paled in comparison to earlier periods. With shrinking rates of honorific commemoration and building, opus doliare industries were no longer as prolific as in previous generations. A survey of workshops in central Italy showed not only decreasing production among opus doliare workshops but also fewer workshops manufacturing dolia (Figure 1.4).[62] With fewer production centers in operation, a shortage of dolia, amphorae, and other opus doliare products was not inconceivable and potentially problematic. Archaeological evidence for dolium production drops significantly in the fourth and fifth centuries with only a few workshops identified. As Chapter 2 discussed, because the risks (and profits) of dolium production were balanced by brick and tile production in opus doliare workshops, a decline in the opus doliare industries was probably a major setback for dolium production. Because the production of these massive jars was difficult, expensive, and time consuming, workshops often buffered and folded in dolium production with the manufacture of other heavy terracotta objects that brought in stable, but perhaps low, income. With the risks and costs of dolium production offset by other terracotta products, however, dolium production relied on the success and operations of opus doliare workshops. The ability to undertake such a specialized type of pottery production was thus limited by a range of conditions. But the highly developed skills that dolium potters cultivated, in building and reinforcing the jars, might have also led to a stagnation in the craft. Once opus doliare workshops supplied their customers, demand might have dropped and led to the shuttering of a formerly lucrative branch of heavy terracotta production. Moreover, the reorganization of the building industry and new taxation system, by which Italian landowners were taxed in kind for the first time, might have resulted in fewer incentives and opportunities for entrepreneurs and craftspeople to engage in opus doliare production.[63] The opus doliare industries never regained traction, and only a handful of workshops manufactured dolia after the third century CE, dropping to numbers comparable to the third century BCE when the craft was in its infancy.

The growing preference for barrels might have also been due to their production and availability. Manufacturing both dolia and barrels required specialized skills and labor, but they diverged in the duration and seasonality of the work. The entire dolium production process required weeks, maybe even months, and was probably pursued only during late spring into early autumn, the dry months of the year, as was other ceramic production. Barrel making (cooperage) was also a specialized craft industry, but with different concerns and constraints.[64]

61. Liebeschuetz 2006, 18.

62. See entries in Olcese 2012.

63. Lancaster 2005, 18–21. The new land tax (*iugatio*) helped secure building material for state architectural projects. See Ward-Perkins 1984, 14–48, for an overview of the building industry and decline of elite civic building programs after the second century and scarcity of building and issues with repairing and restoring older buildings in later periods.

64. Cooperage was similar to ship making, which required making watertight joins between wooden pieces. Howard 1996, 441–444: nineteenth-century coopers on whaling voyages also assembled and repaired other wooden containers (tubs, buckets, bins) and occasionally functioned as the ship's carpenter too.

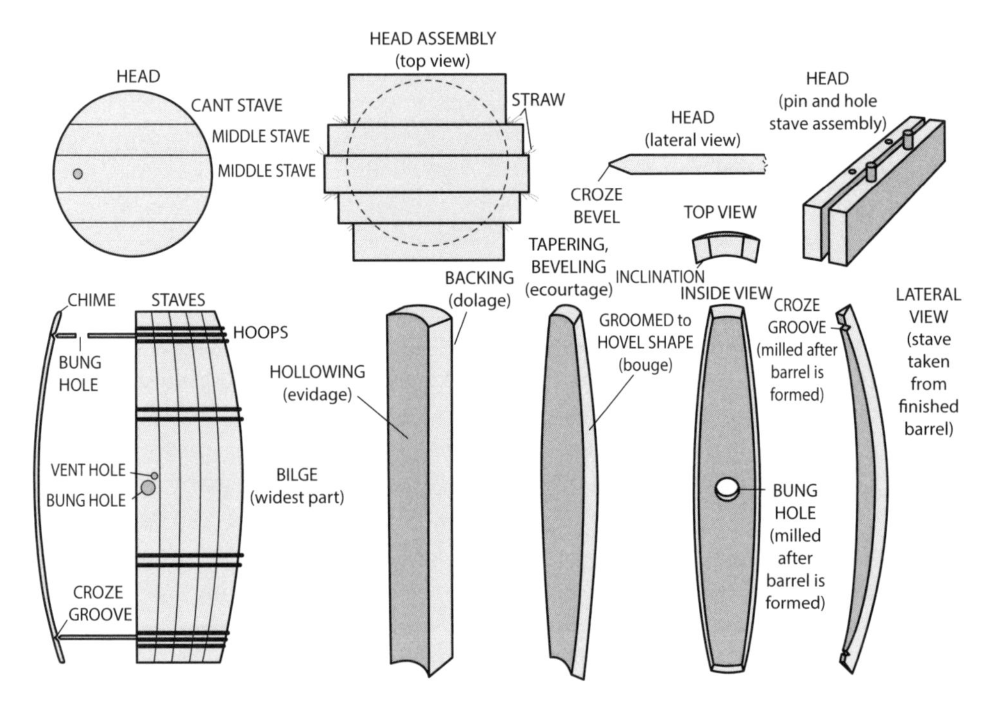

FIGURE 8.6. (*L*) Illustration of a barrel and its parts; (*R*) the process of cooperage. Illustrated by Gina Tibbott.

Barrel makers (coopers) needed the expertise to select and shape staves of the correct length, width, thickness, and quality, as well as join them properly (Figure 8.6L).[65] After selecting the staves, the cooper would shape them (*dressing*) following several steps: taper the staves' edges individually (*listing*); shave and shape the staves to give the exterior surface a convex shape (*backing*) and interior surface a concave one (*hollowing*); and then trim the staves to the barrel's desired width (*tapering, beveling, jointing*), the most challenging step, drawing on the cooper's extensive experience and skill for the accuracy and precision to produce leak-proof joins (Figure 8.6R). The cooper would place the staves together in the correct order in hoops to pull them together (*raising up*), then fabricate and fit the heads with the rest of the barrel (*chiming* and *topping*), and finally drill a bunghole into the strongest stave. There were no blueprints or molds involved, and a single cooper or pair of coopers was responsible for each individual barrel, from selecting the staves to listing them to joining them. Although the work was strenuous, a cooper could produce a barrel in one or even just half a day, and probably any time during the year because the craft was not beholden to weather conditions.[66] Although ceramic production used local natural resources, the duration and seasonal nature limited the

65. See Kilby 1971; Twede 2005b; Work 2014. Kilby 1971, 99–100: "There are no amateur barrel-makers" (15). Kilby 1971; Work 2014; Howard 1996: nineteenth-century British coopers apprenticed for four years; coopers in London, a major hub of whaling activity, apprenticed for seven years; coopers apprenticed in the United States for two or three years.

66. Sixteenth- and seventeenth-century Basque coopers produced barrels for the whale oil trade just before shippers departed for their journey.

products' availability, especially during the winter season and in later periods when production was not widespread. Against the backdrop of a shortage of opus doliare objects from the mid-third century onward, procuring barrels might seem expeditious by comparison.

The repair and general maintenance of dolia and barrels also shed light on differences in labor, skills, and frequency of maintenance. As Chapter 7 discussed, dolia often cracked, but these were expensive vessels, and customers naturally wanted their investments to last. The production process itself was challenging and risky enough that dolium makers anticipated damage and preemptively repaired the pots during their manufacture before the pot was ever even used. Limited opus doliare activity starting in the second to third century was potentially problematic for the stability of the industry. If workshops did not prepare or have the same expertise to create quality reinforcements, dolium users could face serious damage to their vessels. It was not easy to repair a dolium that had been fired either. Due to the chemically transformative process of firing clay, repairers could not take dolia apart, patch them up, and put them together again. Depending on available labor and markets, a dolium user might have also had trouble finding a mender. On the other hand, checking and maintaining barrels was routine work for a cooper. The potential to disassemble and reassemble barrels enhanced not only their portability but also their repairability and upkeep. Barrels were inherently leaky vessels, and evidence from later periods tells us that barrels needed regular upkeep; coopers aboard whaling ships were tasked with barrel maintenance and repair, such as replacing faulty staves and reinforcing hoops and heads. Although this kind of routine maintenance was necessary and expected, with barrels, one could take them apart, tinker with them, and reassemble them. Furthermore, because barrels were made of wooden staves bound together, they did not shatter as easily as a ceramic vessel, and might even bounce, rather than break, if porters dropped them. If a barrel did shatter, coopers could repair it by just replacing the damaged staves more effectively than a tinker could repair and render liquid-tight a broken dolium. Even though menders developed sophisticated repair techniques for dolia, these repairs could probably never have transformed a severely broken dolium into a container capable of holding liquids again.

Barrels and a New Container System

By the late third century, a period considered the "barrel revolution," an inscription recording a tariff from the northeastern part of the Campus Martius in Rome shows that the Roman Empire had adopted barrels as the official containers to carry wine for the treasury (*fiscalia vina*) into the capital's river ports:[67]

To the emptiers—for each barrel:	30 *nummi*
To the clerks—for individual receipts:	20 *nummi*
To the extractor—for each barrel:	10 *nummi*
To the porters, whose job it is to carry the barrels from the Ciconiae to the temple:	[??] *nummi*
To the guards of the barrels:	[?? *nummi*]

67. On the barrel revolution, see Tchernia 1986; Marlière 2001, 2002; McCormick 2012; Bevan 2014; *CIL* 6.1785; Purcell 1985, 12n56; Rougé 1957, 1966; Vera 2006.

Regarding the sample flasks, it has been approved that after the tasting they be returned to the landowner.

To the tax declaration personnel from the
Ciconiae straightaway upon the arrival
of the wine—for each barrel: 120

nummi
Austoribus in **cupa** una numm(is) XXX,
tabulariis in singulis apocis numm(is) XX,
exasciatori in **cupa** una numm(is) X,
falancariis qui de ciconiis ad templum **cupas**
referre consuerunt numṃ(is) [—],
custodibus **cuparum** [—],
d⌜e⌝ ampullis placuit ut post degustatiọ[nem]
possessori reddantur
professionariis de ciconiis statim ut adveneret
vinum in una **cupa** nụmm(is) CXX. (*CIL* 6.1785)

The inscription preserves what is likely a portion of the urban prefect's edict of set prices paid for the consignment of the treasury's wine in a region of the ancient city known as Ciconiae Nixiae (Straining storks) for the *fiscalia vina*, Aurelian's wine ration mentioned in the *Historia Augusta*:

He had also planned to give free wine to the Roman people, in order that, just as they are supplied with free oil and bread and pork, they might be given wine in the same way. . . . The fact that wine belonging to the treasury is stored in the porticos of the Temple of the Sun, disbursed to the people (not free of cost but at a price), proves that Aurelian actually considered this measure, and, indeed that he made arrangements for accomplishing it and even did so to a certain extent.

statuerat et vinum gratuitum populo Romano dare, ut, quemadmodum oleum et panis et porcina gratuita praebentur, sic etiam vinum daretur . . . argumento est id vere Aurelianum cogitasse, immo etiam facere disposuisse vel ex aliqua parte fecisse, quod in porticibus Templi Solis fiscalia vina ponuntur, non gratuita populo eroganda sed pretio. (SHA *Aurel.* 48.2–4)

Scholars have posited, based on the inscription (Ciconiae Nixiae, almost certainly a reference to dockside cranes), the topography of the ancient city, and archaeological remains of moles and other harbor infrastructure, that the river port for wine shipments was located in the area of the Campus Martius along the Tiber River near the Temple of Sol, where fiscal wine was stored.[68] Although the inscription marked a change in wine containers, this system of wine importation and distribution was not new. The use of barrels was grafted onto areas previously designated for wine shipments and transactions, such as the various cellae vinariae in operation by the second century CE, if not earlier, that were abandoned or reused in the third century, perhaps to take in shipments of wine in barrels instead.

68. R.E.A. Palmer 1974, 265; 1978, 237–238; 1980; 1990, 52ff.; Purcell 1985; Rice 2019, 209. See Salzman 2021 for a discussion of the topography of the area during the reign of Aurelian.

TABLE 8.2. Comparison of the logistics of barrels, dolia, amphorae, and cullei.

	Barrels	Dolia	Amphorae	Cullei
Production	Potential year-round production	Seasonal production (late spring to early fall)		Not necessarily seasonal (but could follow slaughter of animals)
	Half a day to one day to make one barrel	Weeks to months to make one dolium	Less than a day to form, a few days for drying and firing	A few days to make
Portability	Relatively light compared with capacity	Heavy compared with capacity		Light compared with capacity
	Rollable and stackable	In most cases immovable	Carried manually	Difficult to maneuver when full
	Easy to transport on carts, boats, and ships	Difficult to transport	Ideal for transport on boats and ships	Could be placed on carts and boats
	Could be disassembled and reassembled	One piece		One piece
Use	Could be used for fermentation, storage, transportation, distribution	Used for fermentation and storage, occasionally storage aboard specialized ships	Used for transport; could be used to store	Used for temporary transport
	Could be used anywhere and for both overland and overseas transport	Used at production sites and storage sites	Used for transporting goods; ideal for overseas transport	Ideal for overland transport
	Requires constant upkeep, frequent repairs	Might not require upkeep or repair (except cleaning)	Probably no upkeep or repair (unless reused)	Some upkeep might be required
	Average life span: eight years	Average life span: decades to centuries	Average life span: unknown	Average life span: unknown

Vintners and merchants used ceramic containers and cullei for centuries, but barrels became attractive options for storage and packaging (Table 8.2). The decreased production time, increased accessibility, and less demanding handling requirements were surely important factors in the barrel's rising popularity throughout the Mediterranean. Once cooperage was an established craft, farmers and merchants had access to a container that could be quickly manufactured and repaired any time of year. Moreover, barrels cut down on labor and time in their transport and use, and porters could move them much more easily both on and between boats and carts without special equipment, with the barrel functioning as a general container capable of moving goods on different modes of transportation without the porter handling or transferring the contents (known as an intermodal container), almost as a preindustrial precursor to steel shipping boxes today.[69] Although barrels, dolia, amphorae, cullei, and their industries were radically different, they emerged from and were responding to similar needs: the ability to store, protect, and deliver more wine more effectively and more quickly between different urban and rural settlements and across land and sea. Barrels had the obvious advantage of functioning as bulk yet portable and lightweight containers.

A great deal more than just the containers had changed, though. The adoption of barrels resulted in new industries, different relationships between specialized craftspeople and consumers, changing labor, and developing tastes. But that is another story. For several possible reasons, the dolium-based container system started to lose its mass appeal at the end of the second century. Dolium production sites and elite villas that continued to use dolia were probably the last gasp of the dolium-based containerization model that had supersized the Roman wine industry for centuries.

69. Commodities stay inside steel shipping containers as they are loaded on ships and trucks; see Levinson 2006, 2020; Klose 2015 for the history and development of shipping containers.

9

Dolia

THE STORAGE CONTAINER OF THE ROMAN EMPIRE

When she was planning the book that ended up as *Three Guineas*, Virginia Woolf wrote a heading in her notebook, "Glossary"; she had thought of reinventing English according to a new plan, in order to tell a different story. One of the entries in this glossary is *heroism*, defined as "botulism." And *hero*, in Woolf's dictionary, is "bottle." The hero as bottle, a stringent reevaluation. I now propose the bottle as hero.

Not just the bottle of gin or wine, but bottle in its older sense of container in general, a thing that holds something else.

"THE CARRIER BAG OF THEORY," URSULA LE GUIN

IN HER ESSAY "The Carrier Bag of Theory," Ursula Le Guin proposes a new theory that the first tool was not a weapon but a container.[1] Whether it was "a bag a sling a sack a bottle a pot a box," a container held oats, transported seeds, and stored vegetables. Containers, for Le Guin and Virginia Woolf, are the unsung heroes that enable people and communities to move, store, and save precious resources, and we should write new histories revolving around the stories of these containers. We need them today—think of all the bottles, cans, takeout boxes, and shipping boxes we use and throw away—just as much as the Romans needed them in antiquity. Containers made, and still make, possible the production, storage, and distribution of a plethora of foods.

Towering among them during the Roman period was the dolium, a supersized strawberry-shaped jar capable of long-term storage that contributed to developing flavors and textures of wine and that, with the right rigging, could ship thousands of liters of wine. When exactly the dolium first appeared on the scene in west-central Italy is murky. By the late third century BCE, dolia were produced in ceramic workshops and, certainly by the second century BCE, they were installed in large numbers in elite Roman villas practicing viticulture, around the same time amphorae exporting wine (Dressel 1, and later Dressel 2/4) were found throughout the Mediterranean. Dolia opened new doors for farmers, merchants, and communities. Their capaciousness supersized every step in the supply

1. Le Guin 1989.

chain, from production to distribution, and their material and design made them ideal long-term storage containers for both the fickle Mediterranean climate and the Roman wine trade. Estate owners who installed these massive vessels greatly expanded their productive power, not only in quantity but also in quality, sometimes installing multiple wine presses and hundreds of dolia, as seen in the western Mediterranean, to maximize profits. With large sets of dolia, vintners had more control over both the amount and types of wines they produced, as they could select batches of wine to age or modify with additives. Merchants and traders also tapped into the dolia's massive size as they moved tens of thousands of liters of wine to distant markets in specialized tanker ships. The ability to move and sell so much wine abroad extended trade networks, with the result that the city of Rome could stock up on a variety of vintages from across the Mediterranean. The installation of storage hardware enabled an increased scale of production, storage, and trade and connected various urban settlements, allowing them to partake in a new scale and form of trade. In towns and cities, shopkeepers and warehouse owners with plentiful storage could turn a considerable profit by strategically storing and selling wine to urban residents. In large cities, such as Ostia and Rome, dolia were packed into storerooms that supported urban living conditions—namely, the dense *insula* residential units and apartments that did not have individual storage spaces. Moreover, though dolia were designed with the primary purpose to ferment and store wine, they were incorporated into not only various storehouses but also towns and cities, especially in the retail spaces so distinctive of urban settlements. If the Vesuvian towns can shed light on wider urban food distribution and retail patterns of the Roman Empire during the first century CE, we can infer that dolia and other large-scale ceramic storage vessels became integral components of these spaces, storing perishable foods that would later be prepared and served in ways that contributed to an urban diet and lifestyle. Dolia in urban settlements fostered an unprecedented level and network of food storage, feeding into a sophisticated apparatus that signaled and allowed rising levels of urban services and quality of life.

This book set out to trace the "life" of the dolium-based storage technology and how it served the imperial capital and made some people rich, from the development of the craft industry in west-central Italy to the proliferation of dolia in the northwestern Mediterranean and ultimately to their disappearance. Along the way we saw how dolia could transform a farmer's fortunes, expand a shopkeeper's reserves, and add to a bar's offerings. Here I take the opportunity to expand on several threads that have been running through the previous chapters, especially concerning the people who drove this complex specialized storage system for several hundred years.

Investors, Workshops, and Personnel

Large dolia capable of holding huge quantities of wine (and other liquids) were the handiwork of highly specialist potters known as *doliarii*, many of whom worked in opus doliare workshops that also produced bricks, tiles, and other heavy terracotta objects for a burgeoning imperial capital. In order to offset the enormous costs and risks in manufacturing these vessels and diversify their portfolios, workshop owners folded in dolium production with the less profitable but stable production of construction material. During the first and second centuries CE in the Tiber River Valley, dolium manufacture reached new heights. Specialist potters were able to make dolia not only bigger but also better as they mastered the techniques and materials to treat potential production flaws. They drew on architectural techniques to reinforce their vessels preemptively, lessening the likelihood that the jar would break. As potters refined the production for this storage container technology, dolia became more reliable and dolium use proliferated.

Large opus doliare workshops might have offered opportunities for craftspeople. Within such a workshop, different workers could learn and oversee the various steps for making a dolium, allowing them to specialize in the construction of these enormous vessels. Learning could even begin with other, simpler opus doliare products—for example, starting with brick molding and then tile production to master control over groggy clay (tiles could easily warp) before advancing to larger objects such as basins and *mortaria*. As Chapters 2 and 7 posited, opus doliare workshops were places conducive for learning and sharing craft skills and knowledge. With their concentration of different skills, they were places where dolium makers refined their skills and formulated new techniques to treat production-based flaws and reinforce their vessels. The frequency and detailed information of dolium stamps indicate that many of the dolium makers and *officinatores* were able to rise in the ranks, perhaps based on their performance and mastery of the craft. A significant number were enslaved and then manumitted and "promoted" to supervisory roles, sometimes with their own slaves, suggesting the incentives in dolium production were substantial and even life changing. As Chapter 2 mentioned, the formerly enslaved Cimber, once manumitted, became another tentacle of the expansive Tossius opus doliare workshops and had at least two slaves of his own working under him.

Dolium production became an important and lucrative industry, and it offered appealing financial and social benefits for investors, entrepreneurs, and even craftspeople, but the rise of opus doliare workshops, and the dolium technology in general, potentially had a dark side.[2] One obvious case is the operation of massive cellae vinariae in and around the imperial capital we saw in Chapter 6. Given the dolia's and cellae vinariae's high installation and operational costs, only the rich and powerful could have a hand in this lucrative activity, further widening the gap between the haves and have-nots. Perhaps less obvious is the effervescence of large opus doliare workshops. Due to economies of scale, independent dolium potters and smaller workshops were unlikely to be able to compete against wealthy opus doliare workshops. With their resources and greater output, large workshops might have pushed out smaller ones that could not afford more equipment, better materials, or labor. Wealthier workshops owned by powerful individuals had a larger and more influential network. Certain families, especially the imperial family, dominated the industry from the first century CE on. Furthermore, the training to become a successful dolium maker was probably restricted to people who had the means to complete a lengthy (and perhaps unpaid) apprenticeship, which was likely out of reach for many.[3] Given the seasonality of ceramic production and agricultural work, having enslaved labor was one way to ensure a steady and reliable labor supply, which was crucial for craft specialization.[4] Specialization might have

2. Fernández-Götz et al. (2020, 1631) cautions against following the "material-cultural turn" in ways that provide "an unbalanced view of the workings of imperialism, marginalising hard power, violence and extreme social hierarchies."

3. Westermann 1914; Wendrich 2012a; Laes 2015; Freu 2016.

4. On skilled and unskilled labor prices in Diocletian's *Price Edict*, see Groen-Vallinga and Tacoma 2017. For time in apprenticeship, see Groen-Vallinga and Tacoma 2017; papers in Wendrich 2012a.

offered advancement opportunities within the workshop, as well as manumission.[5] For example, dolium stamps tell us that the *offinciatores* C. Vibius Fortunatus and C. Vibius Crescens, initially slaves of C. Vibius Donatus, were manumitted.[6] But these are only the success stories. Conditions for the enslaved working in mines, in quarries, and on agricultural estates were difficult, and opus doliare workshops were in line with that kind of dangerous work and environment;[7] some of the most physically demanding tasks would have included mining clay, firing the dolia, and transporting and cleaning them. We can only imagine how many enslaved workers were forced into this line of work and suffered for every Cimber, C. Vibius Fortunatus, or C. Vibius Crescens. And even for all the success stories, we generally only have their names. We do not know what they endured to reach that station within the workshop, or whether they even wanted to practice the craft.[8]

For workshop owners, however, successful dolium production enriched their portfolios as well as their profits and prestige. Ancient literary texts highlight the aversion of wealthy Romans, especially senatorial elites, to risky money-making endeavors.[9] Elite landowners focused on farming and other "proper" enterprises, such as pottery production and quarrying, while trade was considered vulgar.[10] Yet the elite bias in literary texts only distorts our understanding. The dolium storage technology drew on different skills, resources, and labor, offering investors new additions to and possible overlaps in their portfolios, even opportunities for vertical integration.[11] Entrepreneurs in the opus doliare industry could have invested in dolium production for a range of motivations. Many prominent families involved in the industry owned (wine-producing) villas and other properties. In fact, Varro's discussion of agricultural matters encouraged estate owners to exploit their clay pits for ceramics production, which would also secure their supply of dolia and other terracotta materials.[12] Workshops, especially smaller ones, might not have been able to keep up with demand. Their production rate probably could not supply entire wineries, which would explain cellae vinariae's usual assortment of dolia from a range of workshops. With the seasonality, risks, and time- and labor-intensive nature of dolium production, there was no way for most people to guarantee they could acquire a dolium to set up or expand their cellar or to replace a broken vessel. Lacking a single dolium could reduce the capacity of a vineyard to produce or a wine shop to store by a thousand liters of wine, which could bring in a potentially hefty profit. Owners of opus doliare workshops, however, could fulfill their own demands for dolia, whether it was to equip their villas to produce wine or supply their shops or warehouses to store and sell bulk amounts of wine. And in fact, many of the identified dolium production sites have been found on the property of wine-producing estates. Quintus Iulius Primus/Priscus, for example, supplied his vast estate of Saint-Bézard à Aspiran with bricks, tiles, dolia, and amphorae made on-site, enabling the expansion of his villa and economic

5. Hawkins 2016, 2017; Erdkamp 2015. On imperial slaves and freedmen in the opus doliare industry, see Weaver 1998.

6. Bloch 1947, nos. 564, 565.

7. Millar 1984.

8. Benton (2020, 121–140) discusses the widespread practice of employing forced labor in bakeries.

9. On landowners, senators, and trade, see Tchernia 2016, 10–37.

10. Varro *Rust.* 1.

11. Silver 2009, 2013; Broekaert 2012, 2014, 2011.

12. Varro *Rust.* 1.2.22–23.

activities, including large-scale viticulture. Vertical integration could extend across the whole supply chain. The Sestius family, for example, operated a large amphora workshop that packaged central Italian wines to be shipped to different corners of the Mediterranean basin. Yet other felicitous evidence also indicates their involvement in wine production as well as financing ships. The Sestii could have had their hands in multiple related industries, producing the wine that their amphorae bottled and their ships delivered. Similarly, the Piranus family might have been involved in not only dolium production but also viticulture, tanker shipbuilding, and the shipping as well.

Owners and investors of a seemingly humble industry, especially the opus doliare industry, had the potential to reap not only huge profits but also influence. John B. Kelly Sr., father of Grace Kelly, actress and the princess of Monaco, is a prime example of how one could become a millionaire through bricks. In antiquity too, powerful workshop owners could command control over access to fluvial ports and transport, connections to wine merchants, and availability of amphorae. Someone with the resources and portfolio could employ family, freedmen, and other associates to manage their opus doliare workshop, oversee the production of dolia, or manage their wine-producing villas, urban warehouses, shops, bars, or ships, creating an expansive and tightly knit network spanning every aspect of the wine trade. For someone like Domitia Lucilla, her opus doliare workshops could supply the building materials, dolia, and amphorae for her wine-producing *horti* in Rome as well as her villas out in the countryside, including Villa Magna. Folding in dolium manufacture brought other potential advantages too. A workshop that produced dolia could also supply building materials, potentially attracting more clients. If a client was looking to set up a wine cellar on his villa, he might prefer to go with Domitia Lucilla's workshop—because she could supply the bricks, tiles, and dolia—rather than arrange two separate contracts for the sale and transport of the dolia and construction materials. Dangling the dolia could have applied some pressure on the customer too: Domitia Lucilla might not have sold the dolia unless the client arranged to purchase the bricks and tiles as well. Given the challenges in dolium manufacture, a workshop could also point to its superior clay beds and the high mastery of its craftspeople to gain a competitive edge. If they were able to manufacture dolia, which were notoriously difficult vessels to produce, just think of the quality of their bricks, tiles, and amphorae. Yet we cannot forget that the potential power and profit of dolia, and the broader opus doliare industries, might have also motivated more sinister and predatory strategies. Even the emperor confiscated brickyards.[13]

Choosing Container Technologies

With the rise of an economically and politically unified Mediterranean, the wine trade significantly expanded thanks to the dolium, but dolium-based storage was scaffolded by a range of players. To be able to produce and move such large amounts of wine required vast resources and capital, and craftspeople who made these vessels had to refine their techniques and procedures in order to produce and repair these expensive investments effectively. But to design, develop, and refine dolia required particular conditions and skills, and was an ongoing process that lasted for centuries. The highly specialized nature

13. SHA *Hadr.* 15.2, 23.4; Setälä 1977, 160–162.

of dolia—their lengthy and challenging production, the upkeep and repairs, and the workflows, equipment, and vehicles they demanded—also leaned on a variety of physically demanding, and often hazardous, work.

If we take a step back and consider the dolium in context, it becomes obvious how extraordinary the whole industry was: over the course of over four hundred years, dolium makers designed and developed, out of raw material typically used for the manufacture of small items, a vessel that was so big that even a person could live in it. To do so, the dolium industry brought together different crafts, workforces, skills, and knowledge networks in a specialized world to spark innovation. Potters found ways to improve their production, architects and villa owners established conventions regarding their placement, and shipbuilders and merchants engineered new bulk transport systems. As the storage regime for Rome became more sophisticated, the very practices and technologies of storage cast a wider net that drew in more potters, metallurgists, tinkers, architectural workers, farmers, porters, cellar guards, and migrant and seasonal workers to propel the largest premodern wine industry. Dolia, and dolia-laden warehouses and shops, represented and embodied significant potential profits, which in turn could elevate social status. But this system for urban liquid bulk storage was not sustainable in the long term. The container system revolving around the dolium was vast and complex, requiring enormous inputs of labor and mass coordination for the upkeep of the hardware and to transfer the foodstuffs between containers and between settlements. Acquiring and implementing the dolium storage technology was prohibitively expensive and challenging for some places and most people. Dolia were cumbersome vessels: demanding regular maintenance to keep them clean and functional; requiring several people, vehicles and draft animals, and ships to move them; and necessitating transfers of contents and specialized equipment such as cranes, special tanker ships, and heavy-duty freight wagons.

In addition, dolium use within a larger system of other, often ceramic, containers also required transferring the vessels' contents, which drew on a steady supply of labor and containers, ultimately generating a great deal of waste. Remains of the packaging apparatus in Rome in particular highlight the stark consumer nature of the city and the downsides of the ceramic container system. With a population of over one million residents starting in the first century BCE, the ancient capital has been estimated to have consumed 100–250 million liters of wine and 20 million liters of olive oil on an annual basis, almost all imported. If even just a quarter of that wine arrived in amphorae, it would have arrived in at least one million wine amphorae for one year alone.[14] The olive oil trade helps visualize the magnitude of ceramic container waste: the single-usability of oil amphorae led to an accumulation of these vessels, as many as 320,000 Baetican amphorae each year, something that crystallizes most clearly at Monte dei Cocci, an artificial hill thirty-five meters high, made of roughly one million discarded, broken olive oil amphorae, covering approximately twenty thousand square meters, which was just one of several trash mounds in the city.[15] People also tried to clear some of that waste by

14. See Purcell 2007. Assuming a population of 1 million with consumption rates of 145–250 liters per year, and an amphora holding 26.1 liters. James (2020, 67) estimates ca. 1.4–7.2 million amphorae each year for a population of 600,000 to 1,200,000.

15. For the region, see *CIL* 14.20; Peña 1999, 21; Holleran 2012, 76; Lancaster 2005, 58–85; Peña 2007b, 174–178; Plin. *NH* 19.41.142; Rodríguez Almeida 1984, 116–117, 182–184. See Emmerson 2020, 92–124, 121n112; De Caprariis 1999; Aguilera Martín 2002, 215–218; Dey 2011, 187–197.

reusing it in construction, such as a horreum under the modern Nuovo Mercato Testaccio.[16] Such a substantial amount of food coming in required careful organization of labor and management of the waste generated from so many single-use vessels.

Efforts to streamline the storage and packaging apparatus might have emerged early on as merchants used animal-hide containers or dolium tanker ships to save time and resources, while significantly decreasing amphora use and waste. With dolia installed at different points of the supply chain—from vineyards to tanker ships to urban warehouses to shops and bars—traders could sidestep supplying or using amphorae, instead using an ox-hide container to transfer wine from dolia at the vineyard to their urban cella vinaria to their bar. Retailers still looking to buy or sell nicer vintages in amphorae could procure those vessels, but traders engaged in wholesale shipping and sales could bypass the single-use container, and skip the additional costs of commissioning an amphora potter, paying workers or enlisting enslaved workers to transfer contents, or paying the taxes associated with importing amphorae.[17] The bulk shipments between Italy and the northwestern Mediterranean facilitated a new type of trade that helped lower amphora use and waste, saving traders time, money, and a logistical headache, but there was still the problem of transferring wine between containers.

The widespread decision to abandon or dismantle dolia defossa points to the vulnerabilities of and even disadvantages to such a specialized container system, one that had been in place for centuries. Traders and vintners sought new and better solutions for moving bulk amounts of wine quickly and gravitated toward a new container technology, one that combined the capaciousness of dolia and cullei with the portability of amphorae and cullei but eschewed their specialized and labor-intensive handling: barrels. Over the course of a couple of centuries, traders and merchants increasingly turned to barrels to move large quantities of wine. Barrels were so efficient and versatile—they traveled well on ships and wagons, they were rollable and stackable, they did not require specialized equipment, and they functioned in Mediterranean and temperate climates—that the Roman state officially received consignment wine in barrels by the late third century. Overall, the adoption of barrels moved the Roman Empire's food supply system to a more "generalized" version of containerization, from a plethora of specialized forms to fewer forms that were more versatile for transport. While amphora diversity had reached its peak during the first two centuries CE as agricultural production centers manufactured amphora to bottle their wine, olive oil, fish sauces, and other commodities, after the second century, the remarkable diversity of amphora types plunged significantly and never reappeared again, with more general and mixed-use containers being adopted instead.[18]

The move from a more specialized to a more general infrastructure was happening across the board. As Steven Ellis notes, the specialized bar counters so abundant in the first century CE mostly disappeared into the second and third centuries CE.[19] Instead,

16. Sebastiani and Serlorenzi 2011, 84–85: some wine amphorae were also reused (Lamboglia 2 and Dressel 2/4). For other types of amphora reuse in building projects in Rome, see Lancaster 2005, 68–85.

17. *CIL* 6.1016a–c, 31227; Le Gall 1953; De Laet 1949, 347–349; R.E.A. Palmer 1980; Peña 1999, 6.

18. Bevan 2014, figure 4.

19. Ellis 2018.

more homogeneous and standardized shops were set up, blurring the distinctions among their functions and commodities. The shape and face of Roman retail was changing dramatically, from more individual and specialized structures to more generic ones across the empire. Around the same time, we see similar transformations to storage, as the more specialized cellae vinariae were intentionally put out of use. The horrea and cellae of Ostia began to contract during the third century. And while wealthy, elite Romans constructed horrea from the second century BCE into the second century CE, no new horrea were built in the capital after the second century, and the only work on them in the third and fourth centuries was restoration and repair. From the mid-third century on, massive warehouses and other facilities in the capital itself, such as the Nuovo Mercato and Monte dei Cocci in Testaccio, began to fall out of use as well, intensifying during the fourth and fifth centuries with the widespread abandonment of warehouses, not only in the capital but also in other parts of the empire.[20] The warehouses in Rome and Ostia were largely deserted at this time, and many horrea in the provinces, including those constructed for the military in the third and fourth centuries, were also deserted during the fourth through sixth centuries.[21] Portus, Rome's massive artificial harbor and main entry for grain and other bulk shipments, was also abandoned around the sixth century. Overall, the built-up area previously dedicated to urban food storage in the capital had been transformed into open agricultural land, such as in the conversion of Testaccio's warehouse district into a vineyard, and the almost monumental, specialized, and centralized food storage so distinctive of Rome's economic might mostly faded, giving way to a more dispersed, local, and generalized system of storage.[22]

Although changes to the storage and packaging system might seem dramatic, they were gradual. The adoption of barrels happened slowly, and throughout the process many vintners and merchants continued to use the traditional dolia and amphorae. Barrel use might have lagged behind in some places due to the time it took to establish a new craft to such a degree that it could support widespread use. Because their materials were not abundant in the Mediterranean and the skills necessary for cooperage were so highly specialized, barrels were probably not widely available or in high demand early on. Just imagine a new type of container that was hard to find, was expensive, and required specialist labor for routine maintenance. Using barrels was not just about buying them but also about being able to find a cooper to service them with regular upkeep and repairs, especially since a barrel's average life span was eight years, after which it could be rehabilitated or reworked to create new objects, such as writing tablets, well linings, and household utensils. In first-century Londinium, for example, a wooden writing tablet (recycled from an old barrel) documents the presence of a cooper named Junius, who maintained and repaired the barrels in town.[23] The successful implementation of a (new) technology depended on not only its availability but also the infrastructure, labor, and cultural framework.

20. Rizos 2013, 671–679: while horrea in Italy only underwent restoration and repair in the third and fourth centuries, new horrea were constructed in the provinces primarily to supply the military.

21. Rizos 2013, 684–688.

22. Serlorenzi 2010; Sebastiani and Serlorenzi 2011. On viticulture in late antiquity, see Rossiter 2007.

23. Tomlin 2016, 86, WT14: late first-century writing tablet with the text *dabes Iunio cupario/contra Catullu—*. See WT12 (probably addressed to a brewer) and WT72 (beer account) for evidence of beer making

The barrel's potential function as *both* storage equipment and portable packaging, two roles that had traditionally been in the separate realms of dolia, amphorae, and cullei, was unprecedented and posed new questions. Indeed, legal texts disputed the murky status and ownership of barrels: Were they considered viticultural equipment or packaging? To clear up the barrel's status, the third-century jurist Ulpian referenced traditional wine containers:

> If wine has been legated, let us consider whether it is owed along with its containers. Celsus also states that when wine has been legated, the containers also appear to have been legated, even if it [the wine] has not been legated with containers, not because containers are part of the wine, as for instance silver appliqués (on cups or a mirror), but because it is plausible that the testator wanted the amphorae with the wine as an accessory. In this way, he states, we say that we have one thousand amphorae, referring to the amount of wine. When wine is stored in dolia, I do not believe it true that when wine has been legated the dolia are also owed, especially if they have been buried in a wine cellar or they are the sort that it would be difficult to move due to their size. However, when wine is stored in large or small barrels, I think we must allow that these too are owed, unless in a similar fashion they have been so fixed in a farm just as part of that farm's equipment. When wine has been legated, small wineskins are not owed, nor, indeed, I say, are ox hides.
>
> si vinum legatum sit, videamus, an cum vasis debeatur. et Celsus inquit vino legato, etiamsi non sit legatum cum vasis, vasa quoque legata videri, non quia pars sunt vini vasa, quemadmodum emblemata argenti (scyphorum forte vel speculi), sed quia credibile est mentem testantis eam esse, ut voluerit accessioni esse vino amphoras: et sic, inquit, loquimur habere nos amphoras mille, ad mensuram vini referentes. In doliis non puto verum, ut vino legato et dolia debeantur, maxime si depressa in cella vinaria fuerint aut ea sunt, quae per magnitudinem difficile moventur. in cuppis autem sive cuppulis puto admittendum et ea deberi, nisi pari modo immobiles in agro velut instrumentum agri erant. vino legato utres non debebuntur: nec culleos quidem deberi dico. (*Dig.* 33.6.3)

According to Ulpian, wine buyers expected to receive and keep containers such as amphorae, and an amphora would even denote a specific amount of wine. Large containers (dolia and animal skins), however, were not meant to be bought along with the wine. For dolia, the restriction was related to their bulkiness and fixed position, both contributing to their immovability and their status as architectural features and farm equipment.[24] Barrels, however, did not have a set status. Ulpian gives two possibilities. Large and small barrels were generally expected to be part of the wine purchase, except fixed *cupae* used as cellar equipment, as Scaevola also clarifies in the second century (*Dig.* 32.93.4). While some barrels delivered wine and formed part of the transaction, fixed barrels were architectural features, like dolia. The unclear status of barrels posed additional challenges of integrating a new container into traditional supply chains.[25] If a vintner adopted barrels, would he or she decide to install them in the cellar so that they were fixed and immovable, necessitating transfers later? If the barrels

in the area. *Cuparius* is found in several other inscriptions (*CIL* 10.7040, 12.2669, 13.3700), but the dates of the inscriptions are uncertain.

24. See also *Dig.* 33.7.8, 33.7.13, 33.7.21, 33.9.3.11.

25. On the cultural impact of viticulture in the northern provinces, see Baratta 2017.

FIGURE 9.1. Stone epitaph of merchants moving wine, second to third century CE. Musee Calvet, Musee Lapidaire, Avignon.

were not installed, would the vintner stipulate the barrels' return after delivering the wine, or would he or she include the value of the barrel in the wine's price?

Yet archaeological evidence and different texts show a surprising endurance for dolia and amphorae, suggesting merchants and vintners had some access to the two bulk container technologies in the first through fifth centuries. One technology did not necessarily displace another, as related studies of presses and mills have demonstrated, but offered more options to farmers and estate owners, with some preferring the traditional Mediterranean storage jar.[26] Archaeological evidence also shows that merchants, shippers, and food producers continued to use amphorae for centuries, albeit with fewer types, even though barrels were becoming more popular. As studies of (ancient) technology have persuasively demonstrated, technologies could and often did coexist. Just because a new or more efficient technology was available did not mean people would abandon what they already had in place. As Tamara Lewit's recent work on wine presses has articulated so well, adopting a new technology such as the screw press required a calculated decision to designate space for and invest in what was probably an expensive piece of equipment, train staff in new practices, and take new safety precautions.[27] Furthermore, newer did not mean better. Wine made by foot-treading was praised as high-quality wine, while pressing wine would yield a larger quantity of lower-quality wines. One of the questions, then, was between technologies supporting quality or quantity, as was the case with barrels and dolia.

Some wine merchants likely used barrels, amphorae, dolia, and cullei alongside one another in various contexts. After the removal and replacement of dolia defossa with barrels at one cella vinaria at Lattara in the late first century CE, the other cella vinaria maintained its dolia defossa and the two systems coexisted for another two centuries. In the second or third century, a wine merchant in southern Gaul was commemorated by an epitaph depicting two men towing and a third navigating a boat of two wine barrels, while amphorae (and baskets) sit on a shelf (Figure 9.1). Even as late as the early fifth century, the agronomist Palladius describes a cella vinaria as having dolia defossa, with the option of adding barrels

26. A.I. Wilson 2002; Lewit 2020.

27. Lewit 2020.

to expand the storage capacity, as well as the production of both dolia and barrels at villas.[28] The traditional Mediterranean "containerization" system was still in play.

The Legacy of Dolia

Although many farmers, entrepreneurs, and merchants seem to have gravitated toward barrels by the third or fourth century, dolia were not abandoned completely or evenly. Against the backdrop of rapidly changing industries and economic strategies and a more efficient container technology, some communities continued to use dolia to make wine, underscoring the enduring value of the ceramic giants and their association with traditional viticulture. During the third century, for example, the emperor (perhaps Gordian) chose to relocate the site of the Vinalia closer to Rome to the Villa of the Quintilii and had first-century CE dolia installed in a lavish cella vinaria.[29] Diocletian's *Price Edict*, issued in 301 CE, listed dolia among common ceramic commodities and set a maximum price for them, suggesting that people still purchased the ceramic storage jars even when barrels were widely used.[30] Indeed, dolium potters were still practicing their crafts as a handful of ceramic and terracotta workshops continued to produce dolia during the fourth and even into the fifth century, though production was not as prolific as in earlier periods (Figure 1.4). Dolium use also persisted. Although villas in central Italy installed almost no *new* dolia after the second century CE, multiple estates that had already installed dolia in earlier periods continued to use them well into the fourth and fifth centuries. Symmachus, in the late fourth century, mentioned dolia defossa in his cella vinaria, while Macrobius in the early fifth century advised sealing dolia properly to prevent wine from mixing with air, suggesting these discussions were not merely antiquarian topics but reflected contemporary practical knowledge and concerns.[31] In central Italy, occupants of Villa Augustea di Somma Vesuviana in the fourth century and Villa Magna in the sixth century even *reinstalled* their old wine dolia, which were made in the first or second century and had been removed and stored for several centuries.[32] Since dolia were expensive, reliable, and long-term investments, some estates might have preferred to continue using their dolia rather than introduce a different type of vessel that likely affected the wine as well. Wine aged and stored in barrels tasted differently from wine made in dolia, and vintners might have hesitated to produce wine that their customers were not accustomed to drinking. Barrels also were not ideal storage containers, because their leakiness led to some evaporation of their contents, diminishing the quality and ability to age vintages. Estates that continued to employ dolia for fermentation and storage perhaps endeavored to offer a superior wine or reputable vintages that fetched higher prices from wealthier clientele, while wine made in barrels was cheaper and catered to a mass market.

28. Palladius 1.18, 10.11, 1.6: fabricators of dolia and barrels at villas (*doliorum cuparumque factores*).

29. Dodd et al. 2023.

30. Unfortunately, the surviving fragments of the *Price Edict* do not preserve information on barrels.

31. Symm. *Ep.* 3.23.1; Macrob. *Sat.* 7.12.14–15.

32. Aoyagi et al. 2018, 151; De Simone and Russell 2018.

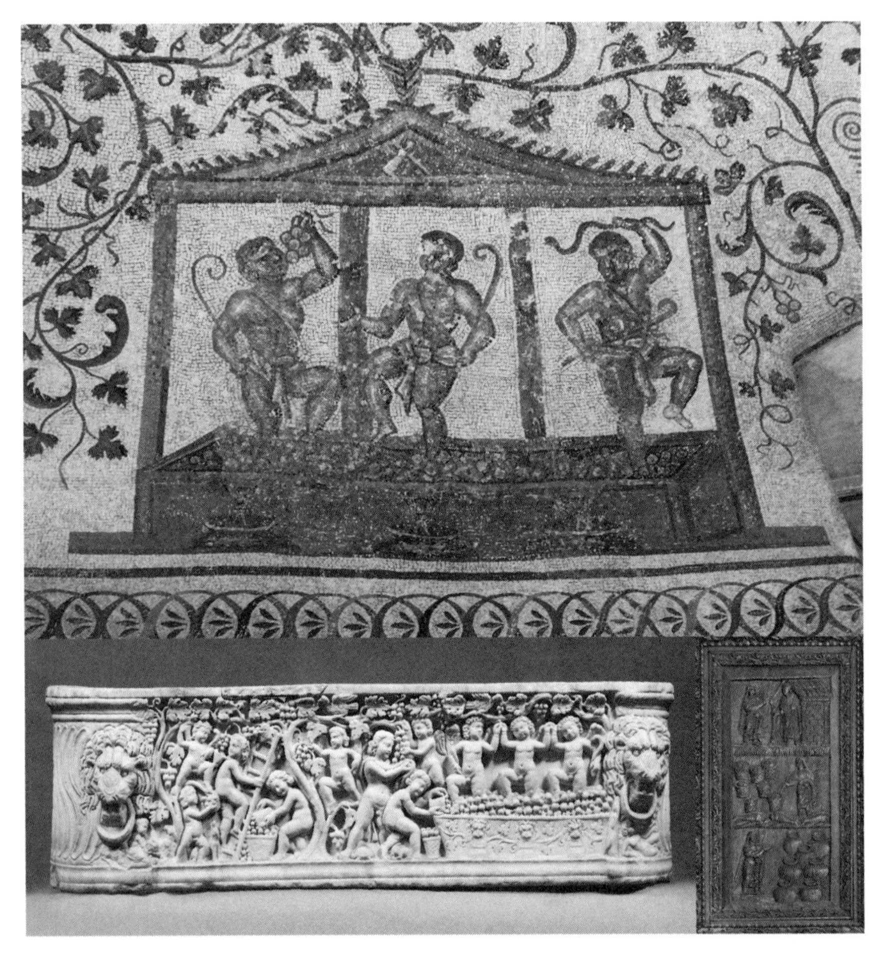

FIGURE 9.2. (*Top*) Wine production, Mausoleum of Santa Costanza. (*Bottom left*) Sarcophagus representing a Dionysiac vintage scene, 290–300 CE. J. Paul Getty Museum, 2008.14. (*Bottom right*) Miracle of Cana, Basilica of Santa Sabina in Rome, by Sailko, 2017.

Dolia might have also retained their popularity because they supported a type of traditional Roman viticulture, one where quality was paramount. Almost half a century after Rome officially began to receive consignment wine in barrels, both the ceiling mosaics and the porphyry sarcophagus in the Mausoleum of Santa Costanza, a funerary monument the emperor Constantine dedicated to his daughter(s), depicted agricultural workers treading grapes by foot, with the juices running into dolia embedded into the ground (Figure 9.2 top; Plate 30 top). The mosaics were no ordinary artworks but imperially funded, lavish pieces that highlighted viticulture and agricultural abundance against a golden backdrop, while the sarcophagus exhibited fine carving on a valuable, exotic stone. This motif was also popular in the private sphere, as depicted on a marble sarcophagus dated to the late third century CE (Figure 9.2 bottom left; Plate 30 bottom left). The legacy of dolia endured as the representation of the Miracle of Cana on the doors of the

FIGURE 9.3. (*L*) Mauro Gandolfi, *Alexander and Diogenes*. Harvard Art Museums/Fogg Museum, Gift of Belinda L. Randall from the collection of John Witt Randall, Photo © President and Fellows of Harvard College, R3579. (*R*) Frontispiece of *Cornelianum dolium*, Rare Book Division, Special Collections, Princeton University Library.

Basilica of Santa Sabina, constructed in the fifth century, which depicted the water that Jesus turned into wine in dolia, not barrels or some other vessel (Figure 9.2 bottom right; Plate 30 bottom right). Although we should be cautious not to read these images at face value, dolia were familiar and anchored enough in the contemporary cultural milieu for viewers to recognize them. Even if these representations of dolia were chosen as rustic containers to represent the past, they were only effective because they were loaded with cultural cachet, signifying traditional wine production and the agricultural abundance of the olden days.

This chapter opened with Ursula Le Guin's call to rewrite history centering on a new hero—the bottle, or the containers. Writing a history of the Roman Empire centered on the dolium has shed light on the impact a single container type could make not only on the wine trade but also on people's businesses, work, and even ideas. Containers, in general, carry cultural values and ideas and shape the world around us, something that the changing depictions of Diogenes the Cynic over time demonstrate. As we saw earlier, ancient authors described Diogenes as so intent to scorn wealth, comfort, and general amenities that he lived in an old, discarded storage jar. Greek authors described it as a pithos in the agora, while the Romans translated Diogenes' ceramic abode into the dolium, sometimes even a repaired one (Figure 7.1). As barrels became more established and common, the concept of a bulk (storage) container, and the representation of Diogenes' humble dwelling, changed. Artists and writers in later periods frequently portrayed Diogenes across different media—paintings, ceramics, stone sculpture—not in a pithos or dolium but in a barrel, the signature bulk container that had become entrenched as *the* bulk container in the day-to-day activities as well as the cultural imagination (Figure 9.3L).

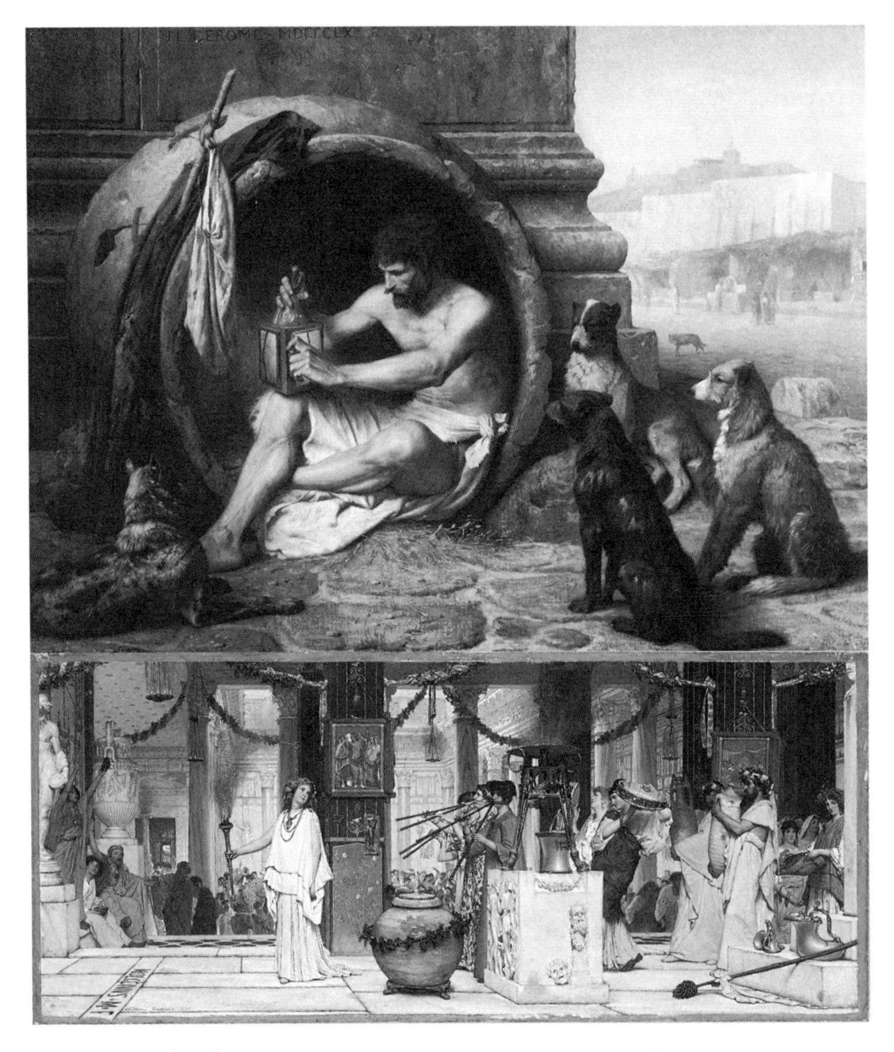

FIGURE 9.4. (*Top*) *Diogenes*, by Jean-Léon Gérôme, 1860. Walters Art Museum. Acquired by William T. Walters, 1872. (*Bottom*) *The Vintage Festival*, by Lawrence Alma-Tadema, 1871, National Gallery of Victoria, Melbourne.

Even when people today know that he lived in a pithos, there is no equivalent term for the vessel, and they instead describe Diogenes as having lived in "a ceramic barrel."[33] Dolia, and their cultural value, were not completely forgotten, though. In the seventeenth century, Thomas Randolph entitled his Latin comedy on a character's quest to treat syphilis *Cornelianum Dolium*, because the character climbs into a sweating tub, called Cornelius' tub (Figure 9.3R).[34] Although the tub did not have the characteristic shape or material of dolia, Randolph used the term *dolium* to highlight the massive container the titular

33. Williams 2018, 1.

34. Kilby (1971, 107) suggests that dolly tubs were related to both barrels and dolia.

character steps into during the play. As interest in ancient material culture grew, some artists studied ancient artifacts more carefully, and some even revived dolia. Jean-Léon Gérôme, for example, placed Diogenes back into a cracked dolium, mended and braced with rushes or lead (Figure 9.4 top; Plate 31 top). Lawrence Alma-Tadema's painting not only showcased a vintage festival in a marble-adorned space with ceremonious music, a marble altar, and a procession to mark the Bacchic revelry but also placed a festooned dolium in the middle of the festival (Figure 9.4 bottom; Plate 31 bottom). Alma-Tadema's choice to center the scene on a dolium underscored its role as a wine vessel, as well as a symbol of agricultural abundance, fecundity, and wealth. Yet we can only understand the significance and legacy of dolia by tracing the development of the storage container technology, uncovering not only the skills and mastery in their production and repair but also the different people and resources that sustained and profited from the Roman Empire's greatest and most precious pot, a feat of clay that made the Roman wine trade possible.

GUIDE TO THE APPENDIXES

THE FOLLOWING appendixes present collected data that inform the book. Appendix 1 consists of tables organized according to chapter and the order in which the data are referenced in the text.

For Chapter 2, Tables A1.1, A1.3, A1.5, and A1.6 contain dimensions recorded for dolia I was able to study closely in person. The dimensions I gathered, when available, include both interior and exterior diameter of the rim, the thickness of the vessel wall, the volume, the height (from bottom of the base to the rim—i.e., the height of the vessel according to outermost points), the depth (from the top of the base to the rim—i.e., the interior height of the vessel), the exterior and interior diameters of the widest point of the vessel ("belly" or shoulder), and the exterior diameter of the base. The dimensions help visualize the range in size and shape of dolia at and across the different sites and detect possible attempts to standardize the vessels. The volumes of the vessels are based on capacity markings found on the vessels or calculated estimates based on the vessels' dimensions. Tables A1.2, A1.4, and A1.7 include stamps found on dolia of the case study sites and combine published material and unpublished data collected during my fieldwork. The tables include the text of the stamp along with available information about the vessel's findspot or ancient location, references and publications, and cross-references to other ancient examples such as other dolia or opus doliare objects. Table A1.8 collects information on identified dolium production sites in west-central Italy.[1] The table includes information about the production site's location (region, province), the type of site (based on Gloria Olcese's categorization of villa, rural, urban, et al.), the date, evidence for its identification, and other terracotta products (bricks, tiles, amphorae, lamps, coarseware, fineware, and other heavy terracotta objects) that were produced there. Altogether, the table provides a sense of when and where dolium production occurred in west-central Italy, and how the craft was part of a larger terracotta production.

For Chapter 3, Table A1.9 includes information about villas and farms in west-central Italy with dolia in storerooms. The table includes any available information about the chronology of these storerooms, the number of dolia, the size of the storeroom, and the contents of the dolia.

For Chapter 6, Table A1.10 compiles all the capacity markings found on dolia in Ostia. This includes the ancient inscription in Roman numerals (units in amphorae) and its conversion to Arabic numerals as well as the total volume in liters.

For Chapter 7, Tables A1.11–17 cover dolium repairs. Table A1.11 reviews the possible dolium menders according to repair type and stage of execution and material.

1. The data are drawn mostly from Olcese 2012.

Tables A1.12–16 provide an overview of the different repaired dolia (and their repair type, likely stage of execution, and material) across the case study sites in central Italy. Table A1.17 gives a preliminary comparison of the dolium repairs at different sites, noting what types of repairs appear at which sites.

Appendix 2 provides detailed descriptions of select individual dolia from Cosa, Pompeii, Ostia, and Rome, including observations about their repairs. The dolia were chosen based on their better state of preservation and their utility for informing our understanding of dolium production, repair, and use. As such, most of them feature some form of repair and several are referred to at least once in the book. The descriptions are arranged according to site and, when possible, brief descriptions about their findspots are provided. Entries are referenced in the book according to the following abbreviation method:

A2 + first letter of site + abbreviation of location (if available) + dolium number

for example, A2 P 1.22 no. 2 for Appendix 2 Pompeii I.22 dolium no. 2, or A2 R NMR no. 3 for Appendix 2 Rome National Museum of Rome dolium no. 3.

APPENDIX 1

Tables

Chapter 2

Volume in liters, all other dimensions in centimeters. Ext. = exterior (measurements taken on the outside of the vessel); int. = interior (measurements taken inside the vessel; does not include vessel walls); height is the total height of the vessel; depth is the interior height of the vessel; pres. = preserved.

TABLE A1.1. Cosa dolium and other storage jar dimensions.

No.	Other ID	Ext. rim. diam.	Int. rim diam.	Wall thickness	Volume	Height	Depth	Ext. belly diam.	Int. belly diam.	Base diam.
1	CD 707	70	48	5.0	—	—	—	—	—	—
2	CD 708	75	50	4.5	—	—	—	—	—	—
3	C65.124	60	40	3.1	—	—	—	—	—	—
4	C66 V D E.21 S 4, 950433	80	50	—	—	—	—	—	—	—
5	C67.177	70	45	—	—	—	—	—	—	—
6	VIII D II I 15	60	40	4.3	—	—	—	—	—	—
7	CD 47	60	40	—	—	—	—	—	—	—
8	CE 984	—	—	3.2	—	—	—	—	—	37
9	CE 1160	23	15	1.8	—	—	—	—	—	—
10	CE 928	67	43	—	—	—	—	—	—	—
11	C14.100	80	55	—	—	—	—	—	—	—
12	CD 576	33	25	—	—	—	—	—	—	—
13	CE 724	80	50	—	—	—	—	—	—	—
14	NA	24	17	—	—	—	—	—	—	—
15	NA	30	—	—	—	—	—	—	—	—
16	CD 371	85	55	—	—	—	—	—	—	—
17	CD 267	65	45	—	—	—	—	—	—	—
18	CD 266	85	60	—	—	—	—	—	—	—
19	213445	76	50	5.4	—	—	—	—	—	—
20	Puteal SU 17009	—	—	4.9	—	—	—	—	—	—
21	Puteal SU 17004	—	—	5.0	—	—	—	—	—	—
22	VIII D 15 III	80	63	—	—	—	—	—	—	—
23	VIII D 24 I 11	75	50	—	—	—	—	—	—	—

(continued)

TABLE A1.1. (*continued*)

No.	Other ID	Ext. rim. diam.	Int. rim diam.	Wall thickness	Volume	Height	Depth	Ext. belly diam.	Int. belly diam.	Base diam.
24	VIII D 16 IIII	80	55	—	—	—	—	—	—	—
25	Puteal 17009 a	—	—	3.6	—	—	—	—	—	—
26	Puteal 17009 b	—	—	4.6	—	—	—	—	—	—
27	Puteal 17009 c	—	—	3.6	—	—	—	—	—	—
28	Puteal SU 17008	80	—	—	—	—	—	—	—	—
29	C 70 V D SG St. 5	—	—	3.6	—	—	—	—	—	25
30	ANS1-GR 950034	—	25	1.6	—	—	—	—	—	—
31	ANS1-GR 950034	38	30	—	—	—	—	—	—	—
32	C67.353	90	60	—	—	—	—	—	—	—
33	PC72-92	65	40	—	—	—	—	—	—	—
34	Horreum frag.	80	55	—	—	—	—	—	—	—
35	82.8	60	45	—	—	—	—	—	—	—
36	C70 VD SH MGT Garden	60	40	—	—	—	—	—	—	—
37	79	60	45	2.1	—	—	—	—	—	—
38	2016 22011	—	—	4.8	—	—	—	—	—	—
39	C 70–81	80	65	4.3	—	—	—	—	—	—
40	2016 SU 23003	—	—	3.1	—	—	—	—	—	—
41	2013 SU 5002	—	—	4.8	—	—	—	—	—	—
42	C65.337	80	55	—	—	—	—	—	—	—
Avg.		66	45	3.9	—	—	—	—	—	31

TABLE A1.2. Dolium stamps from Cosa.

Dolium	Stamp text	Reference
Cosa 34, "horreum"	H (in triangular border)	Unpublished
Cosa 19, find spot unknown	C·TVRI	Unpublished
Cosa 42, Temple of Jupiter	L·REMIO·C·F	Bace 1984, 172 D1

TABLE A1.3. Pompeii dolium dimensions.

Property	Dolium no.	Ext. rim diam.	Int. rim diam.	Wall thickness	Volume	Height	Pres. height	Depth	Pres. depth	Ext. belly diam.	Int. belly diam.	Base diam.
I.8.8	12	47	32	—	—	—	—	63	—	—	75	—
I.8.8	13	53	37	—	—	—	—	—	78	—	86	—
I.8.15	1	—	—	3.00	—	—	—	—	70	82	77	—
I.8.15	2	48	31	—	156.0[a]	—	—	—	76	—	50	—
I.9.4	3	48	31	—	—	—	—	—	—	—	—	—
I.9.4	4	49	30	—	—	—	—	—	—	—	—	—
I.13.13	1	—	—	1.80	—	—	—	—	—	—	—	—
I.20.1	4	43	28	—	—	—	—	—	—	—	—	—
I.20.5	1	—	—	3.10	—	—	—	—	—	90	—	—
I.20.5	2	62	39	—	—	—	30	—	85	95	—	—
I.20.5	3	64	36	—	—	—	96	—	—	110	—	—
I.20.5	4	—	—	2.60	—	—	—	—	—	90	—	—
I.21.2	1	—	—	3.30	—	—	76	—	76	84	—	—
I.21.2	2	53	31	—	—	—	83	—	83	93	—	—
I.21.2	3	57	37	—	—	—	77	—	77	93	—	—
I.22	1	68	42	—	—	—	90	—	—	90	—	—
I.22	2	—	—	—	—	—	67	—	—	—	—	—
I.22	3	—	—	3.30	—	—	110	—	—	—	—	21
I.22	4	—	—	—	—	—	52	—	—	—	—	20
I.22	5	72	44	3.60	732.0[b]	132	—	—	—	—	—	21
I.22	6	—	—	4.25	—	—	108	—	—	—	—	—
I.22	7	47	28	—	192.0[b]	94	—			75	69	22
I.22	8	—	37	3.00	229.0[b]	86	—	—	—	86	—	18
I.22	9	48	29	3.50	133.0[b]	84	—	—	—	77	—	19
I.22	14	—	—	3.10	—	—	—	—	—	—	—	—
I.22	15	—	—	4.00	—	—	—	—	—	—	—	18

(*continued*)

TABLE A1.3. (*continued*)

Property	Dolium no.	Ext. rim diam.	Int. rim diam.	Wall thickness	Volume	Height	Pres. height	Depth	Pres. depth	Ext. belly diam.	Int. belly diam.	Base diam.
I.22	16	—	—	3.20	—	—	—	—	—	—	—	—
I.22	17	—	—	3.80	—	—	—	—	—	—	—	—
I.22	18	—	—	4.00	—	—	—	—	—	—	—	—
I.22	19	—	—	3.70	—	—	—	—	—	—	—	—
I.22	20	—	—	3.30	—	—	—	—	—	—	—	—
I.22	21	—	—	3.30	—	—	—	—	—	—	—	—
II.1.8–9	1	50	30	3.00	—	—	65	—	—	80	—	—
II.5.5	1	61	43	—	550.0[a]	—	—	105	—	—	120	—
II.5.5	2	—	—	3.60	300.0[a]	—	—	80	—	—	85	—
II.5.5	3	69	44	—	—	—	—	115	—	—	117	—
II.5.5	4	71	45	—	600.0[a]	—	—	104	—	—	117	—
II.5.5	5	70	45	—	600.0[a]	—	—	104	—	—	120	—
II.5.5	6	65	42	—	—	—	—	105	—	—	110	—
II.5.5	7	71	45	—	770.0[a]	—	—	106	—	—	120	—
II.5.5	8	72	46	4.70	630.0[a]	—	—	102	—	—	110	—
II.5.5	9	—	—	4.60	300.0[a]	—	—	80	—	—	84	—
II.5.5	10	77	49	—	750.0[a]	—	—	116	—	—	120	—
II.8.6	1	72	44	—	—	125	—	—	—	117	—	—
II.8.6	2	66	41	—	—	—	90	—	—	90	—	—
II.8.6	3	49	30	—	—	—	—	—	—	80	—	—
V.4.6–7	2	53	38	—	—	—	—	—	—	—	—	—
V.4.6–7	3	64	40	—	—	—	—	—	—	100	—	—
V.4.6–7	5	60	38	—	—	—	—	—	—	100	—	—
VI.9.2	1	64	39	—	—	—	115	—	—	100	—	—
VI.9.10	1	64	40	—	—	—	75	—	—	115	—	—
VI.14.27	1	50	31	—	—	—	—	—	—	—	—	—
VI.14.27	2	44	26	—	136.0[a]	—	—	—	—	75	—	—
VI.14.27	3	—	—	2.70	—	—	—	—	—	82	—	—
VI.14.27	Frag 1	—	—	—	—	—	—	—	—	—	—	—

VI.15.13–15	1	43	28	—	—	—	—	—	—	60	—	—
VI.15.13–15	3	41	26	—	100.0[a]	—	—	—	64	—	56	—
VI.15.13–15	5	65	42	—	—	—	—	—	50	—	45	—
VI.15.13–15	Frag 1	—	—	—	—	—	—	—	—	—	—	—
VI.15.13–15	Frag 2	—	—	3.20	—	—	—	—	—	—	—	—
VI.15.16	2	—	—	—	—	—	—	—	—	—	—	—
VI.14.36	4	44	27	—	—	—	—	—	47	76	66	—
VI.16.40	1	47	27	—	—	—	45	—	67	68	—	—
VII.4.58	1	34	24	—	—	—	—	—	—	55	—	—
VII.5.21	1	78	45	—	—	—	—	—	—	140	—	—
VII.6.15	1	49	32	—	—	—	77	—	—	80	—	—
VII.9.54	1	68	42	3.70	463.0[a]	—	—	—	93	—	117	—
VII.9.54	2	63	43	—	—	—	—	—	92	—	107	—
VII.9.54	4	—	—	3.50	—	—	—	—	90	—	103	—
VII.13.20	1	46	30	4.10	150.0[a]	—	—	—	68	—	64	—
VII.16.4	1	46	30	—	—	—	—	—	70	70	—	—
VII.16.4	1	48	32	—	—	—	55	—	65	65	—	—
VII.16.6	1	42	26	—	—	—	66	—	56	62	54	—
VII.16.7	1	39	25	—	—	—	45	—	47	62	51	—
VII.16.7	2	—	—	2.60	—	—	—	—	—	60	54	—
VII.16.7	3	42	26	—	—	—	—	—	—	—	—	—
IX.9.10	1	53	32	3.50	425.0[c]	—	—	—	80	95	—	—
V Mysteries	1	68	41	—	—	—	—	—	116	114	110	—
V Mysteries	2	—	—	3.80	—	—	—	—	—	110	—	—
V Mysteries	3	73	46	—	—	—	—	—	—	—	—	—
V Mysteries	4	76	48	—	—	—	—	—	—	120	—	—
Unknown	21425	—	—	3.80	—	—	—	—	—	—	—	—
Unknown	47003	80	50	—	—	—	—	—	—	—	—	—
Unknown	?	55	40	—	—	—	—	—	—	—	—	—
Unknown	47004	65	50	—	—	—	—	—	—	—	—	—
Unknown	47006	55	40	—	—	—	—	—	—	—	—	—

(continued)

TABLE A1.3. *(continued)*

Property	Dolium no.	Ext. rim diam.	Int. rim diam.	Wall thickness	Volume	Height	Pres. height	Depth	Pres. depth	Ext. belly diam.	Int. belly diam.	Base diam.
Unknown	47005	80	50	—	—	—	—	—	—	—	—	—
Unknown	?	78	58	—	—	117	—	—	—	115	—	—
V Regina	1	—	43	—	695.0[c]	—	—	—	—	—	—	—
V Regina	2	—	43	—	522.0[c]	—	—	—	—	—	—	—
V Regina	3	—	43	—	522.4[d]	110	—	—	—	110	—	—
V Regina	4	—	43	—	695.0[c]	—	—	—	—	—	—	—
V Regina	5	—	—	—	522.0[c]	—	—	—	—	—	—	—
V Regina	6	—	—	—	—	—	—	—	—	—	—	—
V Regina	7	—	42	—	522.0[c]	—	—	—	—	—	—	—
V Regina	8	—	—	—	—	—	—	—	—	—	—	—
V Regina	9	—	45	—	480.8[d]	113	—	—	—	107	—	—
V Regina	10	—	43	—	695.3[d]	122	—	—	—	121	—	—
V Regina	11	—	45	—	566.3[d]	122	113	—	—	—	—	—
V Regina	12	—	41	—	522.4[d]	111	110	—	—	—	—	—
V Regina	13	—	—	—	—	—	—	—	—	—	—	—
V Regina	14	—	43	—	581.8[d]	118	114	—	—	—	—	—
V Regina	15	—	34	—	392.5[d]	115	100	—	—	—	—	—
V Regina	16	—	44	—	712.7[d]	131	122	—	—	—	—	—
V Regina	17	—	44	—	581.5[d]	122	114	—	—	—	—	—
V Regina	18	—	35	—	216.4[d]	91	82	—	—	—	—	—
Lowest value		34	24	1.3	100.0	84	30	63	47	55	45	18
Mean		58	38	3.4	478.4	112	83	98	73	90	88	19
Median		57	40	3.5	522.2	116	82	104	76	90	85	20
Highest value		80	58	4.7	770.0	132	122	116	116	140	120	22

[a] 3-D scanning.

[b] Calculated with mathematical computation.

[c] Estimate based on similar vessels with volume determined.

[d] Ancient inscription of volume on vessel.

TABLE A1.4. Dolium stamps from Pompeii.

Property	Stamp text	Reference	Miscellaneous information
VII.15.15	ANTEROTIS *litt. cavis pulchr.* GALLICI	*CIL* 10.8047, 2	
VI.8.9	(*anulo impressa*) A·APPVLEI (*anulo impressa*) (*uva vel folium*) HILARIONIS (*uva vel folium*)	*CIL* 10.8047, 3	Also found on dolium at Villa B in Gragnano; also found at Stabiae
	FIRMVS·FEC		
Unknown	(*anulo impressa*) A·APPVLEI (*anulo impressa*) (*uva vel folium*) HILARIONIS (*uva vel folium*)	*CIL* 10.8047, 3	In Naples Museum
VII.4.11	(*anulo impressum pomnum*) A· APPVLEI/QVIETI (*anulo impressum pomnum*) *litteris cavis*	*CIL* 10.8047, 4	Also found at Stabiae
VII.12.9	ASCL·PONTI *Ascl(epiadis) Ponti.*	*CIL* 10.8047, 5	*CIL* X 8042 76: Q MCI ASLEPIAD appears on at least seven tiles
VII.7.19	(*caduceus altus*) D·F·C·CLVENTI/AMPLIATI (*caduceus altus*)	*CIL* 10.8047, 7	
	(*palmae a ramus insertus coronae; supra luna crescens*) CORINTHVS·S·F (*palmae a ramus insertus coronae; supra luna crescens*)		
	D(e?) f(gilinis?) C. Cluenti Ampliati. Corinthus s(ervus) f(ecit)		
I.3.2	(*ramus palmae*) C·CLVENTI/AMPLIATI (*ramus palmae*) *litteris cavis pulchris*	*CIL* 10.8047, 6	Also found *a ziro di mattone* at Pompeii and on four ceramic vases
I.3.20 VI.14.36 Two dolia of unknown provenance	(*sigillum vaccum*) L·CORNELI RUFION (*sigillum vaccum*)	*CIL* 10.8047, 8	Four dolia

(*continued*)

TABLE A1.4. *(continued)*

Property	Stamp text	Reference	Miscellaneous information
IX.5.11	PHILEROS M·FULVI·SER	*CIL* 10.8047, 15	Phileros appears on a stamp in Rome
VI.8.8	LAVRINI PINNIAES	*CIL* 10.8047, 9	On two dolia, both of which have incised Roman numerals for capacities: Ↄ·XVI ↃXIX·P·C//
VII.4.11, VII.4.14	(*anulo impressum: folium uvae*) M·LVCCEI QVARTIONIS (*anulo impressum: folium uvae*) *litt. cavis pulchris*	*CIL* 10.8047, 10a	Incised with Roman numeral for capacity: XLIII
VII.4.11, VII.4.14	(*anulo impressum: guttus*) M·LVCCEI QVARTIONIS (*anulo impressum: guttus*) *litt. cavis pulchr.*	*CIL* 10.8047, 10b–c	Two dolia, one incised with Roman numerals for capacity: XLVI
VI.14.36	M·LVCCEI *litt. cavis pulchr.* QVARTIONSI *sic*	*CIL* 10.8047, 10d	
	AVKIOY (*litt. cavis*)		
NA	C·N·V C·NA*evi vital*IS C·N	*CIL* 10.8047, 11a	Also produced tiles, *CIL* X 8042 81. C·N·V C·N·V/C·NAEVI VITALIS C·N·V
VII.2.48	*sigillum: folium?* (*idem sigillum*) C·NAEVI VITALIS (*idem sigillum*) *litt. cavis pulchr.*	*CIL* 10.8047, 11b	Also produced tiles, *CIL* X 8042 81. C·NAEVI VITALIS
NA	SEX·OBINI·SALVI (*litt. cavis*)	*CIL* 10.8047, 12	Also found at Castellammare di Stabia
VII.4.11, VII.4.14	M·PACCI·HILARI (*litt. cavis*) (*anulo impressum: vasculum*)	*CIL* 10.8047 13	Also found at Castellammare di Stabia
I.I.2	M·PACCI SEC (*litt. cavis*)	*CIL* 10.8047 14	

VII.2.48	(*sigillum detritum*) L·SAGINI (*sigillum detritum*)	*CIL* 10.8047, 16	Also produced tiles, *CIL* 10.8042 90. L·SAGINI; L·SAGINI PRODMI appears on many tiles, *CIL* 10.8042 91.
VII.7.21	C SATRINI COMMUNIS MARCIAN *sic*	*CIL* 10.8047, 17	Also produced "urban" tiles, *CIL* 10.8042 93. C·SATRINI/COMMUNIS MARCIAN *sic*
VII.4.11, VII.4.14	L·TITI·T·F·PAP	*CIL* 10.8047, 18	On at least three dolia; one with post-cocturam incision "P CXI," one with post-cocturam incision "P CI"
VII.4.56	(*sigillum: vasculum?*) M·VIBI LIBERALIS (*idem sigillum*)	*CIL* 10.8047, 19a–b	On two dolia
VI.2.5	M·VIBI LIBERALIS (*sigillum: folium*) (*litt. cavis*)	*CIL* 10.8047, 19c	
VII.2.32/VII.3.3	MV·A·P (*litteris pulchris*)	*CIL* 10.8047, 1a–b	On two dolia
I.2.1	MVA////// ////////// *litt. cavis pulchr.*	*CIL* 10.8047, 1c	
IX.1.3	(*sigillum incertum*) VITALIS GALLICI (*sigillum incertum*) (*litt. cavis*)	*CIL* 10.8047, 20	
IX.2.7	L////VORVM LVCC///V·S	*CIL* 10.8047, 21	
I.5.5	EROTICV[	Unpublished?	
I.5.5	]..MICVI . . .	Unpublished?	
Dolium lid, inv. no. 17464	A·PLAVTI EVTACTI	Unpublished?	Also on lid of a dolium from Gragnano "Contrada Messigno" (Della Corte 1923, 274) and "APE" on dolium lid at Gragnagno "Contrada Carita" and another villa at Gragnano (Della Corte 1932, 278)

TABLE A1.5. Ostia dolium dimensions.

Property	Dolium no.	Ext. rim diam.	Int. rim diam.	Wall thickness	Volume	Height	Pres. height	Depth	Pres. depth	Ext. belly diam.	Int. belly diam.	Base diam.
I.4.5	1	104	73	—	1,102.04	—	—	—	—	130	—	—
I.4.5	2	99	64	—	1,061.10	—	—	—	—	130	—	—
I.4.5	3	—	—	5.5	—	—	—	—	—	133	—	—
I.4.5	4	—	—	5.5	1,010.34	—	—	—	—	134	—	—
I.4.5	5	—	—	5.8	—	—	—	—	—	130	—	—
I.4.5	6	—	—	4.4	—	—	—	—	—	130	—	—
I.4.5	7	—	—	—	—	—	—	—	—	140	—	—
I.4.5	8	—	—	6.0	—	—	—	—	—	135	—	—
I.4.5	9	—	—	5.2	—	—	—	—	—	135	—	—
I.4.5	10	—	—	4.8	746.70 or 1,008.70	—	—	—	—	135	—	—
I.4.5	11	—	—	—	1,231.40	—	—	—	—	140	—	—
I.4.5	12	99	67	—	1,101.49	—	—	—	—	145	—	—
I.4.5	13	—	—	4.0	—	—	—	—	—	130	—	—
I.4.5	14	—	—	5.3	—	—	—	—	—	137	—	—
I.4.5	15	—	—	5.1	—	—	—	—	—	132	—	—
I.4.5	16	104	70	—	1,061.10	—	—	—	—	130	—	—
I.4.5	17	96	63	—	1,100.40	—	—	—	—	134	—	—
I.4.5	18	—	—	4.8	—	—	—	—	—	125	—	—
I.4.5	19	—	—	—	957.39	—	—	—	—	135	—	—
I.4.5	20	—	—	—	1,036.00	—	—	—	—	133	—	—
I.4.5	21	—	—	5.1	—	—	—	—	—	127	—	—
I.4.5	22	—	—	—	877.70	—	—	—	—	128	—	—
I.4.5	23	—	—	4.8	1,061.10	—	—	—	—	123	—	—
I.4.5	24	—	—	4.5	930.10	—	—	—	—	120	—	—
I.4.5	25	—	—	4.9	1,139.70	—	—	—	—	136	—	—
I.4.5	26	—	—	—	1,036.00	—	—	—	—	125	—	—
I.4.5	27	—	—	—	851.50	—	—	—	—	110	—	—
I.4.5	28	—	—	—	774.00	—	—	—	—	110	—	—
I.4.5	29	96	65	—	1,074.20	—	—	—	—	118	—	—
I.4.5	30	—	—	5.0	1,126.60	—	—	—	—	130	—	—

I.4.5	31	—	—	4.5	1,139.70	—	—	—	—	135	—	—
I.4.5	32	101	66	—	1,113.50	—	—	—	—	130	—	—
I.4.5	33	100	64	—	—	—	—	—	—	—	—	—
I.4.5	34	102	71	—	1,036.50	—	—	—	—	135	—	—
I.4.5	35	—	—	4.1	—	—	—	—	—	125	—	—
I.4.5	36	102	72	—	1,087.30	—	—	—	—	130	—	—
III.14.3	1	83	55	—	—	—	—	—	—	—	—	—
III.14.3	2	99	71	—	—	—	—	—	—	—	—	—
III.14.3	3	93	64	—	—	—	—	—	—	—	—	—
III.14.3	4	—	66	—	—	—	—	—	—	—	—	—
III.14.3	5	93	62	—	—	—	—	—	—	—	—	—
III.14.3	6	87	60	—	—	—	—	—	—	—	—	—
III.14.3	7	85	58	—	—	—	—	—	—	—	—	—
III.14.3	8	80	57	—	—	—	—	—	—	—	—	—
III.14.3	9	90	60	—	—	—	—	—	—	—	—	—
III.14.3	10	83	55	—	—	—	—	—	—	—	—	—
III.14.3	11	78	51	—	—	—	—	—	—	—	—	—
III.14.3	12	85	57	—	—	—	—	—	—	—	—	—
III.14.3	13	81	53	—	—	—	—	—	—	—	—	—
III.14.3	14	—	—	—	—	—	—	—	—	—	—	—
III.14.3	15	—	—	—	—	—	—	—	—	—	—	—
III.14.3	16	85	59	—	—	—	—	—	—	—	—	—
III.14.3	17	73	47	—	—	—	—	—	—	—	—	—
III.14.3	18	87	60	—	—	—	—	—	—	—	—	—
III.14.3	19	—	—	—	—	—	—	—	—	—	—	—
III.14.3	20	93	66	—	—	—	—	—	—	—	—	—
V.11.5	1	—	—	3.9	—	—	—	—	—	124	117	—
V.11.5	2	—	—	5.6	—	—	—	—	—	133	125	—
V.11.5	3	—	—	5.1	—	—	—	—	—	136	124	—
V.11.5	4	—	—	5.3	—	—	—	—	—	131	122	—
V.11.5	5	—	—	4.7	—	—	—	—	—	140	127	—
V.11.5	6	—	—	5.0	—	—	—	—	—	146	139	—
V.11.5	7	—	—	5.3	—	—	—	—	—	137	125	—
V.11.5	8	—	—	5.3	—	—	—	—	—	142	132	—

(*continued*)

TABLE A1.5. (*continued*)

Property	Dolium no.	Ext. rim diam.	Int. rim diam.	Wall thickness	Volume	Height	Pres. height	Depth	Pres. depth	Ext. belly diam.	Int. belly diam.	Base diam.
V.11.5	9	—	—	4.4	—	—	—	—	—	140	132	—
V.11.5	10	—	—	5.2	—	—	—	—	—	121	110	—
V.11.5	11	—	—	5.6	—	—	—	—	—	125	117	—
V.11.5	12	—	—	5.4	—	—	—	—	—	116	105	—
V.11.5	13	—	—	4.3	—	—	—	—	—	123	116	—
V.11.5	14	—	—	4.7	—	—	—	—	—	134	127	—
V.11.5	15	—	—	4.7	—	—	—	—	—	138	122	—
V.11.5	16	—	—	4.5	—	—	—	—	—	139	129	—
V.11.5	17	—	—	4.5	—	—	—	—	—	136	126	—
V.11.5	18	—	—	4.1	—	—	—	—	—	128	120	—
V.11.5	19	—	—	4.9	—	—	—	—	—	125	116	—
V.11.5	22	—	—	4.9	—	—	—	—	—	131	124	—
V.11.5	23	—	—	4.9	—	—	—	—	—	137	128	—
V.11.5	24	—	—	4.8	—	—	—	—	—	130	120	—
V.11.5	26	—	—	—	—	—	—	—	—	—	—	—
V.11.5	27	—	—	4.8	—	—	—	—	—	129	120	—
V.11.5	28	—	—	4.8	—	—	—	—	—	124.5	115	—
V.11.5	29	—	—	4.4	—	—	—	—	—	129	118	—
V.11.5	30	—	—	4.6	—	—	—	—	—	124	115	—
V.11.5	31	—	—	4.8	—	—	—	—	—	121	113	—
V.11.5	32	—	—	4.8	—	—	—	—	—	127	119	—
V.11.5	33	—	—	5.0	—	—	—	—	—	138	128	—
V.11.5	34	—	—	5.1	—	—	—	—	—	137	130	—
V.11.5	35	—	—	4.4	—	—	—	—	—	—	—	—
V.11.5	36	—	—	5.2	—	—	—	—	—	136	131	—
V.11.5	37	—	—	4.2	—	—	—	—	—	128	118	—
V.11.5	38	—	—	4.2	—	—	—	—	—	—	—	—
V.11.5	39	—	—	4.6	—	—	—	—	—	135	124	—
V.11.5	40	—	—	4.2	—	—	—	—	—	—	—	—
V.11.5	41	—	—	—	—	—	—	—	—	—	—	—
V.11.5	42	—	—	5.0	—	—	—	—	—	139	130	—

V.11.5	43	—	—	4.6	—	—	—	—	—	135	126	—
V.11.5	44	—	—	4.1	—	—	—	—	—	122	114	—
V.11.5	45	—	—	5.0	—	—	—	—	—	123	113	—
V.11.5	46	—	—	4.1	—	—	—	—	—	134	128	—
V.11.5	47	—	—	4.9	—	—	—	—	—	136	125	—
V.11.5	48	—	—	3.0	—	—	—	—	—	129	123	—
V.11.5	49	—	—	4.8	—	—	—	—	—	121.5	113	—
V.11.5	50	—	—	4.7	—	—	—	—	—	144	136	—
V.11.5	51	—	—	4.6	—	—	—	—	—	129	122	—
V.11.5	52	—	—	6.7	—	—	—	—	—	142	129	—
V.11.5	53	—	—	5.2	—	—	—	—	—	140	133	—
V.11.5	54	—	—	4.7	—	—	—	—	—	141	132	—
V.11.5	55	—	—	5.0	—	—	—	—	—	134	125	—
V.11.5	56	—	—	4.9	—	—	—	—	—	132	122	—
V.11.5	57	—	—	4.7	—	—	—	—	—	132	122	—
V.11.5	58	—	—	5.0	—	—	—	—	—	136	127	—
V.11.5	59	—	—	5.3	—	—	—	—	—	140	129	—
V.11.5	60	—	—	6.4	—	—	—	—	—	137	126	—
V.11.5	61	—	—	4.6	—	—	—	—	—	141	132	—
V.11.5	62	—	—	4.6	—	—	—	—	—	109	100	—
V.11.5	63	—	—	4.3	—	—	—	—	—	—	—	—
V.11.5	64	—	—	4.3	—	—	—	—	—	—	—	—
V.11.5	65	—	—	4.9	—	—	—	—	—	—	—	—
V.11.5	66	—	—	4.1	—	—	—	—	—	—	—	—
V.11.5	67	88	56	—	997.24	—	83	—	—	140	—	—
V.11.5	68	—	—	—	—	—	—	—	—	—	—	—
V.11.5	BF1	—	—	—	—	—	—	—	—	—	—	—
V.11.5	BF2	—	—	—	—	—	—	—	—	—	—	—
V.11.5	BF3	—	—	—	—	—	—	—	—	—	—	—
V.11.5	BF4	—	—	—	—	—	—	—	—	—	—	—
Lowest value		73	47	3.0	774.00	—	83	—	—	109	100	—
Mean		91	61	4.9	1,016.30 or 1,027.20	—	83	—	—	131	123	—
Median		93	63	4.8	1,061.10	—	83	—	—	132	124	—
Highest value		104	73	6.4	1,139.70	—	83	—	—	146	136	—

Note: Volume determined by ancient incisions in unit of amphorae.

TABLE A1.6. Rome dolium dimensions.

Property	Dolium no.	Ext. rim diam.	Int. rim diam.	Wall thickness	Volume	Height	Pres. height	Depth	Pres. depth	Ext. belly diam.	Int. belly diam.	Base diam.
CapMus	1	97	60	—		135	—	—	—	130	—	35
CapMus	2	97	63	—		130	—	—	—	130	—	35
CapMus	3	92	57	—	—	140	—	—	—	125	—	37
MusNaz	1	92	67	—		150	—	—	—	135	—	25
MusNaz	2	93	61	—	—	—	—	—	—	135	—	—
MusNaz	3	100	74	—	—	—	120	—	—	130	—	—
MusNaz	4	—	—	—	—	—	135	—	—	110	—	34
MusNaz	5	—	—	—	—	140	—	—	—	140	—	35
MusNaz	6	80	—	—	—	165	—	—	—	140	—	24
MusNaz	7	—	—	—	—	—	—	—	—	135	—	—
MusNaz	8	—	—	—		165	—	—	—	140	—	20
MusNaz	9	—	52	5		135	—	—	—	130	—	—
MusNaz	10	92	65	—		—	115	—	—	130		—
MusNaz	11	—	—	—		—		—	—	—	—	—
MusNaz	12	—	—	—		—		—	—	—	—	—
Lowest value		92	52	5		130	115	—	—	110	—	20
Mean		92	60	5		145	123	—	—	130	—	30
Median		92	61	5		140	120	—	—	130	—	34
Highest value		100	74	5		165	135	—	—	140	—	37

Note: Volume determined by ancient incisions in units of amphorae.

TABLE A1.7. Dolium stamps from Ostia.

Property	Stamp text	Reference	Notes
III.14.3	*cornucopiae* L·AVTRONI XANTHI *bipennis*	Bloch 1948, 96 no. 470	
III.14.3, no. 12	*bucranium* L·C[A]ECILIVS PROCLVS *signum detritum*	Bloch 1948, 96 no. 471	
cella prope thermas adhuc extantibus	M [F]VRI· VINDICIS *alterum sigillum in labro eiusdem dolii impressum praeter ornamenta totum evanuit*	*CIL* 15.2447a = *CIL* 14.4093, 12	
nel fondo la Torretta	M·FVRI VINDICIS	*CIL* 15.2447b	Information from catalog; dolium said no longer to exist
I.4.5	GENIALIS RASINI PONTICI SER FE	Gatti 1903, 202	Same stamp found on dolium from Rome, *CIL* 15.2449
III.14.3, no. 6	*bucranium* C IVLI RVFI *bucranium* L·ARISTAEVS RESTITVTVS **FF** *caput bovis caput bovis caput bovis caput bovis*	Bloch 1948, 101 no. 509	
Ostia Museum	L·LVRIVS *ramus palmae* VERECVN FE *L. Lurius Verecun(dus) fe(cit)*	*CIL* 15.2459	
cella prope thermas adhuc extantibus	Q·OCI/////	*CIL* 15.2475	Stamped on rims of two dolia?

(*continued*)

TABLE A1.7. *(continued)*

Property	Stamp text	Reference	Notes
cella prope thermas adhuc extantibus	*ramus palmae* *corona* FAVSTVS·FEC *corona* *corona* L·PETRONI FVSCI FECIT FAVSTVS·SER *corona* *ramus palmae*	*CIL* 15.2479 = *CIL* 14.4093, 3	Two separate stamps
III.14.3, no. 1	PYRAMI ENCOLPI AVG DISP·ARCARI AMPLIATVS·VIC·F *Pyrami, Encolpi Aug(usti) disp(ensatoris) arcari; Ampliatus vic(arius) f(ecit)*	Bloch 1948, 136 no. 537	Pyramus could have been T. Flavius Pyramus in Bloch (1948, 101 no. 506): T· FLAVI· PYRAMI ADIVTOR· SER· FEC, on dolium in Rome[a]
Ostia surroundings	*ramus palmae* L RVFEN·PROCVLI LEO·SER·FEC *caduceus* *ramus palmae* *L. Rufen(i) Proculi; Leo ser(vus) fec(it)*	*CIL* 15.2488 = *CIL* 14.4093.6; Lanciani 1885, 77	Another slave from same estate appears on a dolium in Rome, *CIL* 15.2487: L· RVFENI· PROCVL/ COGIT*a*TVS· SER· F
III.14.3	*caput bovis caput bovis caput bovis* II·RVFENORVM CELERIS ET POLLIO *(Duorum) Rufenorum Celeris et Pollio(nis)*	Bloch 1948, 106 no. 539	
cella prope thermas adhuc extantibus	C·TITIENI C F FLORI *caput bovis infulatum* REPENTINVS F *caput bovis infulatum*	*CIL* 15.2500a = XIV 4093, 7	Two separate stamps on same dolium; identical set on dolium found on Equiline in Rome, *CIL* 15.2500b

cella prope thermas adhuc extantibus	*thyrsus vittis exornatus* *caput bovis caput bovis* Q· TOSSIVS· PROCVLVS· F *caput bovis caput bovis caput bovis* *ramus palmae*	*CIL* 15.2507	Different members of the Tossius family appear on many opus doliare products from the late Republic period onward, typically in southern Etruria and northern Latium[b]
III.14.3	*caput bovis* C VIBI FORTVNATI C VIBI CRESCENTIS *caput bovis*	Bloch 1948, 111n565	Bloch: "C. Vibius Crescens was undoubtedly a slave of Vibius Donatus before his manumission"
III.14.3	*caput bovis* C· VIBIVṢ *caput bovis* F̣ORTVNẠṬ· FEC.	Bloch 1948, 111n564	Bloch: "C. Vibius Fortunatus is the *Fortunatus ser(vus)* of 2512 after his manumission by C. Vibius Donatus."
I.4.5	RHODINVS SER· FEC	Bloch 1948, 106n538	

[a] Bloch 1948, 106n537: "Pyramus was the *arcarius* of the imperial *dipensator* Encolpus, Ampliatus his *vicarius*. An *Aug(usti) disp(ensator) arcar(ius) regn(i) Noric(i)* occurs in *ILS* 1506 (Virunum = Klagenfurt); an *Aug(ustorum) n(ostrum) dispensatoris arkarius* in *ILS* 1661 (Caesarea, Cappadocia); a *dispensatoris fisci castrenis arcarius* in *ILS* 1660 from Rome (cf. O. Hirschfeld, *Kaiserl. Verwaltungsbeamte*2, 401 n. 3 and 461 n. 3). Encolpus may have been connected with the *annona*."

[b] Taglietti (2015) discusses the activities and chronology of the Tossius family. Carrato (2017, 619) includes a dolium in Gaul with the stamp of Q. Tossius Priscus, who has been attested on dolia in Rome.

TABLE A1.8. Dolium production sites in west-central Italy (from Olcese 2012).

Region	Province	Site	Context	Date	Indication of production	Bricks	Tiles	Amphorae	Lamps	Coarseware	Other heavy terracotta objects	Fineware
Toscana	Arezzo	Ossaia	Villa	100 BCE–5th c. CE; 2nd–3rd c. CE (dolium production)	Structures (kilns, paved areas, vats, channels); instruments (molds); discard	X	X		X	X		
Toscana	Arezzo	Pain di Scò	Settlement	1st c. BCE to early 1st c. CE	Three kilns; wasters; discard			X		X		
Toscana	Firenze	Scandicci—Vingone	Productive area	20 BCE–20 CE	Kilns; discard	X	X			X	X	X
Toscana	Firenze	Scarperia—Monte Calvi, Fonte Laterina	Not defined	2nd–3rd c. CE	Discard; misfired; wasters	X	X			X	X	
Toscana	Livorno	Cecina—Podere Canciana	Farm	2nd c. BCE–5th c. CE	Spacers; discard; wasters; misfired	X	X	X		X		
Toscana	Livorno	Cecina—Podere del Pozzo	Settlement	1st–2nd c. CE	Spacers; discard	X	X	X		X		
Toscana	Livorno	Quercianella, Gorgo	Not defined	NA	Discard	X	X	X		X		
Toscana	Livorno	Rosignano Marittimo—Galafone	Farm	Augustan–4th c. CE	Spacers; discard; misfired	X	X	X		X		
Toscana	Livorno	Rosignano Marittimo—Poggio Fiori	Settlement	End of 1st c. BCE–2nd/3rd c. CE	Kilns; discard; spacers; misfired	X	X	X		X		
Toscana	Pisa	Pontedera—Fossa Nuova	Not defined	Republican period–late antiquity	Discard; misfired; wasters			X		X		
Toscana	Pistoia	Larciano—Case Belriposo	Settlement	NA	Discard	X	X	X				
Toscana	Pistoia	Larciano—Poggio Bagnolo, Brugnana	Not defined	Imperial period	Discard	X	X					
Toscana	Siena	Castellina in Chianti—S. Iacopo	Farm	2nd c. CE	Kiln; discard	X	X	X		X		
Toscana	Siena	Castelnuovo Berardenga—Gaggiola	Farm	End of 1st–2nd c. CE	Kiln; discard	X	X	X		X		

Toscana	Siena	Murlo—Vescovado	Settlement	End of 7th–mid-3rd c. BCE	Discard					X		
Toscana	Siena	Poggibonsi—Castelluccio	Not defined	2nd–1st c. BCE	Discard; wasters; misfired	X	X					
Lazio	Roma	Allumiere—Macchia di Freddara	Villa	3rd c. BCE–early imperial	Structures (kilns, *tettoia*/canopies); waste	X	X			X	X	X
Lazio	Roma	Guidonia	Not defined	Late Republican–imperial period	Discard; wasters	X	X				X	X
Lazio	Viterbo	Bolsena—Poggio Moscini, cistern 5	Urban	End of 2nd c. BCE–first half of 1st c. CE	Spacers; discard; kilns			X	X	X	X	X
Lazio	Viterbo	Bomarzo	Productive area	Trajanic–Caracalla	Discard; wasters	X	X				X	
Lazio	Viterbo	Mugnano in Teverina (Gasperoni 2003, sito 57)	Productive area	Imperial period	Discard; wasters						X	
Lazio	Viterbo	Mugnano in Teverina (Gasperoni 2003, sito 7)	Productive area	Imperial period	Discard; wasters	X	X					
Lazio	Vitbero	Mugnano in Teverina—Rota Rio (Gasperoni 2003, sito 41)	Productive area	Beginning of 1st c. CE–Commodus	Structures (kilns, rooms, walls); discard; wasters	X	X				X	
Lazio	Viterbo	Mugnano in Teverina—S. Liberato (Gasperoni 2003, sito 56)	Productive area	From reign of Caligula to Diocletian	Structures (kilns?); discard; wasters	X	X				X	
Lazio		Pontine Plain	Rural site									
Lazio		Pontine Plain	Rural site									
Campania	Naples	Pompeii—Casa dei Fiori (VI.5.9–19, 10)	Urban	3rd c. BCE	Misfirings		X					
Campania	Naples	Sorrento—Massa Lubrense, Arorella	Villa	Roman	Wasters; discard	X	X					
Campania	Naples	Sorrento—Piano di Sorrento, Trinità S. Massimo	Urban	3rd–1st c. BCE	Kilns	X	X					X
Campania	Caserta	Ruviano—S. Martino	Settlement	Roman		X	X			X		
Campania	Avellino	Lioni—Gasdotto	Not defined	Late Republic–2nd c. CE			X					
Umbria		Scoppieto						X			X	
Total						22	24	12	2	16	10	5

TABLE A1.9. Villas and farms with dolia in west-central Italy.

Town	Villa name	ID no.	Date(s)	Number of dolia	Size of storeroom (area m^2)	Dolium contents
Anguillara Sabazia	Mura di S. Stefano	L11	2nd–early 4th c. CE		225	Wine
Astura	Colle Falcone	L10	Late 1st c. BCE/early 1st c. CE			Wine
Astura	La Saracca, Villa	L23	7th c. BCE–?			
Bassano di Sutri	Castellina	L26	? (starting in late Republican?)			
Boville (Bovillae)	Casal Morena	L42	1st c. BCE–4th c. CE			
Casperia	Paranzano	L56	1st–2nd c. CE			
Castel di Guido (Lorium?)	Colonnacce	L60	2nd/1st c. BCE–3rd c. CE			Wine
Circeo	Monte Circeo	L79	Republican–1st c. CE at least			?
Cottanello (ager Foronovanus)	Cottanello	L96	End of 2nd c. BCE–6th c. CE			Wine?
Fiano Romano	Volusii Saturnini Villa	L106	50 BCE–4th c. CE			Wine
Ischia di Castro	La Selvicciola	L120	Late 3rd c. BCE–5th c. CE			Olive oil?
Licenza	Cerri Hill	L126	?			?
Licenza	Mandela	L128	2nd c. BCE–?			?
Licenza	Via Licinese	L135	?			?
Montopoli	Bocchignano, Caravilla	L144	?			?
Ostia	Dragoncello, Villa	L151	2nd/1st c. BCE–4th c. CE			Wine?
Poggio Catino	Vigna Paleani	L158	?			Preserved birds
Poggio Mirteto	S. Maria in Turano	L160	?			?
Poggio Mirteto	S. Savino	L161	2nd c. BCE–imperial period			?
Procoio Nuovo	Procoio Nuovo	L172	Late 2nd/early 1st c. BCE–4th c. CE	2		Wine
S. Marinella	Castello Odescalchi	L177	1st–3rd c. CE			?
S. Palomba	Palazzo	L186	3rd c. BCE–4th c. CE	5	9	?
Suburbium	Auditorium	L195	6th c. BCE–2nd c. CE			Olive oil?
Suburbium	Borgata Ottavia	L197	Late 2nd c. BCE–at least 1st c. CE			Wine
Suburbium	Casale di Aguzzano	L201	50 BCE–at least 1st c. CE			?
Suburbium	Casale Ghella	L202	1st c. BCE–late imperial period	4		Wine; olive oil
Suburbium	Castel Giubileo	L205	Late 1st c. BCE–at least 2nd c. CE		105	Wine
Suburbium	Castel Giubileo	L206	Late Republican period–late 5th c. CE			Wine?

Suburbium	Cecchignola	L207	2nd/1st c. BCE–at least 2nd c. CE			?
Suburbium	Cinecittà, Quaietta	L211	2nd c. CE			Wine?
Suburbium	Cinquina	L213	Late 1st c. BCE–4th c. CE	7	196	Wine?
Suburbium	Fosso Lombardo	L218	Late 2nd c. BCE–4th c. CE	2	0.5	?
Suburbium	Fosso Santa Maura, Tor Vergata	L219	2nd c. BCE–at least 2nd c. CE	3	48	Wine
Suburbium	Grotte di Cervara	L222	2nd–1st c. BCE		17.5	Olive oil?
Suburbium	Muracciola	L227	1st c. CE at least	12	144	Wine
Suburbium	Podere Anna/Podere Rosa	L229	1st c. BCE–5th c. CE			?
Suburbium	Prima Porta, Cimitero Flaminio	L231	Late Republican period–at least 6th c. CE	11	140	Wine
Suburbium	Prima Porta, Via Tiberina	L234	1st c. BCE–Julio-Claudian			Wine?
Suburbium	S. Alessandro	L237	1st–5th c. CE			Wine?
Suburbium	Tor Carbone, "Domus Marmeniae"	L243	1st–5th c. CE	10		Wine
Suburbium	Via Capobianco	L251	Late Republican period–at least 2nd c. CE			Wine?
Suburbium	Via Casalotti	L253	2nd–4th c. CE			Wine and olive oil?
Suburbium	Via Ripa Mammea	L257	2nd c. BCE–at least 2nd c. CE	9		Wine?
Suburbium	Via Togliatti	L258	2nd c. BCE–6th c. CE	15	88	Wine and olive oil?
Suburbium	Via Togliatti	L258	5th c. CE	4	8	
Suburbium	Via Vigne Nuove, Val Melaina	L259	Late Republican period–3rd c. CE			Wine? Oil?
Suburbium	Via Vigne Nuove, Val Melaina	L260	1st c. BCE	1		Wine?
Tarquinia	Portaccia	L265	?			?
Tivoli	Colle Lecinone	L273	2nd c. BCE–1st c. CE			
Tivoli	Guidonia	L288	1st c. CE		12.5	
Tivoli	Lecinone, Scalzacane	L290	Late Republican period–at least 2nd c. CE			?
Tusculum	Cornufelle	L330	1st c. BCE–early 3rd c. CE			?
Tusculum	Villa dei Furii	L350	Republican period–at least 1st c. CE	2		Wine
Vescovio (Forum Novum)	Vescovio	L363	Imperial period			?
Viterbo	Asinello	L375	Mid-1st c. BCE–4th c. CE	15	33.3	Wine or olive oil
Anagni	Villa Magna		2nd c. CE			
Rome	Villa dei Quintili		2nd c. CE			
Ansedonia	Settefinestre	T3	1st c. BCE	58	176	
Ansedonia	La Tagliata or Torre Tagliata	T4	1st–5th c. CE			?
La Befa	Carcerelle	T6	Imperial period			?

(*continued*)

TABLE A1.9. (*continued*)

Town	Villa name	ID no.	Date(s)	Number of dolia	Size of storeroom (area m^2)	Dolium contents
Scansano	Fattoria Pomonte	T45	200 BCE–600 CE			?
Talamone	Poggio Mulinaccio	T49	?			?
Volterra	Pieve Vecchia di Casale Maritimo	T50	Imperial period			?
Penna in Taverina	Pennevecchia	U11	Imperial period	9	45.5	Wine
Perugia	Deruta (Perugia Vecchia)	U13	1st–4th c. CE			?
Perugia	Valcaprara	U17	Imperial period	1		Wine
Perugia	Colle Preiano	U18	3rd–4th c. CE (maybe earlier)			Wine
Perugia	Fossalto	U19	1st c. CE			?
S. Giustino	Colle Plinio	U23	2nd c. BCE–4th c. CE		160	Wine
S. Giustino	Pitigliano	U24	?			?
Stabiae	Oliaro	C2	1st c. CE		33	
Stabiae	Casa di Miri	C5	1st c. CE		445.3	
Stabiae	Sassole	C7	1st c. CE		32.4	Wine
Stabiae	Belvedere	C9	1st c. CE	7	260.3	Wine
Stabiae	i Medici	C10	1st c. CE		31.6	Wine
Boscoreale	Pisanella	C13	1st c. CE	90	195	Wine
Boscoreale	Giuliana	C14	1st c. CE	4	90	Wine
Scafati	Spinelli	C19	1st c. CE	6	4.8	Wine
Boscoreale	Civita-Giuliana	C25	1st c. CE	5	54	Wine
Boscoreale	stazione ferroviaria	C28	1st c. CE		17.5	
Boscoreale	Pisanella (N. Popidius Florus)	C29	1st c. BCE	7	70	Wine
Scafati	S. Abbondio	C32	1st c. CE			Wine
Gragnano	Messigno	C33	1st c. CE	35	90	Wine
Gragnano	Carità	C34	1st c. CE	11	169	Wine
Scafati	Spinelli	C35	1st c. CE	12	88	
Boscoreale	Villa Regina	C39	1st c. CE	18	55.9	Wine
Terzigno	Boccia al Mauro, Villa 1	C42	1st c. CE	42	132	Wine
Terzigno	Boccia al Mauro, Villa 2	C43	1st c. CE		110	
Ponticelli	Villa of C Olius Ampliatus	C45	1st c. CE	31	152	Wine
Gragnano	Carmiano		1st c. CE	12		
Somma Vesuviana	Villa "of Augustus"		3rd–4th c. CE	64		Wine

Chapter 6

TABLE A1.10. Volume incisions on dolia at Ostia (from Gatti 1903 and Carcopino 1909).

Property and dolium	Ancient volume inscription (units of amphora)	Number of amphorae; number of *sextarii*	Volume (liters)
I.4.5 no. 1	XLII ƆIII	42.0; 3	1,102.0
I.4.5 no. 2	XLS	40.5	1,061.1
I.4.5 no. 4	XXXVIIIS ƆIII	38.5; 3	1,010.3
I.4.5 no. 10	XXVIIIS (XXXVIIIS)[a]	28.5 (38.5)[a]	746.7 (1,008.7)[a]
I.4.5 no. 11	XLVII	47.0	1,231.4
I.4.5 no. 12	XLII ƆII	42.0; 2	1,101.5
I.4.5 no. 16	XLS	40.5	1,061.1
I.4.5 no. 17	XLII	42.0	1,100.4
I.4.5 no. 19	XXXVIS ƆII	36.5; 2	957.4
I.4.5 no. 20	XXXIXS ƆII	39.5; 2	1,036.0
I.4.5 no. 22	XXXIIIS	33.5	877.7
I.4.5 no. 23	XLS	40.5	1,061.1
I.4.5 no. 24	XXXVS	35.5	930.1
I.4.5 no. 25	XLIIIS	43.5	1,139.7
I.4.5 no. 26	XXXIXS ƆII	39.5; 2	1,036.0
I.4.5 no. 27	XXXIIS	32.5	851.5
I.4.5 no. 28	XXIXS ƆII	29.5; 2	774.0
I.4.5 no. 29	XLI	41.0	1,074.2
I.4.5 no. 30	XLIII	43.0	1,126.6
I.4.5 no. 31	XLIIIS	43.5	1,139.7
I.4.5 no. 32	XLIIS	42.5	1,113.5
I.4.5 no. 33	XXXIXS ƆIII	39.5; 3	1,036.5
I.4.5 no. 35	XLIS	41.5	1,087.3
V.11.5 no. 68	XXXVIII ƆIII[b]	38.0; 3[b]	997.2[b]
I.19 no. 3	XXIII	23.0	602.6
I.19 no. 4	XXXV	35.0	917.0
I.19 no. 5	XXVI or XXXVI	26.0 or 36.0	681.2 or 943.2
I.19 no. 6	XXXIV	34.0	890.8
I.19 no. 7	XXXIIS or XXXIIƆ	32.5 or 32.0; 2	851.5 or 839.5
I.19 no. 8	XXXII	32.0	838.4
I.19 no. 14	XXXVIII	38.0	995.6
I.19 no. 15	XXXIV	34.0	890.8
I.19 no. 16	XXIX	29.0	759.8
I.19 no. 21	XLVS	45.5	1,192.1

[a] Indicates a volume incision I recorded that differed from Gatti (1903).

[b] Indicates a volume incision that was unpublished.

TABLE A1.11. Possible craftspeople of dolium repairs, according to stage of execution.

Repair	Stage	Tools, equipment	Location	Possible repairer(s)
Fills	Production	Lead and open flame; lead and other metal (for lead alloy) and furnace; tool for applying metal into crack	Dolium workshop	(a) Member of the dolium production site: dolium maker, specialist repairer (b) Outsider: specialist repairer, another craftsperson, metal worker (especially for lead alloys)
	Use	Lead and open flame; tool for applying metal into crack	Place of use	(a) User of vessel: member of farm, shop, or warehouse (b) Outsider: tinker, specialist craftsperson, metallurgist
Lead or lead alloy staples or clamps	Production	Lead and open flame; lead and other metal (for lead alloy) and furnace; drill	Dolium workshop	(a) Member of the dolium production site: dolium maker, specialist repairperson (b) Outsider: specialist repairperson, another craftsperson, metal worker (especially for lead alloys)
Lead staples or clamps	Use	Lead; open flame; mold for crosspiece; drill	Place of use	(a) User of vessel: member of farm, shop, or warehouse (b) Outsider: pottery mender, specialist craftsperson, tinker, metal worker
Iron and lead staples or clamps	Use	Iron pins and crosspieces; lead; open flame; tool for applying metal; drill	Place of use	(a) User of vessel: member of farm, shop, or warehouse (b) Outsider: tinker, metalworker, architectural craftsperson
Double dovetail or double dovetail tenons	Production	Lead and open flame; lead and other metal (for lead alloy) and furnace; tools for cutting mortises and applying and securing metal tenon into mortises	Dolium workshop	(a) Member of the dolium production site: dolium maker, specialist repairer (b) Outsider: architectural craftsperson, specialist repairperson, another craftsperson, metal worker (especially for lead alloys)
Hybrid (staple or clamp + mortise and tenon)	Use	Lead and open flame; drill; tools for cutting mortises and applying and securing metal tenon	Place of use	(a) User of vessel: member of farm, shop, or warehouse (b) Outsider: architectural craftsperson, pottery mender, tinker, specialist repairer

TABLE A1.12. Dolium repairs at Cosa.

Vessel	Production phase	Use-life	Unknown stage
Cosa no. TC	—	Clamp	—
Cosa no. 19	—	Hybrid mortise-and-tenon clamp; hybrid mortise-and-tenon double dovetail	—
Cosa no. 29	—	Hybrid mortise-and-tenon double dovetail	—
Cosa no. 11	—	Clamp	Double dovetail
Cosa no. 1	—	Clamp	—

TABLE A1.13. Types of dolium repairs at Pompeii.

	Count	% of repaired dolia	% of all repaired jars
Repaired	30	—	73.2
Production	21	70.0	51.2
Use	7	23.3	17.1
Production and use	4	13.3	9.8
Lead	18	60.0	43.9
Lead alloy	6	20.0	14.6
Fill(s)	3	10.0	7.3
Staple(s)	4	13.3	9.8
Clamp(s)	11	36.7	26.8
Hybrid	5	16.7	12.2
Double dovetail	16	53.3	39.0
Double dovetail tenon	14	46.7	34.1
One technique	17	56.7	41.5
Two or more techniques	13	43.3	31.7

TABLE A1.14. Dolium repairs at Pompeii.

Vessel	Production phase	Use-life	Unknown stage
I.22 no. 2	—	—	Lead fill
I.22 no.3	Lead alloy double dovetails and double dovetail tenons	—	—
I.22 no.5	Double dovetail	—	—
I.22 no. 6	Lead alloy double dovetail tenon	—	—
I.22 no. 7	—	Lead fill; hybrid double dovetail tenon staple (unknown material)	—
I.22 no. 14	Double dovetail	—	—
I.22 no. 17	Double dovetail and double dovetail tenon	—	—
I.22 no. 18	Lead clamp	—	—
I.22 no. 19	—	Clamp	—
I.22 no. 20	Double dovetail	—	—
I.8.8 no. 13	Lead double dovetail	—	—
I.21.2 no. 1	Lead alloy double dovetail and double dovetail tenon	—	—
I.21.2 no. 2	—	Lead staple and clamp	—
II.8.6 no. 1	Lead alloy double dovetail and hybrid double dovetail tenon clamp	Lead clamps	—
II.8.6 no. 2	Lead double dovetail tenon	—	—
VII.5.21 no. 1	Lead alloy double dovetail and double dovetail tenon	—	—
II.1.8–9 no. 1	—	—	Lead staple and clamp
VI.15.13 no. 5	Clamp	—	—
VI.14.2 fragment 1	Clamp	—	—
VII.4.58 no. 1	—	Lead double dovetail	—
II.5.5 no. 1	Lead double dovetail and double dovetail tenon	—	Lead clamp
II.5.5 no. 4	Double dovetail	—	—
II.5.5 no. 5	Lead double dovetail tenon	—	—
II.5.5 no. 8	Double dovetail tenon	Lead hybrid staple mortise and tenon	Lead clamp
II.5.5 no. 9	Lead clamp	—	—
II.5.5 no. 10	Lead double dovetail and double dovetail tenon	—	Lead fill
VM n. 2	Lead double dovetail	—	—
VM n. 3	Lead double dovetail and double dovetail tenon	Lead hybrid double dovetail staple	—
VM n. 4	Lead double dovetail and double dovetail tenon	—	—
VI.16.40 no. 1	—	—	Lead clamp

TABLE A1.15. Types of dolium repairs at Ostia.

Repairs	Total (124)	Repaired (50)
Type of repair = raw quantity of dolia	% of total	% of repaired
Number of repaired dolia = 50	40.3	(100)
Dolia repaired with lead = 32	25.8	64
Dolia repaired with lead alloy = 19	15.3	38
Dolia repaired with lead and lead alloy = 9	7.3	18
Dolia repaired with one technique = 40	28.2	80
Dolia repaired with two techniques = 8	10.5	16
Dolia repaired during production phase, with certainty = 24	19.4	48
Fills on interior surface = 18	14.4	36
Lead fills on interior surface made in lead = 13	10.5	26
Lead alloy fills on interior surface = 2	1.6	4
Lead and lead alloy fills on interior surface = 1	0.8	2
Fills on exterior surface = 16	12.9	32
Lead fills on exterior surface = 9	7.3	18
Lead alloy fills on exterior surface = 3	2.4	6
Double dovetails on exterior surface = 9	7.3	18
Lead double dovetails on exterior surface = 6	4.8	12
Lead alloy double dovetails on exterior surface = 1	0.8	2
Lead and lead alloy double dovetails on exterior surface = 1	0.8	2
Lead alloy double dovetail tenons on interior surface = 1	0.8	2
Double dovetail tenons on exterior surface = 21	16.9	42
Lead double dovetail tenons on exterior surface = 6	4.8	12
Lead alloy double dovetail tenons on exterior surface = 8	6.5	16
Lead and lead alloy double dovetail tenons on exterior surface = 1	0.8	2

TABLE A1.16. Dolium repairs at Ostia.

Vessel	Production phase	Use-life	Unknown stage
I.4.5 no. 1	Lead double dovetail	—	—
I.4.5 no. 2	—	—	Lead fill
I.4.5 no. 3	—	—	Lead fill
I.4.5 no. 5	Double dovetail	—	—
I.4.5 no. 6	Double dovetail tenon	—	—
I.4.5 no. 11	Lead alloy double dovetail tenon; fill	—	Lead fill (probably use)
I.4.5 no. 12	Lead alloy double dovetail tenon; double dovetail	—	—
I.4.5 no. 17	Lead alloy double dovetail tenon	—	Fill
I.4.5 no. 23	Double dovetail	—	—
I.4.5 no. 26	—	—	Fill
I.4.5 no. 27	Double dovetail tenon	—	—
I.4.5 n. 28	Lead double dovetail tenon	—	—
I.4.5 n. 30	Lead alloy double dovetail; double dovetail tenon	—	—
I.4.5 no. 34	—	—	Lead fill
I.4.5 no. 35	Lead alloy double dovetail tenon	—	—
I.4.5 no. 36	Double dovetail tenon	—	—
III.14.3 no. 1	Lead double dovetail	—	Lead fill
III.14.3 no. 6	—	—	Lead fill
V.11.5 no. 2	Lead alloy double dovetail tenon	—	Lead fill
V.11.5 no. 3	Lead alloy fill	—	—
V.11.5 no. 4	—	—	Fill
V.11.5 no. 5	Lead double dovetail tenon	—	—
V.11.5 no. 6	—	—	Fill
V.11.5 no. 7	Lead double dovetail tenon	—	—
V.11.5 no. 8	Lead alloy fill	Lead fill	—
V.11.5 no. 17	Lead alloy double dovetail tenon	—	Lead fill
V.11.5 no. 22	Double dovetail tenon	—	Lead fill
V.11.5 no. 26	Lead double dovetail tenon	—	—
V.11.5 no. 32	Lead alloy double dovetail tenon	—	—
V.11.5 no. 37	Lead and lead alloy double dovetails	—	—
V.11.5 no. 42	Lead double dovetail tenon	—	—
V.11.5 no. 46	Double dovetail tenon	—	—
V.11.5 no. 47	—	—	Lead fill
V.11.5 no. 48	Double dovetail tenon	—	—
V.11.5 no. 52	Lead alloy fill	Lead fill	—
V.11.5 no. 53	Lead and lead alloy double dovetail tenons	—	—
V.11.5 no. 54	—	—	Lead fill
V.11.5 no. 56	—	—	Lead alloy double dovetail tenon; lead fill
V.11.5 no. 58	Lead alloy fill	—	—
V.11.5 no. 59	—	—	Lead fill
V.11.5 no. 61	—	—	Lead fill

TABLE A1.16. (*continued*)

Vessel	Production phase	Use-life	Unknown stage
V.11.5 no. 62	Lead double dovetail tenon	—	—
V.11.5 no. 64	Lead double dovetail	—	—
V.11.5 no. 65	—	—	Lead fill
V.11.5 no. 67	Lead alloy fill	—	—
V.11.5 no. 68	Lead double dovetail	—	—
V.11.5 BF 1	—	—	Lead fill
V.11.5 BF 2	Lead double dovetail	—	—
V.11.5 BF 3	Lead double dovetail	—	—
V.11.5 BF 4	—	—	Lead fill

Notes: Metal noted when preserved; lead alloy considered a *production* repair material; lead grouped in unknown stage except when dolium also has lead alloy filler repairs.

TABLE A1.17. Dolium repairs in Rome.

Vessel	Production phase	Use-life	Unknown stage
RMN no. 1	—	Lead fill	—
RMN no. 2	—	Lead fill	—
RMN no. 4	—	—	Lead (alloy) dovetail and double dovetail tenon[a]
RMN no. 5	—	Lead fill	—
RMN no. 6	Lead (alloy) double dovetail tenon	—	—
RMN no. 8	Lead (alloy) double dovetail tenon	—	—
RMN no. 10	Lead (alloy) double dovetail tenon	—	—
RMN no. 1	—	—	Lead (alloy) dovetail and double dovetail tenons[a]
AAR no. 1	Lead double dovetail	—	—

[a] Probably production phase.

TABLE A1.18. Comparison of dolium repair types.

Repair technique	Phase	Material(s)	On pottery	Sites
Fill	Production and/or use	Lead, lead alloy	Rare	Ostia, Pompeii
Staple	Production and/or use	Lead	Common	Pompeii
Clamp	Production and/or use	Lead	Common	Pompeii, Cosa
Hybrid mortise-and-tenon staple	Use	Lead	Rare	Pompeii
Hybrid mortise-and-tenon clamp	Use	Lead	Rare	Cosa
Hybrid mortise-and-tenon double dovetail	Use	Lead	Rare	Cosa, Pompeii
Double dovetail	Production (ideally)	Lead, lead alloy	Rare	Ostia, Pompeii, Cosa
Double dovetail tenon	Production (ideally)	Lead, lead alloy	None	Ostia, Pompeii

APPENDIX 2

Descriptions of Select Dolia

Cosa

Cosa no. 1

Description: Rim fragment from large dolium; three iron screws were drilled into the upper surface of the rim, one of which was connected to a thin strip of lead shallowly engraved into the vessel surface.

ID number: CD 707, 550. Cosa Catalog Card: Incorrectly identified as rim of a mortar. Found in square VIII D Bas. ENW 2 Level II.

Date: Last quarter of the second century BCE to first quarter of the first century BCE (reused).

Repairs: Three iron (?) nails drilled into the upper face of the rim; dimensions of nail 1 were 2.2 × 1.5 cm, nail 2 were 2.1 × 1.5 cm, and nail 3 were 2.3 × 1.4 cm. One nail (at the end) was connected to a thin strip of lead placed in a shallow cutting in the vessel surface; metal strip length 1.642 cm, width 0.4–0.48 cm, depth 0.32 cm.

Cosa no. 11

Description: Rim fragment of large dolium featuring two repairs.

ID number: C14.1000; 2014 Western Castellum Sounding 2, SU 10003 (puteal).

Date: First or second century CE (reused for construction).

Repairs: Two different repairs on ends of fragment. One is the preserved cutting for half a double dovetail with small traces of lead; unclear whether this was made during the production phase or use-life. The other features the preserved half of a drill hole through rim, which was probably executed during use-life to form modifications like Cosa no. 1.

Cosa no. 19

Description: Joining fragments of large dolium rim, shoulder, and middle wall; repairs on dolium shoulder.

ID number: 213445.

Date: First century CE.

Repairs: Three hybrid clamp and mortise-and-tenon repairs of lead arranged in horizontal alignment on upper shoulder of vessel where crack formed between two coils. Settings for the clamps on the exterior wall were carved into the vessel wall, whereas the crossbars of the clamps on the interior wall of the vessel lay on the surface of the vessel. Two of the clamps were similar in execution, featuring a stylized border for the clamp surface on the exterior wall. The third clamp is smaller, with somewhat asymmetrical form and rougher edges and no stylized border. The three clamps were made during the vessel's use-life; the form and execution indicate that two of the three clamps were made at the same time and probably by the same person, whereas one clamp was probably made at a different time and by a different person.

Stamp: A small (4.7 × 1.8 cm) stamp, *C.TVRI*, preserved on rim.

Incisions: A symbol (8.4 × 8.2 cm, 0.2 cm deep), possibly of an anchor or stylized phallus, incised onto shoulder of vessel.

Cosa no. 29

Description: Base fragment of medium or small dolium preserving 16 percent of the base and a portion of the lower wall; repaired.

ID number: C 70 V D SH St. S. 5, M L 0.I K Ware.

Date: First half of the first century BCE.

Repairs: Lead hybrid double dovetail tenon and clamp repair on base and lower wall, almost certainly made during use-life of vessel. On exterior surface, base had double dovetail connected to a tenon that extended from the base onto the lower wall of the well. The base was drilled through to insert a pin that connected the double dovetail on the underside of the base to a clamp on the interior surface of the base. Double dovetail and clamp made of lead; traces of lead visible on tenon.

Cosa TC

Description: Fragment of terracotta object (dolium or architectural terracotta), repaired.

ID number: CB 1176.

Date: First century BCE.

Repairs: Large clamp made of very dark material (possibly iron with lead), added during use-life. One side of clamp flat, other highly irregular. The clamp was uneven and not neatly executed, with large space left between pins. Cracks radiate from the drill holes.

Pompeii

I.21.2

This is the house connected to the Garden of the Fugitives. The room connected to the garden had an area with a treading vat and three dolia for wine production.

I.21.2 no. 1

Description: Nearly complete and intact dolium, missing entire rim and almost all of rim core. The dolium features two sets of repairs made during the production phase; one set consisted of at least three double dovetails on the shoulder and the second set is a vertical double dovetail tenon on the shoulder.

Repairs: On one area of vessel shoulder are (1) three lead double dovetails; on another area of vessel shoulder is part of a (2) vertical lead double dovetail tenon with three double dovetails (lead mostly missing); both sets of repairs were made during the production phase. The repairs were all neatly and consistently executed and were probably cut (less than 1 cm deep) into the vessel surface when the dolium was leather-hard. The (1) double dovetails are ca. 8–9 cm long × ca. 3 cm wide, with a width at the middle of ca. 1.5 cm. The (2) double dovetail tenon probably extended from the rim (now missing) to the middle wall; 35 cm of its length and three double dovetails (ca. 9–10 cm × 3 cm) are preserved.

I.21.2 no. 2

Description: Large intact dolium, complete except for small chip off rim lip. Dolium features two sets of repairs on vertical dunting cracks.
Repairs: Two vertical sets of lead (alloy?) repairs opposite each other, extending from upper face of rim to upper vessel wall. One set consists of a hybrid mortise-and-tenon staple on rim face, with two clamps followed by a triangular clamp on upper wall. The second set has a hybrid mortise-and-tenon staple on rim face, a clamp on juncture between rim and shoulder, and two clamps on upper wall. All repairs, except for the clamp on the juncture between rim and shoulder, were probably made at the same time, perhaps in the workshop during the production phase. The clamps (ca. 9–10 cm × 1 cm) on the upper wall might have been drilled when the vessel was leather-hard, or after firing. The hybrid mortise-and-tenon staples on the rim were an attempt to repair dunting cracks that had formed during firing and were probably formed in the workshop. They are the same shape and filled with the same material; the minor differences between their dimensions are probably the result of difficulty in chiseling into an already-fired ceramic surface (9.4 × 2.0 cm; 11.7 × 1.7 cm). The lead clamp (13 × 1 cm) on the juncture between the rim and shoulder was likely added later, during use-life.

I.22

The House of Stabianus was excavated several times during the twentieth century. Wilhelmina Jashemski's excavations revealed a large olive tree orchard. Excavations in the 1980s uncovered several dolia. Because many of the dolia are below the layer of lapilli where several Pompeians died during the eruption, the dolia remain buried to preserve the plaster casts of the bodies in situ. At the time of the eruption, the House of Stabianus was undergoing renovations, and the dolia were being reused to hold construction materials.

I.22 no. 1

Description: Large dolium buried in lapilli, only one small area with a repair visible.
Repairs: Single, thin vertical lead fill on dolium upper wall.

I.22 no. 3

Description: Large dolium, partly buried in lapilli; the base and most of the vessel wall are visible, but rim and shoulder are under lapilli. The dolium was repaired in several areas.

Repairs: Numerous production phase repairs on vessel; metal is lead alloy. Two sets of vertical double dovetail tenons visible on vessel middle and upper walls (likely extended from rim to middle wall); the double dovetail tenons repaired vertical dunting cracks. Single double dovetail on vessel shoulder to repair crack between coils; horizontal set of double dovetails on middle wall, with additional clay smeared between double dovetails on emerging crack.

I.22 no. 5

Description: Large dolium, partly buried in lapilli; most of the vessel is visible, but part of the upper wall is still buried in lapilli. The dolium was probably repaired.

Repairs: Faint remains of cutting for double dovetail preserved on upper wall.

I.22 no. 7

Description: Medium-sized dolium, repaired. A large horizontal crack formed on the vessel's middle wall and was repaired.

Repairs: Large hybrid repair on vessel's middle wall made during the vessel's use-life either at one time or in two stages. The repairer drilled three or four sets of holes on either side of the break and chiseled cuttings for double dovetail tenon. The repairer added lead to the crack and drill holes and added a bar. The repairer then added a dark metal substance (lead alloy?) to fill the cutting for the double dovetail tenon. This use-life repair was probably an attempt to mimic production phase double dovetail tenons.

I.22 no. 9

Description: Medium dolium, rim broken (ca. 45 percent preserved).

I.22 no. 18

Description: Body fragment of dolium repaired with clamps.

Repairs: Preserved on the dolium wall are (1) half a drill hole and (2) a drill hole filled with a lead pin, crossbar not preserved. The holes are preserved halves of two clamps. The repairer drilled through the vessel wall to form clamps during the dolium's use-life. Cracks formed around the (2) drill hole, and the dolium broke halfway through the (1) drill hole, suggesting that the repairs were not effective or further damaged the vessel.

II.5.5

This large property was a vineyard with a wine press, cella vinaria, and large masonry triclinium. The wine press room was connected to the cella vinaria, and at least four lead pipes directed the must from the pressed grapes into some of the dolia. The cella vinaria was a long, narrow room with five dolia installed on either side of a path through the room. The dolia are all buried up to the shoulder or rim.

II.5.5 no. 10

Description: Complete, intact large dolium, repaired.
Repairs: The rim was heavily repaired with lead, almost all made during production. Three single double dovetails and three double dovetail tenons were neatly formed on the rim to repair dunting cracks. A large area of the rim surface (ca. 5 × 20 cm) had been abraded or damaged, and was filled with lead.

II.8.6. Garden of Hercules

The Garden of Hercules featured a commercial flower garden, likely for a perfume production facility, and was installed with several dolia to collect and store rainwater for irrigation.

II.8.6 no. 1

Description: Large dolium with many repairs; the dolium was placed on a supportive base mostly made of stone, with one inverted fragment of a dolium rim.
Repairs: Numerous repairs were executed on different parts of the vessel, mostly made during the production phase of the vessel. Two lead alloy double dovetails on upper surface of rim, placed on opposite sides of each other, to repair dunting cracks that formed on rim. Also placed on opposite sides of each other were two vertical sets of hybrid clamp mortise-and-tenon repairs on dolium body, almost certainly made during the production phase. A small set of horizontal lead alloy clamps were placed on the middle dolium wall and another set near the base. Repairers regularized the emerging vertical cracks and drilled holes on either side; after firing, they added lead alloy to the regularized crack and formed clamps. In one area, repairers chiseled a double dovetail (after firing) to add to the hybrid clamp repair.

II.8.6 no. 2

Description: Medium dolium, cracked, with repairs; the dolium was partly buried in the corner of the garden to collect rainwater.
Repairs: The dolium shoulder and belly were mended during the production phase by two horizontal lead double dovetail tenons, connected by a short vertical double dovetail tenon.

VI.14.27. House of Memmius Auctus

This was a small, narrow house belonging to M. Memmius Auctus (*CIL* 10.8058, 50), who is believed to have been a *vinarius* (wine dealer or vintner). In the back of the house was a small room where four dolia were buried, now only three visible.

VI.14.27 frag. 1

Description: Body fragment with repair.
Repairs: Single drill hole through wall of body fragment, made during the production phase before the vessel was fired in the kiln. The drill hole was for a clamp.

VI.14.36. Caupona of Salvius

This street-corner caupona featured a room on the street with a masonry counter, in which two cylindrical jars were installed, and a cylindrical jar buried in the ground by the counter and a dolium buried in the ground by one of the entrances.

VI.14.36 no. 4

Description: Complete, well-preserved medium dolium in northwest corner. General surface abrasion on rim surface, some gouges on exterior wall of belly, three dunting cracks forming on rim into middle wall. Rim stamped during production phase.
Stamps: 5.6 × 2.0, letter size 0.7 cm. Two registers C.NAEVI/VITALIS
Surrounded by oval seal 1.8 × 1.3 cm
CIL 10.8047, 11b

VII.6.15

This multiroom bar featured a room on the street with a masonry counter. The room currently has one dolium and one cylindrical jar.

VII.6.15 no. 1 Villa of the Mysteries

The Villa of the Mysteries had a *pars rustica* that included a wine press room with one dolium defossum (no. 1) and a cella vinaria with three dolia defossa (nos. 2–4).

Villa of the Mysteries no. 3

Description: Large dolium defossum outside wine press room, features many repairs.
Repairs: Numerous repairs on the rim, shoulder, and upper wall of the vessel, extending into the middle wall (and probably farther below what is visible). On the rim, four dovetail-type repairs were preserved, a fifth broken. Three of the double dovetails on the rim were neatly executed during production. Two of the double dovetail repairs were hybrid double dovetail staples or clamps made after firing. All five dovetails were part of a vertically aligned set of double dovetails to correct dunting cracks, and they usually had two or three double dovetails on the upper wall or shoulder; one was above a double dovetail tenon. One long horizontal double dovetail tenon, with at least six double dovetails, was on the shoulder for a horizontal crack; materials were mostly lead with maybe a small portion of stronger metal. Very shallow 0.5–0.6 cm deep and consistent in form and size.

Ostia

I.4.5. Caseggiato dei Doli

The storeroom, located across from the museum, was built and installed with thirty-five dolia during the first quarter of the second century CE (Figure 6.7D). The dolia were probably used to store wine but fell out of use by the end of the second or beginning of the third century CE. The dolia are numbered starting from the right-hand side (the northeastern most dolium) in columns.

I.4.5 no. 1

Description: Large, intact dolium with a repair.
Repairs: One production phase lead double dovetail on the upper face of the rim to repair dunting crack. Length 12.5 cm, width of ends 3.4 cm and 3.2 cm, width in middle 1.7 cm.
Inscriptions: Two inscriptions visible, one on rim, one on upper shoulder: XLII ⊃III.

I.4.5 no. 12

Description: Large, intact dolium with repaired dunting cracks.
Repairs: One vertical double dovetail tenon from rim to shoulder (or beyond) of vessel, unevenly preserved. On the rim upper surface was the impression of a double dovetail, ca. 1.0 cm deep, metal not preserved. Under the rim lip was a vertical tenon, mostly lead, terminating in a double dovetail; ca. 1.2 cm deep and 1.5 cm wide. Repair made during the production phase.
Inscriptions: One inscription on upper shoulder: XLII⊃III.

I.4.5 no. 16

Description: Large, intact dolium.
Repairs: None apparent.
Inscriptions: One inscription on upper shoulder: XLS.

I.4.5 no. 17

Description: Large intact dolium, dunting cracks repaired. Stamp on dolium rim.
Repairs: Two lead alloy double dovetail tenons on rim and shoulder. One double dovetail tenon reinforced with a lead alloy strip. One end of the strip latched onto the underside of the rim lip and the other end of the strip latched onto the inner surface of the rim. On the exterior upper wall near one double dovetail tenon were two short vertical lead alloy fills.
Inscriptions: Inscription XLII that appears on upper shoulder and on rim.
Stamp: GENIALIS RASINI/PONTICI SER FE.

I.4.5 no. 18

Description: Large dolium, broken, missing rim and most of shoulder, repaired.
Repairs: A shallow lead alloy fill (14 × 1.5 cm) on exterior wall, diagonally oriented.

I.4.5 no. 28

Description: Repaired dolium, smaller than other dolia found at Ostia, with a different fabric (redder, fewer inclusions). Small part of the rim (ca. 25 percent) missing.
Repairs: Horizontal lead double dovetail tenon across entire shoulder, connected to vertical double dovetail tenons, placed on opposite sides of the vessel, that extended the repair from the shoulder of the vessel to just under the rim. On rim at break, above one vertical double dovetail, is preserved half of a lead double dovetail.
Inscriptions: One on upper shoulder at eleven o'clock: XXIXSC/⊃II.

III.14.3. Magazzino dei Doli

This storeroom was originally connected to the House of Annius, a residence with multiple working areas and a shop (Figure 6.7C). The storeroom and dolia were installed during the first quarter of the second century CE; the dolia were likely not fully buried when originally installed (the ground level has been raised significantly in later periods). On the facade of the house were several terracotta plaques. Three spelled out OMNIA FELICIA ANNI. Two terracotta plaques featured reliefs: one of a man depicted between dolia and the other depicting a boat with dolia. If the owner of this house is the same Annius as Annius Serapiodus, an attested oil-lamp producer of Ostia, it is likely that the dolia of the storeroom contained oil.[1] The dolia are numbered from left to right, starting with the top row.

III.14.3 no. 1

Description: Large, intact dolium, repaired.
Repairs: Lead double dovetail executed on rim surface in workshop during production phase, in anticipation of dunting crack.
Stamp: Two stamps placed next to each other on rim upper surface: (1) PYRAMI ENCOLPI/AVG DISP ARCARCI, (2) AMPLIATVS VIC F.

III.14.3 no. 6

Description: Large, intact dolium, stamped and repaired.
Repairs: One shallow lead alloy fill on interior surface of rim that extends to upper shoulder to repair vertical dunting crack that formed during firing.
Stamp: One very faint stamp with text, text illegible now.

V.11.5. Magazzino Annonario

Large, trapezoidal-shaped storeroom across the decumanus from the theater (Figure 6.7B). This storeroom was excavated hastily in 1939, with only one or two pages of notes recorded in the *Giornale degli Scavi*. The excavations

1. Ceci 2001, 2003; Gianfrotta 2008; Meiggs 1973, 275. See ostia-antica.org for images.

recovered a dolium lid with a stamp (*CIL* 1063?), dating the dolia and warehouse to the first quarter of the second century CE. There were approximately one hundred dolia installed in this storeroom. Every dolium, except one (no. 67), was broken and is still partly buried, with only part of the middle wall visible today. The storehouse is overgrown, with up to one-third of the dolia inaccessible for study. The dolia are numbered in rows from the left to right in two groups, starting with the left half of the property.

V.11.5 no. 8

Description: Large dolium, broken, with repairs.
Repairs: Fills on both interior and exterior walls. Two dark lead alloy vertical fills next to each other on interior wall, likely added in the workshop during production phase, postfiring. One vertical lead fill on exterior wall, might have been added during use-life or production.

V.11.5 no. 52

Description: Large dolium, broken, repaired.
Repairs: Lead alloy fill and double dovetail tenon on interior surface. The double dovetail tenon is unusual because it is on the inner wall, which had an additional layer of clay. These are likely production phase repairs.

V.11.5 no. 61

Description: Large dolium, broken, repaired.
Repairs: One vertical lead fill on interior surface. Fragment of lead fill (possibly part of double dovetail tenon) on exterior wall, continues below topsoil.

V.11.5 no. 67

Description: Large complete dolium with major cracks and several repairs.
Repairs: Four dark lead alloy fills and a lead alloy "plug" on exterior surface. Most repairs are in one area, likely made in the workshop during the production phase, postfiring.
Inscriptions: Two sets of identical incisions on shoulder of vessel, nearly 180 degrees from each other: XXXVIII ⊃III.

Rome

All dolia in Rome are without provenance but likely found in the city or immediate hinterland.

Capitoline Museum

Dolium without inventory number: Large complete and intact dolium, stamped.
Two stamps on the rim: (1) FELIX, (2) CORNELIVS.
Dolium inv. 3905/A: Large complete and intact dolium, stamped.

Stamp on rim: Illegible.
Dolium inv. 3895/A: Large complete and intact dolium.

National Museum of Rome, Diocletian's Baths

Several dolia, without provenance, located in the gardens of Diocletian's Baths as part of the National Museum of Rome.

NMR Dolium no. 4

Description: Large dolium, rim missing. It was extensively repaired with lead double dovetail tenons extending from the base of the vessel onto different areas of the vessel walls. The metal elements are likely lead. The repairs are not as neatly formed as other production phase repairs, yet they expand across much larger areas of the dolium than most use-life repairs. The stage during which the vessel was repaired is inconclusive, but the metal elements feature traces of a tool that seems to have been rolled along the edges to tap down and regularize the metal elements.

NMR Dolium no. 8

Description: Large, complete dolium with production phase lead double dovetail tenons on lower wall and near base. The metal elements are made of lead.

NMR Dolium no. 10

Description: Large, mostly complete dolium that had been extensively repaired with production phase double dovetails and double dovetail tenons, using lead alloy. The metal elements were decorated and embellished using a tool to create a border of small squares that enclosed rosette-shaped decorations.

REFERENCES

Adam, J.P. 2005. *Roman Building: Materials and Techniques*. London: Routledge.

Aguilera Martín, A. 2002. *El monte Testaccio y la llanura subaventina*. Madrid: Consejo Superior de Investigaciones Científicas.

Albert, K. 2012. "Ceramic Rivet Repair: History, Technology, and Conservation Approaches." *Studies in Conservation* 57, Supplement 1: S1–S8.

Alcock, S.E., J.F. Cherry, and J.L. Davis. 1994. "Intensive Survey, Agricultural Practice, and the Classical Landscape of Greece." In *Classical Greece: Ancient Histories and Modern Archaeologies*, edited by I. Morris, 135–168. Cambridge: Cambridge University Press.

Amarger, M.P. 2009. "'Le meilleur et le pire serviteur de l'humanité': Fer, forges et forgerons à Pompéi." In *Artisanats antiques d'Italie et de Gaule: Mélanges offerts à Maria Francesca Buonaiuto*, edited by J.-P. Brun, 135–168. Naples: Publications du Centre Jean Bérard.

Amarger, M.P., and J.-P. Brun. 2007. "La forge dell'insula I, 6, 1 de Pompéi." *Quaderni di studi pompeiani* 1: 147–168.

Amelung, W. 1927. "Notes on Representations of Socrates and of Diogenes and Other Cynics." *American Journal of Archaeology* 31.3: 281–296.

Angelicoussis, E. 2009. "The Funerary Relief of a Vintner from Ince Blundell Hall." *Bonner Jahrbücher* 209: 95–107.

Annecchino, M. 1982. "Suppellettile fittile per uso agricolo in Pompei e nell'agro vesuviano." In *La regione sotterrata dal Vesuvio: Studi e prospettive: Atti del Convegno Internazionale, 11–15 novembre 1979*, 753–767. Naples: Università degli Studi di Napoli.

Aoyagi, M., A. De Simone, and G.F. De Simone. 2018. "The 'Villa of Augustus' at Somma Vesuviana." In *The Roman Villa in the Mediterranean Basin: Late Republic to Late Antiquity*, edited by A. Marzano and G. Métraux, 141–156. Cambridge: Cambridge University Press.

Appadurai, A., ed. 1986. *The Social Life of Things: Commodities in Cultural Perspective*. Cambridge: Cambridge University Press.

Arangio-Ruiz, V., and G.P. Carratelli. 1954. *Tabulae herculanenses*. Naples: Macchiaroli.

Arce, J., and B. Goffaux, eds. 2011. *Horrea d'Hispanie et de la Méditerranée romaine*. Madrid: Casa de Velázquez.

Artenova. n.d. "Terracotta and Wine." https://jars.terracotta-artenova.com/production/.

Attema, P., and M. van Leusen. 2004. "Intra-regional and Inter-regional Comparison of Occupation Histories in Three Italian Regions: The RPC Project." In *Side-by-Side Survey: Comparative Regional Studies in the Mediterranean World*, edited by S.E. Alcock and J.F. Cherry, 86–100. Havertown, PA: Oxbow Books.

Aubert, J.J. 1994. *Business Managers in Ancient Rome: A Social and Economic Study of Institores, 200 B.C.–A.D. 250*. Columbia Studies in the Classical Tradition 21. New York: E.J. Brill.

Bace, E.J. 1984. "Cosa: Inscriptions on Stone and Brick-Stamps." PhD dissertation, University of Michigan.

Bain, R. 1937. "Technology and State Government." *American Sociological Review* 2.6: 860–874.

Bakels, C., and S. Jacomet. 2003. "Access to Luxury Foods in Central Europe during the Roman Period: The Archaeobotanical Evidence." *World Archaeology* 34.3: 542–557.

Bakker, J.T., ed. 1999. *The Mills-Bakeries of Ostia: Description and Interpretation*. Amsterdam: J.C. Gieben.

Bang, P.F. 2007. "Trade and Empire—In Search of Organizing Concepts for the Roman Economy." *Past and Present* 195.1: 3–54.

Baratta, G. 1994a. "Bolli su botti." In *Epigrafia della produzione e della distribuzione*, 555–565. Collection de l'Ecole française de Rome 193. Rome: Università di Roma La Sapienza, École Française.

Baratta, G. 1994b. "Circa Alpes ligneis vasis condunt circulisque cingunt." *Archeologia Classica* 46: 233–260.

Baratta, G. 1997. "Le botti: Dati e questioni." In *Techniques et economie antiques et médiévales: Le temps de l'innovation*, edited by D. Meeks and D. Garcia, 109–112. Paris: Errance.

Baratta, G. 2001. "Un'alternativa all'anfora: La botte." In *Actas del I simposio de la Asociación Internacional de Historia y Civilización de la Vid y del Vino*, edited by J. Maldonado Rosso, 1:149–155. El Puerto de Santa Maria: Ayuntamiento de El Puerto de Santa María.

Baratta, G. 2004. "La diffusione delle botti: Un dato negativo: Il rifornimento dell'esercito durante il tardo impero e l'assenza di botti." In *L'armée romaine de Dioclétien à Valentinien Ier*, edited by Y. Le Bohec and C. Wolff, 489–492. Paris: de Boccard.

Baratta, G. 2005. "*Ponte itaque cupis facto* . . . (Max. 22, 4): I ponti nella *Historia Augusta*: Il caso particolare di Aquileia." In *Atti "Historiae Augustae" colloquium Barcinonense*, edited by G. Bonamente and M. Meyer, 47–66. Bari: Edipuglia.

Baratta, G. 2006. "Misurare per mestiere." In *Misurare il tempo, misurare lo spazio*, edited by M. Angeli Bertineli and A. Donati, 233–260. Epigrafia e antichità 25. Faenza: Fratelli Lega.

Baratta, G. 2017. "Commercio e identità culturale: Il caso delle *cupae*." In *Insularity, Identity, and Epigraphy in the Roman World*, edited by J. Velaza, 93–107. Newcastle upon Tyne: Cambridge Scholars.

Baratta, G. 2020. "Ipotesi su una serie di piombi monetiformi con la raffigurazione di una botte." In *Ex Baetica Romam: Homenaje a José Remesal Rodríguez*, edited by V. Revilla Calvo, A. Aguilera Martín, L. Pons Pujol, and M. García Sánchez, 563–576. Barcelona: Universitat de Barcelona.

Barisashvili, G. 2011. *Making Wine in Qvevri—A Unique Georgian Tradition*. Tbilisi: Biological Farming Association, Elkana.

Barret, J.C., and P. Halstead, eds. 2004. *The Emergence of Civilisation Revisited*. Havertown, PA: Oxbow Books.

Barrett, A.A. 2002. *Livia: First Lady of Imperial Rome*. New Haven, CT: Yale University Press.

Bassani, M., and F. Romana Berno. 2019. "The Porticus Liviae in Ovid's *Fasti* (6.637–648)." In *The Cultural History of Augustan Rome: Texts, Monuments, and Topography*, edited by M.P. Loar, S.C. Murray, and S. Rebeggiani, 103–125. Cambridge: Cambridge University Press.

Becker, C. 1986. "Note sur un lot d'amphores regionales du 1er siècle ap. J.-C. à Lyon (Fouille de L'îlot 24)." *Figlina* 7: 147–163.

Beltrán, M. 1990. *Guía de la cerámica romana*. Zaragoza: Libros Pórtico.

Beltrán de Heredia Bercero, J., ed. 2001a. *De* Barcino *a* Barcinona *(siglos I–VII): Los restos arqueológicos de la plaza del Rey del Barcelona*. Barcelona: Museu d'Història de la Ciutat, Institut de Cultura, Ajuntament de Barcelona.

Beltrán de Heredia Bercero, J. 2001b. "Una factoría de *garum* e salazón de pescado en *Barcino*." In *De* Barcino *a* Barcinona *(siglos I–VII): Los restos arqueológicos de la plaza del Rey del Barcelona*, edited by J. Beltrán de Heredia Bercero, 58–65. Barcelona: Museu d'Història de la Ciutat, Institut de Cultura, Ajuntament de Barcelona.

Beltrán de Heredia Bercero, J. 2001c. "Uva y vino a través de los restos arqueológicos: La producción de vino en *Barcino*." In *De* Barcino *a* Barcinona *(siglos I–VII): Los restos arqueológicos de la plaza del Rey del Barcelona*, edited by J. Beltrán de Heredia Bercero, 66–73. Barcelona: Museu d'Història de la Ciutat, Institut de Cultura, Ajuntament de Barcelona.

Bene, Z., M. Kállay, B.O. Horváth, and D. Nyitrai-Sárdy. 2019. "Comparison of Selected Phenolic Components of White Qvevri Wines." *Mitteilungen Klosterneuburg, Rebe und Wein, Obstbau und Früchteverwertung* 69.2: 76–82.

Bennett, J. 2009. *Vibrant Matter: A Political Ecology of Things*. Durham, NC: Duke University Press.

Benôit, F. 1961. *Lépave du Grand Congloué a Marseille: Gallia*. Suppl. 14. Paris: Centre National de la Recherche Scientifique.

Benton, J. 2020. *The Bread Makers: The Social and Professional Lives of Bakers in the Western Roman Empire*. London: Palgrave Macmillan.

Bentz, M., and U. Kästner, eds. 2006. *Konservieren oder restaurieren: Die Restaurierung griechischer Vasen von der Antike bis heute*. Beihefte zum Corpus vasorum antiquorum 3. Munich: Beck.

Beresford, J. 2013. *The Ancient Sailing Season*. Boston: Brill.

Berg, R. 2019. "Dress, Identity, Cultural Memory: Copa and Ancilla Cauponae in Context." In *Gender, Memory, and Identity in the Roman World*, edited by J. Rantala, 203–237. Amsterdam: Amsterdam University Press.

Bergamini, M., ed. 2007. *Scoppieto 1: Il territorio e i materiali*. Florence: All'Insegna del Giglio.

Bernal-Casasola, D. 2015. "What Contents Do We Characterise in Roman Amphorae? Methodological and Archaeological Thoughts on a 'Trending Topic.'" In *Archaeoanalytics: Chromatography and DNA Analysis in Archaeology*, edited by C. Oliveira, R.M. Lopes de Sousa Morais, and Á. Morillo Cerdán, 61–83. Esposende: Município de Esposende.

Bernal Casasola, D., P. Berni Millet, and I. Navarro Luengo. 2021. "Quintus Valerius Arator en Baetica: A propósito de un singular dolium itálico de Estepona." *Boletín Ex Officina Hispana* 12: 51–56.

Bernard, S., J. McConnell, F. Di Rita, F. Michelangeli, D. Magri, L. Sadori, A. Masi, et al. 2023. "An Environmental and Climate History of the Roman Expansion in Italy." *Journal of Interdisciplinary History* 54.1: 1–41.

Bevan, A. 2014. "Mediterranean Containerization." *Current Anthropology* 55.4: 387–418.

Bevan, A. 2018. "Pandora's Pithos." *History and Anthropology* 29.1: 7–14.

Bevan, A. 2020. "A Stored-Products Revolution in the 1st Millennium BC." *Archaeology International* 22.1: 127–144.

Bianchi, E. 2016. "L'industria del laterizio e i bolli doliari a Roma in età imperial." In *Made in Roma: Marchi di produzione e di possesso nella società antica*, edited by L. Ungaro, M. Milella, and S. Pastore, 28–30. Rome: Gangemi Editore spa.

Bilde, P.G., and S. Handberg. 2012. "Ancient Repairs on Pottery from Olbia Pontica." *American Journal of Archaeology* 116.3: 461–481.

Bintliff, J. 1997. "Regional Survey, Demography, and the Rise of Complex Societies in the Ancient Aegean: Core-Periphery, Neo-Malthusian, and Other Interpretive Models." *Journal of Field Archaeology* 24: 1–38.

Blitzer, H. 1990. "ΚΟΡΩ ΝΕΪΚΑ: Storage Jar Production and Trade in the Traditional Aegean." *Hesperia* 59.4: 675–711.

Bloch, H. 1947. *I bolli laterizi e la storia edilizia romana: Contributi all'archeologia e alla storia di Roma*. Rome: Comune di Roma, Ripartizione antichità e belle arti.

Bloch, H. 1948. *The Roman Brick Stamps Not Published in Volume XV 1 of the "Corpus Inscriptionum Latinarum."* Cambridge, MA: Harvard Studies in Classical Philology.

Bloch, H. 1961. "A New Edition of the Marble Plan of Ancient Rome." *Journal of Roman Studies* 51: 143–152.

Bodel, J.P. 1983. *Roman Brick Stamps in the Kelsey Museum*. Ann Arbor: University of Michigan Press.

Boetto G., E. Bukowiecki, N. Monteix, and C. Rousse. 2016. "Les grandi horrea d'Ostie." In *Entrepôts et trafics annonaires en Méditerranée: Antiquité-temps modernes*, edited by B. Marin and C. Virlouvet, 177–226. Rome: École française de Rome.

Bond, S.E. 2016. *Trade and Taboo: Disreputable Professions in the Roman Mediterranean*. Ann Arbor: University of Michigan Press.

Bonifay, M. 2004. *Etudes sur la céramique romaine tardive d'Afrique*. Oxford: BAR.

Borowski, O. 1998. *Every Living Thing: Daily Use of Animals in Ancient Israel*. Walnut Creek, CA: AltaMira.

Bos, J.E.M.F. 2000. "Jar Stoppers and Seals." In *Berenike 1998: Report of the 1998 Excavations at Berenike and the Survey of the Eastern Desert, Including Excavations in Wadi Kalalat*, edited by S.E. Sidebotham and W.Z. Wendrich, 275–303. Leiden: Universiteit Leiden.

Bos, J.E.M.F. 2007. "Jar Stoppers, Seals, and Lids, 1999 Season." In *Berenike 1999/2000: Report of the Excavations at Berenike, Including Excavations in Wadi Kalalat and Siket, and the Survey of the Mons Smaragdus Region*, edited by S.E. Sidebotham and W.Z. Wendrich, 258–269. Los Angeles: Cotsen Institute of Archaeology.

Bos, J.E.M.F., and C. Helms. 2000. "Jar Stoppers and Seals." In *Report of the 1998 Excavations at Berenike and the Survey of the Egyptian Eastern Desert, Including the Excavations in Wadi Kalalat*, edited by S.E. Sidebotham and W.Z. Wendrich, 275–303. Leiden: Universiteit Leiden.

Bossard, S. 2019. "Évolution du stockage agricole dans le moitié septentrionale de la France à l'âge du fer." In *Rural Granaries in Northern Gaul (6th Century BCE–4th Century CE): From Archaeology to Economic History*, edited by S. Martin, 51–72. Radboud Studies in Humanities, Vol. 8. Leiden: Brill.

Bowes, K., ed. 2021a. *The Roman Peasant Project 2009–2014: Excavating the Roman Rural Poor*. Philadelphia: University of Pennsylvania Press.

Bowes, K. 2021b. "Tracking Consumption at Pompeii: The Graffiti Lists." *Journal of Roman Archaeology* 34.2: 552–584.

Bowes, K. 2021c. "When Kuznets Went to Rome: Roman Economic Well-Being and the Reframing of Roman History." *Capitalism: A Journal of History and Economics* 2.1: 7–40.

Bowes, K., A.M. Mercuri, E. Rattigheri, R. Rinaldi, A. Arnoldus-Huyzendveld, M. Ghisleni, C. Grey, M. MacKinnon, and E. Vaccaro. 2017. "Peasant Agricultural Strategies in Southern Tuscany: Convertible Agriculture and the Importance of Pasture." In *The Economic Integration of Roman Italy: Rural Communities in a Globalising World*, edited by T.C.A. de Haas and G. Tol, 170–199. Leiden: Brill.

Bowman, A., and A. Wilson, eds. 2013. *The Roman Agricultural Economy: Organization, Investment, and Production*. Oxford: Oxford University Press.

Boyce, G.K. 1937. "Corpus of the Lararia of Pompeii." *Memoirs of the American Academy in Rome* 14: 5–112.

Braconi, P., and D. Lanzi. 2020. "Produzioni ceramiche a Pompei tra III e II sec. a.C.: Le fornaci delle insulae VI 5 e VII 15." In *Fecisti cretaria: Dal frammento al contesto: Studi sul vasellame ceramico del territorio vesuviano*, edited by M. Osanna and L. Toniolo, 13–22. Studi e Ricerche del Parco Archeologico di Pompei 40. Rome: L'erma di Bretschneider.

Bradley, K. 1987. "On the Roman Slave Supply and Slave Breeding." *Slavery and Abolition* 8.1: 42–64.

Bragatini, I., M. de Vos, and F.P. Badoni. 1981. *Pitture e pavimenti di Pompei: Regioni I, II, III*. Rome: Ministero per i beni culturali e ambientali, Istituto centrale per il catalogo e la documentazione.

Braito, S. 2020. *L'imprenditoria al femminile nell'Italia romana: Le produttrici di opus doliare*. Rome: Scienze e Lettere.

Brenni, G.M.R. 1985. "The Dolia and the Sea-borne Commerce of Imperial Rome." MA thesis, Texas A&M University.

Breton, E. 1870. *Pompéi décrite et dessinée suivie d'une notice sur Herculaneum*. Paris: Gide et J. Baudry.

Brizio, E. 1868. "Descrizione dei nuovi scavi. Domus D. Caprasii Primi (Reg. VII. Ins. II. n. 48)." *Giornale degli Scavi di Pompei* 1.4: 8–93.

Broekaert, W. 2011. "Partners in Business: Roman Merchants and the Potential Advantages of Being a *Collegiatus*." *Ancient Society* 41: 221–256.

Broekaert, W. 2012. "Vertical Integration in the Roman Economy: A Response to Morris Silver." *Ancient Society* 42: 109–125.

Broekaert, W. 2014. "Vertical Integration Again. Another Reply to Morris Silver." *Ancient Society* 44: 343–346.

Broekaert, W. 2019. "Wine and Other Beverages." In *The Routledge Handbook of Diet and Nutrition in the Roman World*, edited by P. Erdkamp and C. Holleran, 140–149. London: Routledge.

Brown, F.E. 1951. *Cosa I: History and Topography*. Memoirs of the American Academy in Rome, Vol. 20. Ann Arbor: University of Michigan Press.

Brown, F.E. 1980. *Cosa: The Makings of a Roman Town*. Ann Arbor: University of Michigan Press.

Brown, F.E. 1984. "The Northwest Gate of Cosa and Its Environs (1972–1976)." In *Studi di antichità in onore di Guglielmo Maetzke*, edited by M.G. Marzi Costagli and L. Tamagno Perna, 493–498. Rome: Giorgio Bretschneider Editore.

Brown, F.E., E. Richardson, and L. Richardson. 1993. *Cosa III: The Buildings of the Forum: Colony, Municipium, and Village*. Memoirs of the American Academy in Rome 37. University Park: Pennsylvania State University Press.

Brown, N.G. 2018. "The Country under the City: The Symbolic Topography of the Rustic Past in Late Republican and Early Imperial Rome." PhD dissertation, Princeton University.

Brughmans, T., and A. Pecci. 2020. "An Inconvenient Truth: Evaluating the Impact of Amphora Reuse through Computational Simulation Modelling." In *Recycling and Reuse in the Roman Economy*, edited by C.N. Duckworth and A.I. Wilson, 191–234. Oxford: Oxford University Press.

Brun, J.-P. 1993. "La discrimination entre las installations oléicoles et vinicoles." *Bulletin de Correspondance Hellenique*, Supplement 6: 511–537.

Brun, J.-P. 1997. "La production du vin et de l'huile en Lusitanie romaine." *Conimbriga* 36: 45–72.

Brun, J.-P. 2003. *Le vin et l'huile dans la Méditerranée antique: Viticulture, oléiculture et procédés de transformation*. Paris: Errance.

Brun, J.-P. 2004. *Archéologie du vin et de l'huile dans l'Empire romain*. Collection des Hespérides. Paris: Errance.

Brun, J.-P., M. Cullin-Mingaud, I. Figueiral, N. Monteix, M. Pernot, B. Chiaretti, and V. Monaco. 2005. "Pompéi, Herculanum (Naples) et Saepinum (Molise): Recherches sur l'artisanat antique." *Mélanges de l'École française de Rome—Antiquité* 117.5: 317–339.

Brun, J.-P., and F. Laubenheimer, eds. 2001. *La viticulture en Gaule*. Gallia 58. Paris: CNRS.

Brun, J.-P., and A. Tchernia. 2020. "Conclusions: *Dolia* bien visibles et tonneaux invisibles!" In *Nouvelles recherches sur les dolia: L'exemple de la Méditerranée nord-occidentale à l'époque romaine (Ier s. av. J.-C.–IIIe s. ap. J.-C.)*, edited by C. Carrato and F. Cibecchini, 275–285. Montpellier: Editions de l'Association de la Revue archéologique de Narbonnaise.

Bruno, D. 2010. "Un mercato sulla casa del pontefice massimo." In *Le case del potere nell'antica Roma*, edited by A. Carandini, 138–143. Rome: GLF editori Laterza.

Bruno, D. 2012. "Regione X, Palatium." In *Atlante di Roma antica*, illustrated ed., edited by A. Carandini, 215–280. Milan: Mondadori Electa.

Bruno, V.J., and R.T. Scott. 1993. *Cosa IV: The Houses*. Memoirs of the American Academy in Rome. Ann Arbor: University of Michigan Press.

Bruun, C., ed. 2005. *Interpretare i bolli laterizi di Roma e della Valle del Tevere: Produzione, storia economica e topografia: Atti del convegno all'École française de Rome e all'Institutum Romanum Finlandiae, 31 marzo e 1 aprile 2000*. Acta Instituti Romani Finlandiae 32. Rome: Institutum Romanum Finlandiae.

Buffat, L. 2018. "Villas in South and Southwestern Gaul." In *The Roman Villa in the Mediterranean Basin: Late Republic to Late Antiquity*, edited by A. Marzano and G.P.R. Métraux, 220–234. Cambridge: Cambridge University Press.

Bukowiecki, E., N. Monteix, and C. Rousse. 2008. "Ostia Antica: Entrepôts d'Ostie et de Portus: Les Grandi Horrea d'Ostie." *Mélanges de l'Ecole française de Rome—Antiquité* 120.1: 211–216.

Bustamante Álvarez, M., and T. Cordero Ruiz. 2013. "Une exportation viticole à Mérida? Considération sur la production locale d'amphores de style Haltern 70." In *Patrimonio cultural de la vid y el vino 2: Comunicaciones aceptadas*, edited by S. Celestino Pérez and J. Blánquez Pérez, 81–93. Madrid: Universidad Autónoma de Madrid.

Caillaud, C. 2020. "Pour une meilleure compréhension des vinifications antiques en *dolia*: Approches expérimentales et ethnographiques." In *Nouvelles recherches sur les dolia: L'exemple de la Méditerranée nord-occidentale à l'époque romaine (Ier s. av. J.-C.–IIIe s. ap. J.-C.)*, edited by C. Carrato and F. Cibecchini, 141–156. Montpellier: Editions de l'Association de la Revue archéologique de Narbonnaise.

Capelli, C., and F. Cibecchini. 2020. "La contribution des analyses pétrographiques à l'étude des épaves à *dolia*: Les *dolia* de l'épave *Ouest Giraglia 2* et de quelques autres épaves et sites terrestres italiens et français." In *Nouvelles recherches sur les dolia: L'exemple de la Méditerranée nord-occidentale à l'époque romaine (Ier s. av. J.-C.–IIIe s. ap. J.-C.)*, edited by C. Carrato and F. Cibecchini, 227–240. Montpellier: Editions de l'Association de la Revue archéologique de Narbonnaise.

Carafa, P., and P. Pacchiarotti. 2012. "Regione XVI. Transtiberim." In *Atlante di Roma antica: Biografia e ritratti della città, 1. Testi e immagini*, illustrated ed., edited by A. Carandini, 549–582. Milan: Mondadori Electa.

Carandini, A., ed. 1985. *Setttefinestre: una villa schiavistica nell'Etruria romana*. Modena: Panini.

Carandini, A. 2014. *La Roma di Augusto in 100 Monumenti*. Turin: UTET.

Carandini, A., F. Cambi, M. Celuzza, E. Fentress, and I. Attolini, eds. 2002. *Paesaggi d'Etruria: Valle dell'Albegna, Valle d'Oro, Valle del Chiarone, Valle del Tafone: Progetto di ricerca italo-britannico seguito allo scavo di Settefinestre*. Rome: Edizioni di storia e letteratura.

Carandini, A., M.T. D'Alessio, and H. di Giuseppe, eds. 2006. *La fattoria e la villa dell'Auditorium nel quartiere Flaminio di Roma*. Rome: L'Erma di Bretschneider.

Carcopino, J. 1909. "Ostiensia—I. Glanures épigraphiques." *Mélanges d'archéologie et d'histoire* 29: 341–364.

Carey, R. 2017, July. "Benefits of Egg-Shaped Wine Tanks: Winery Trail Compares Half-Ton Bins and Plastic Eggs." *Wines and Vines*. https://winesvinesanalytics.com/features/article/182255/Benefits-of-Egg-Shaped-Wine-Tanks.

Carpentieri, A., and C. Cheung. 2023. "Chemical Analyses of Dolia at Ostia." Unpublished manuscript.

Carrato, C. 2012. "Le four 3 de l'atelier de potier de Saint-Bézard et ses productions (Aspiran, Hérault): Contribution à la connaissance de l'artisanat potier en Gaule Narbonnaise à la fin de l'époque augustéenne." *Revue archéologique de Narbonnaise Année* 45: 39–73.

Carrato, C. 2013. "Les *dolia* dans la péninsule ibérique à l'époque romaine: État de la question." In *Atti del congreso Internacional sobre Estudios Cerámicos: Homenaje a Mercedes Vegas, Cadiz, Ier–5 novembre 2010, Cadiz*, edited by L. Girón Anguiozar, M. Lazarich González, and M. da Conceiçao Lopes, 1173–1200. Cádiz: Universidad de Cádiz, Servicio de Publicaciones.

Carrato, C. 2017. *Le* dolium *en Gaule Narbonnaise (Ier a.C.–IIIe S. p.C): Contribution à l'histoire socio-économique de la Méditerranée nord-occidentale*. Mémoires 46. Bordeaux: Ausonius Edition.

Carrato, C. 2020. "Les typologies régionales du *dolium* en Méditerranée nord-occidentale à l'époque romaine (fin du Ier s. av. J.-C.–IIIe s. ap. J.-C.)." In *Nouvelles recherches sur les dolia: L'exemple de la Méditerranée nord-occidentale à l'époque romaine (Ier s. av. J.-C.–IIIe s. ap. J.-C.)*, edited by C. Carrato and F. Cibecchini, 23–42. Montpellier: Editions de l'Association de la Revue archéologique de Narbonnaise.

Carrato, C. 2022a. "Les couvercles de Marseille." Paper read at "Bateaux et entrepôts à dolia à l'époque romaine: Problèmes de chronologie et d'évolution du transport maritime en vrac," March 28, Paris.

Carrato, C. 2022b. "Les entrepôts portuaires à *dolia*." Paper read at "Bateaux et entrepôts à dolia à l'époque romaine: Problèmes de chronologie et d'évolution du transport maritime en vrac," March 28, Paris.

Carrato, C., and F. Cibecchini, eds. 2020. *Nouvelles recherches sur les dolia: L'exemple de la Méditerranée nord-occidentale à l'époque romaine (Ier s. av. J.-C.–IIIe s. ap. J.-C.)*. Montpellier: Editions de l'Association de la Revue archéologique de Narbonnaise.

Carrato, C., V. Martínez Ferreras, J.-M. Dautria, and M. Bois. 2019. "The Biggest *Opus Doliare* Production in Narbonese Gaul Revealed by Archaeometry (First to Second Centuries A.D.)." *ArcheoSciences* 43.1: 69–82.

Carre, M.-B. 2020. "Les propriétaires des ateliers de *dolia* de Minturnes." In *Nouvelles recherches sur les dolia: L'exemple de la Méditerranée nord-occidentale à l'époque romaine (Ier s. av. J.-C.–IIIe s. ap. J.-C.)*, edited by C. Carrato and F. Cibecchini, 197–206. Montpellier: Editions de l'Association de la Revue archéologique de Narbonnaise.

Carre, M.-B., and F. Cibecchini. 2020. "Tableau de synthèse des timbres sur *dolia* maritimes." In *Nouvelles recherches sur les dolia: L'exemple de la Méditerranée nord-occidentale à l'époque romaine (Ier s. av. J.-C.–IIIe s.

ap. J.-C.), edited by C. Carrato and F. Cibecchini, 159–162. Montpellier: Editions de l'Association de la Revue archéologique de Narbonnaise.

Carre, M.-B., and R. Roman. 2008. "Hypothèse de restitution d'un navire à dolia." *Archaeonautica* 15: 176–192.

Carreras Monfort, C. 2013. "Evolució de les terrisseries del Baix de Llobregat a partir de les seves marques i els seus derelicts." In *Barcino II: Marques i terrisseries d'àmfores al Baix Llobregat*, edited by J. Guitart and A. López, 323–346. Barcelona: Institut d'Estudis Catalans, Institut Català d'Arqueologia Clàssica.

Carroll, M., ed. 2022a. *The Making of a Roman Imperial Estate: Archaeology in the Vicus at Vagnari, Puglia*. Oxford: Archaeopress.

Carroll, M. 2022b. "Viticulture, Opus Doliare, and the Patrimonium Caesaris at the Roman Imperial Estate at Vagnari (Puglia)." *Journal of Roman Archaeology* 35.1: 221–246.

Carroll, M., G. Montana, L. Randazzo, D. Bara, and B. Stern. 2022. "The *Dolia Defossa* and Viticulture at Vagnari." In *The Making of a Roman Imperial Estate: Archaeology in the Vicus at Vagnari, Puglia*, edited by M. Carroll, 82–91. Oxford: Archaeopress.

Casson, L. 1965. "Harbour and River Boats of Ancient Rome." *Journal of Roman Studies* 55: 31–39.

Casson, L. 1971. *Ships and Seamanship in the Ancient World*. Princeton, NJ: Princeton University Press.

Castagnoli, F. 1980. "Installazioni portuali a Roma." *Memoirs of the American Academy in Rome* 36: 35–42.

Castanyer Masoliver, P., and J. Tremoleda Trilla. 2007. "El paisatge agrari a l'Empordà en temps del romans: L'exemple de la vil·la de la Font del Vilar (Avinyonet de Puigventós)." In *Actes del Congrés: El Paisatge, element vertebrador de la identitat empordanesa*, 1:275–290. Figueres: Institut d'Estudies Empordanesos.

Ceccarelli, L. 2017. "Production and Trade in Central Italy in the Roman Period: The Amphora Workshop of Montelabate in Umbria." *Papers of the British School at Rome* 85: 109–141.

Ceci, M. 2001. "La production des lampes a huile: L'exemple de l'atelier d'Annius Serapiodorus." In *Ostia. Port et porte de la Rome antique*, edited by J.-P. Descoeudres, 192–195. Genève: Ausonius.

Ceci, M. 2003. "L'officina di Annius Serapiodorus ad Ostia." *Rei Cretariae Romanae Fautorum Acta* 38: 73–76.

Celuzza, M.G. 1985. "Un insediamento di contadini: La fattoria di Giardino." In *La romanizzazione del'Etruria: Il territorio di Vulci*, edited by A. Carandini, 106–107. Milan: Electra.

Chankowski, V., X. Lafon, and C. Virlouvet, eds. 2018. *Entrepôts et circuits de distribution en Méditerranée antique. Bulletin de Correspondance Hellenique* Supplement 58. Athens: École française d'Athènes.

Chausson, F. 2005. "Des femmes, des hommes, des briques: Prosopographie sénatoriale et figlinae alimentant le marché urbain." *Archeologia Classica* 56: 225–267.

Chausson, F., and A. Buonopane. 2010. "Una fonte della ricchezza delle Augustae—Le figlinae urbane." In *Augustae: Machtbewusste Frauen am römischen Kaiserhof?*, edited by A. Kolb, 91–110. Berlin: Akademie Verlag.

Cherubini, L., and A. Del Rio. 1997. "Officine ceramiche di età romana del'Etruria settentrionale costiera: Impianti, produzione. Altrezzature." *Rei Cretariae Romanae Fautorum Acta* 35: 133–141.

Cheung, C. 2020. "Managing Food Storage in the Roman Empire." *Quaternary International* 597: 63–75.

Cheung, C. 2021a. "Born Roman between a Beet and a Cabbage." *American Journal of Philology* 142.4: 659–697.

Cheung, C. 2021b. "Precious Pots: Making and Repairing Dolia." In *The Value of Making: Theory and Practice in Ancient Craft Production*, edited by H. Hochscheid and B. Russell, 171–188. Turnhout: Brepols.

Cheung, C., S. Chang, and G. Tibbott. 2022. "Calculating Dolium Capacities and Material Use." *Archaeometry* 64.3: 798–814.

Cheung, C., and G. Tibbott. 2020. "The Dolia of Regio I, Insula 22: Evidence for the Production and Repair of Dolia." In *Fecisti cretaria: Produzione e circolazione ceramica a Pompei: Stato degli studi e prospettive di ricerca, atti del convegno*, edited by M. Osanna and L. Toniolo, 165–175. Studi e Ricerche del Parco Archeologico di Pompei 39. Rome: L'Erma di Bretschneider.

Chiaramonte Treré, C., ed. 1986. *Nuovi contributi sulle fortificazioni pompeiane*. Quaderni di Annali della Facoltà di Lettere e Filosofia dell'Università degli Studi di Milano 6. Milan: Cisalpino-Goliardica.

Chic García, G., and E. García Vargas. 2004. "Alfares y producciones cerámicas en la provincial de Sevilla: Balance y perspectivas." In *Figlinae Baeticae: Talleres alfareros y producciones cerámicas en la Bética Romana (ss. II a.C–VII d.C.)*, edited by D. Bernal Casasola and E. García Vargas, 1:279–348. BAR International Series 1266. Oxford: BAR.

Chioffi, L. 1993. "Cella Civiciana." In *Lexicon Topographicum Urbis Romae I (A–C)*, edited by E.M. Steinby, 256. Rome: Quasar.

Christakis, K.S. 2005. *Cretan Bronze Age Pithoi: Traditions and Trends in the Production and Consumption of Storage Containers in Bronze Age Crete*. Philadelphia: INSTAP Academic Press.

Christakis, K.S. 2008. *The Politics of Storage: Storage and Sociopolitical Complexity in Neopalatial Crete*. Philadelphia: INSTAP Academic Press.

Churchill, J.E. 1983. *The Complete Book of Tanning Skins and Furs*. Harrisburg, PA: Stackpole Books.

Cibecchini, F. 2020. "Pour une nouvelle carte des épaves à *dolia*: *Ouest Giraglia 2, Diano Marina* et le commerce du vin en vrac en Méditerranée occidentale." In *Nouvelles recherches sur les dolia: L'exemple de la Méditerranée nord-occidentale à l'époque romaine (Ier s. av. J.-C.–IIIe s. ap. J.-C.)*, edited by C. Carrato and F. Cibecchini, 163–196. Montpellier: Editions de l'Association de la Revue archéologique de Narbonnaise.

Claridge, A. 1990. "Ancient Techniques of Making Joins in Marble Statuary." In *Marble: Art Historical and Scientific Perspectives on Ancient Sculpture*, edited by M. True and J. Podany, 135–162. Malibu: J. Paul Getty Museum.

Clarke, J.R., and N.K. Muntasser. 2014. *Oplontis: Villa A ("of Poppaea") at Torre Annunziata, Italy*, vol. 1, *The Ancient Setting and Modern Rediscovery*. New York: ACLS Humanities E-Book.

Coarelli, F. 1984. *Roma Sepolta*. Rome: Armando Curcio Editore.

Coarelli, F. 1996. "Il forum vinarium di Ostia." In *"Roman Ostia" Revisited: Archaeological and Historical Papers in Memory of Russell Meiggs*, edited by A. Gallina Zevi and A. Claridge, 105–113. London: British School at Rome in collaboration with the Soprintendenza Archeologica di Ostia.

Coarelli, F. 1999. "Portus vinarius." In *Lexicon Topographicum Urbis Romae IV (P–S)*, edited by E.M. Steinby, 156. Rome: Quasar.

Cockle, H. 1981. "Pottery Production in Roman Egypt: A New Papyrus." *Journal of Roman Studies* 71: 87–97.

Comodi, P., S. Nazzareni, and D. Perugini. 2007. "Dolia e mortaria: Analisi archeometriche." In *Scoppieto 1: Il territorio e i materiali*, edited by M. Bergamini, 187–198. Florence: All'Insegna del Giglio.

Conison, A. 2012. "The Organization of Rome's Wine Trade." PhD dissertation, University of Michigan.

Cooley, A.E., and M.G.L. Cooley. 2014. *Pompeii and Herculaneum: A Sourcebook*. 2nd ed. London: Routledge.

Cooper, F.A. 2008. "Greek Engineering and Construction." In *The Oxford Handbook of Engineering and Technology in the Classical World*, edited by J.P. Oleson, 225–255. Oxford: Oxford University Press.

Corbeill, A. 1994. "Cyclical Metaphors and the Politics of Horace, *Odes* 1.4." *Classical World* 88.2: 91–106.

Corsi-Sciallano, M., and B. Liou. 1985. "Les épaves de Tarraconaise à chargement d'amphores Dressel 2–4." *Archaeonautica* 5: 5–178.

Costin, C. 1991. "Craft Specialization: Issues in Defining, Documenting, and Explaining the Organization of Production." In *Archaeological Method and Theory*, edited by M. Schiffer, 1–56. Tucson: University of Arizona Press.

Coulston, J., and H. Dodge, eds. 2000. *Ancient Rome: The Archaeology of the Ancient City*. Oxford: Oxford University School of Archaeology.

Creese, J.L. 2012. "Social Contexts of Learning and Individual Motor Performance." In *Archaeology and Apprenticeship: Body Knowledge, Identity, and Communities of Practice*, edited by W. Wendrich, 43–60. Tucson: University of Arizona Press.

Csikszentmihalyi, M. 1996. *Creativity: Flow and the Psychology of Discovery and Invention*. New York: HarperCollins.

Curtis, R.I. 2001. *Ancient Food Technology*. Leiden: Brill.

Curtis, R.I. 2015. "Storage and Transport." In *A Companion to Food in the Ancient World*, edited by J. Wilkins and R. Nadeau, 173–182. Chichester: Wiley-Blackwell.

Curtis, R.I. 2016. "Food Storage Technology." In *A Companion to Science, Technology, and Medicine in Ancient Greece and Rome*, edited by G.L. Irby, 587–604. Chichester: Wiley-Blackwell.

D'Altroy, T.N., and T.K. Earle. 1985. "Staple Finance, Wealth Finance, and Storage in the Inka Political Economy." *Current Anthropology* 26.2: 187–206.

Dangreaux, B., A. Desbat, M. Picon, and A. Schmitt. 1992. "La production des amphores à Lyon." In *Les amphores en Gaule: Production et circulation*, edited by F. Laubenheimer, 37–50. Besançon: Université de Besançon; Paris: Diffusé par Les Belles Lettres.

Davis, J.R., ed. 2000. *Corrosion: Understanding the Basics*. Materials Park, OH: ASM International.

Davoli, P. 2005. *Oggetti in argilla dall'area templare di Bakchias (El-Fayyum, Egitto): Catalogo dei rinvenimenti delle campagne di scavo 1996–2002*. Pisa: Giardini.

de Angelis, F. 2011. "Playful Workers. The Cupid Frieze in the Casa dei Vettii." In *Pompeii: Art, Industry and Infrastructure*, edited by E. Poehler, M. Flohr, and K. Cole, 62–73. Oxford: Oxbow Books.

De Caprariis, F. 1999. "I Porti della Città nel IV e V Secolo d.C." In *The Transformations of Urbs Roma in Late Antiquity*, edited by W.V. Harris, 217–234. Portsmouth, RI: Journal of Roman Archaeology.

De Caro, S. 1994. *La villa rustica in località Villa Regina a Boscoreale*. Pubblicazioni scientifiche del Centro di studi della Magna Grecia dell'Università degli Studi di Napoli Federico II. Rome: L'Erma di Bretschneider.

De Giorgi, A.U. 2018. "Sustainable Practices? A Story from Roman Cosa (Central Italy)." *Journal of Mediterranean Archaeology* 31.1: 3–26.

Degrassi, A. 1963. *Inscriptiones Italiae*. Vol. 13, *Fasti et elogia*. Fasc. 2, *Fasti anni numani et iuliani*. Rome: Istituto Poligrafico dello Stato.

De Grummond, N.T. 2020. *Cetamura del Chianti*. Cities and Communities of the Etruscans. Austin: University of Texas Press.

De Juan, C. 2020. "Barcos con *dolia* en las costas de a la *Tarraconensis*." In *Nouvelles recherches sur les dolia: L'exemple de la Méditerranée nord-occidentale à l'époque romaine (Ier s. av. J.-C.–IIIe s. ap. J.-C.)*, edited by C. Carrato and F. Cibecchini, 241–250. Montpellier: Editions de l'Association de la Revue archéologique de Narbonnaise.

De Laet, S.J. 1949. *Portorium: étude sur l'organisation douanière chez les Romains, surtout à l'époque du Haut-Empire*. Vol. 10. Brugge: De Tempel.

DeLaine, J. 1997. *The Baths of Caracalla: A Study in the Design, Construction, and Economics of Large-Scale Building Projects in Imperial Rome*. Portsmouth, RI: Journal of Roman Archaeology Supplement Series.

DeLaine, J. 2004. "Designing for a Market: '*Medianum*' Apartments at Ostia." *Journal of Roman Archaeology* 17: 146–176.

DeLaine, J. 2005. "The Commercial Landscape of Ostia." In *Roman Working Lives and Urban Living*, edited by A. Mac Mahon and J. Price, 29–47. Oxford: Oxbow Books.

DeLaine, J. 2010. "Structural Experimentation: The Lintel Arch, Corbel and Tie in Western Roman Architecture." *World Archaeology* 21.3: 407–424.

DeLaine, J. 2018. "The Construction Industry." In *A Companion to the City of Rome*, edited by C. Holleran and A. Claridge, 473–490. Chichester: Wiley Blackwell.

Della Corte, M. 1923. "Pompei: Scavi eseguiti da privati nel territorio Pompeiano." *Notizie degli scavi di Antichità*: 271–287.

Della Corte, M. 1932. "Somma Vesuviana. Ruderi romani." *Notizie degli scavi di Antichità* 10: 309–310.

Della Corte, M. 1936. "Pompei—nuove scoperte epigrafiche." *Notizie degli Scavi di Antichità* 14: 299–352.

Della Corte, M. 1965. *Case e abitanti di Pompei*. Naples: Faustino Fiorentino.

Dell'Amico, P., and F. Pallarés. 2005. "Il relitto di Diano Marina e le navi a 'dolia': nuove considerazioni." In *De Triremibus, Festschrift in honour of Joseph Muscat, Malta*, edited by T. Cortis and T. Gambin, 67–114. San Gwann (Malta): Publishers Enterprises Group.

Dell'Amico, P., and F. Pallarés. 2011. "Appunti sui relitti a dolia." *Archaeologia maritima mediterranea: International Journal on Underwater Archaeology* 8: 47–135.

Deman, A. 2002. "Avec les utriculaires sur les sentiers muletiers de la Gaule romaine." *Cahiers du Centre Gustave Glotz* 13.1: 233–246.

Denecker, E., and K. Vandorpe. 2007. "Sealed Amphora Stoppers and Tradesmen in Greco-Roman Egypt: Archaeological, Papyrological, and Inscriptional Evidence." *BABESCH* 82: 115–128.

Desbat, A. 1987. "Note sur la production d'amphores à Lyon au debut de l'empire." *Société Française de l'Étude de la Céramique Antique de la Gaule: Actes du Congrès de Caen*: 159–165.

Desbat, A. 2003. "Amphorae from Lyon and the Question of Gaulish Imitations of Amphorae." *Journal of Roman Pottery Studies* 10: 45–49.

De Sena, E.C. 2005. "An Assessment of Wine and Oil Production in Rome's Hinterland: Ceramic, Literary, Art Historical and Modern Evidence." In *Roman Villas around the Urbs: Interaction with Landscape and Environment: Proceedings of a Conference Held at the Swedish Institute in Rome, September 17–18 2004*, edited by B. Santillo Frizell and A. Klynne, 135–149. Rome: Swedish Institute in Rome.

De Sena, E.C., and J.P. Ikäheimo. 2003. "The Supply of Amphora-Borne Commodities and Domestic Pottery in Pompeii 150 BC–AD 79: Preliminary Evidence from the House of the Vestals." *European Journal of Archaeology* 6.3: 301–321.

De Simone, G.F. 2017. "The Agricultural Economy of Pompeii: Surplus and Dependence." In *The Economy of Pompeii*, edited by M. Flohr and A. Wilson, 23–52. Oxford: Oxford University Press.

De Simone, G.F., and B. Russell. 2018. "New Work at Aeclanum (Comune di Mirabella Eclano, Provincia di Avellino, Regione Campania)." *Papers of the British School at Rome* 86: 298–301.

Desmond, W. 2008. *Cynics.* Berkeley: University of California Press.

De Spagnolis, M.C. 2002. *La villa "N. Popidi Narcissi Maioris": In Scafati suburbio orientale di Pompei.* Rome: L'Erma di Bretschneider.

Dey, H. W. 2011. *The Aurelian Wall and the Refashioning of Rome.* Cambridge: Cambridge University Press.

Díaz, C., V. F. Laurie, A.M. Molina, M. Bücking, and R. Fischer. 2013. "Characterization of Selected Organic and Mineral Components of Qvevri Wines." *American Journal of Enology and Viticulture* 64.4: 532–537.

Dicus, K. 2014. "Resurrecting Refuse at Pompeii: The Use-Value of Urban Refuse and Its Implications for Interpreting Archaeological Assemblages." In *TRAC 2013: Proceedings of the Twenty-Third Theoretical Roman Archaeology Conference*, edited by H. Platts, J. Pearce, C. Barron, J. Lundock, and J. Yoo, 56–69. Oxford: Oxbow Books.

Dietler, M. 1997. "The Iron Age in Mediterranean France: Colonial Encounters, Entanglements, and Transformations." *Journal of World Prehistory* 11: 269–357.

Dietler, M. 2005. *Consumption and Colonial Encounters in the Rhône Basin of France: A Study of Early Iron Age Political Economy*. Monographies d'Archéologie Méditerranéenne 21. Lattes: CNRS.

Dietler, M. 2010a. *Archaeologies of Colonialism: Consumption, Entanglement, and Violence in Ancient Mediterranean France.* Berkeley: University of California Press.

Dietler, M. 2010b. "Consumption." In *The Oxford Handbook of Material Culture Studies*, edited by D. Hicks and M.C. Beaudry, 209–228. Oxford: Oxford University Press.

Dietler, M., and I. Herbich. 2011. "Feasts and Labor Mobilization: Dissecting a Fundamental Economic Practice." In *Feasts: Archaeological and Ethnographic Perspectives on Food, Politics, and Power*, edited by M. Dietler and B. Hayden, 240–264. Washington, DC: Smithsonian Institution Press.

Diggory, E. 2018, February 2. "Why Georgian Qvevri Are So Efficient at Making Natural Wine." The Buyer. http://www.the-buyer.net/insight/buyer-road-unique-qvevri-georgia/.

Dinsmoor, W.B. 1922. "Structural Iron in Greek Architecture." *American Journal of Archaeology* 26.2: 148–158.

Dinsmoor, W.B. 1933. "The Temple of Apollo at Bassae." *Metropolitan Museum Studies* 4.2: 204–227.

Djaoui, D. 2020. "Le transport en vrac à l'époque romaine: *Dolia* ou tonneaux?" In *Nouvelles recherches sur les dolia: L'exemple de la Méditerranée nord-occidentale à l'époque romaine (Ier s. av. J.-C.–IIIe s. ap. J.-C.)*, edited by C. Carrato and F. Cibecchini, 261–274. Montpellier: Editions de l'Association de la Revue archéologique de Narbonnaise.

Djaoui, D., and N. Tran. 2014. "Une cruche du port d'Arles et l'usage d'échantillons dans le commerce de vin romain." *Mélanges de l'École française de Rome* 126.2. http://journals.openedition.org/mefra/2549.

Dobres, M.-A. 1999. "Technology's Links and Chaînes: The Processual Unfolding of Technique and Technician." In *The Social Dynamics of Technology: Practice, Politics, and World Views*, edited by M.-A. Dobres and C. Hoffman, 124–146. Washington, DC: Smithsonian Institution Press.

Dobres, M.-A., and C. Hoffman, eds. 1999. *The Social Dynamics of Technology: Practice, Politics, and World Views*. Washington, DC: Smithsonian Institution Press.

Dodd, E. 2020. *Roman and Late Antique Wine Production in the Eastern Mediterranean: A Comparative Archaeological Study at Antiochia ad Cragum (Turkey) and Delos (Greece)*. Oxford: Archaeopress.

Dodd, E. 2022. "The Archaeology of Wine Production in Roman and Pre-Roman Italy." *American Journal of Archaeology* 126.3: 443–480.

Dodd, E. 2023. "Visualising the Glocal through Roman Agricultural Production." Paper read at the workshop "Visualizing the Global, Local, and Glocal in Roman Archaeology," February 3, Royal Netherlands Institute in Rome.

Dodd, E., G. Galli, and R. Frontoni. 2023. "The Spectacle of Production: A Roman Imperial Winery at the Villa of the Quintilii, Rome." *Antiquity* 97.392: 436–453.

Dodd, E., and D. Van Limbergen. Forthcoming. "The 'Place' of Urban Wineries and Oileries in the Greek and Roman World." *Journal of Urban Archaeology*, Special Issue 9: Making Place in the Ancient City.

Dooijes, R., and O.P. Nieuwenhuyse. 2006. "Ancient Repairs: Techniques and Social Meaning." In *Konservieren oder restaurieren: Die Restaurierung griechischer Vasen von der Antike bis heute*, edited by M. Bentz and U. Kästner, 15–20. Beihefte zum Corpus vasorum antiquorum 3. Munich: Beck.

Dubois, P. 1988. *Sowing the Body: Psychoanalysis and Ancient Representations of Women*. Chicago: University of Chicago Press.

Dunbabin, K.M.D. 2008. "Nec grave nec infacetum: The Imagery of Convivial Entertainment." In *Das römische Bankett im Spiegel der Altertumswissenschaften, Internationales Kolloquium 5./6. Oktober 2005, Schloß Mickeln, Düsseldorf*, edited by K. Vössing, 13–26. Stuttgart: Franz Steiner Verlag.

Duncan-Jones, R.P. 1974. *The Economy of the Roman Empire: Quantitative Studies*. Cambridge: Cambridge University Press.

Dyson, S.L. 1976. *Cosa: The Utilitarian Pottery*. Memoirs of the American Academy in Rome. Ann Arbor: University of Michigan Press.

Dyson, S.L. 1978. "Settlement Patterns in the Ager Cosanus: The Wesleyan University Survey, 1974–1976." *Journal of Field Archaeology* 5.3: 251–268.

Earle, T.K. 2002. "Political Economies of Chiefdoms and Agrarian States." In *Bronze Age Economics: The Beginnings of Political Economies*, edited by T.K. Earle, 1–18. Boulder, CO: Westview.

Edlin, H.L. 1973. *Woodland Crafts in Britain: An Account of the Traditional Uses of Trees and Timbers in the British Countryside*. Newton Abbot, UK: David and Charles.

Edmonson, J., T. Nogales Basarrate, and W. Trillmich. 2011. *Imagen y memoria: Monumentos funerarios con retratos en la colonia Augusta Emerita*. Mérida: Real Academia de la Historia, Museo Nacional de Arte Romano.

Ehmig, U., and R. Haensch. 2021. "Die erste bildliche Darstellung eines Römischen Warenetiketts." *Archäologisches Korrespondenzblatt* 51.2: 245–252.

Ekwall, E. 1936. "The Etymology of the Word *Tinker*." *English Studies* 18: 63–67.

Elia, O. 1975. "La scultura pompeiana in tufo." *Cronache Pompeiane* 1: 118–143.

Ellis, S.J. 2004. "The Distribution of Bars at Pompeii: Archaeological, Spatial and Viewshed Analyses." *Journal of Roman Archaeology* 17: 371–384.

Ellis, S.J. 2018. *The Roman Retail Revolution: The Socio-economic World of the* Taberna. Oxford: Oxford University Press.

Ellis, S.J., A.L. Emmerson, A.K. Pavlick, D. Kevin, and G. Tibbott. 2012. "The 2011 Field Season at I. 1.1–10, Pompeii: Preliminary Report on the Excavations." *FastiOnline Documents and Research* 261: 1–26.

Emmerson, A.L.C. 2020. *Life and Death in the Roman Suburb*. Oxford: Oxford University Press.

Erdkamp, P. 1999. "Agriculture, Underemployment, and the Cost of Rural Labour in the Roman World." *Classical Quarterly* 49: 556–572.

Erdkamp, P. 2001. "Beyond the Limits of the 'Consumer City': A Model of the Urban and Rural Economy in the Roman World." *Historia: Zeitschrift für Alte Geschichte* 50.3: 332–356.

Erdkamp, P. 2005. *The Grain Market in the Roman Empire: A Social, Political, and Economic Study*. Cambridge: Cambridge University Press.

Erdkamp, P., ed. 2013. *The Cambridge Companion to Ancient Rome*. Cambridge: Cambridge University Press.

Erdkamp, P. 2015. "Agriculture, Division of Labour, and the Paths to Economic Growth." In *Ownership and Exploitation of Land and Natural Resources in the Roman World*, edited by P. Erdkamp, K. Verboven, and A. Zuiderhoek, 18–39. Oxford: Oxford University Press.

Excoffon, P. 2020. "Le bâtiment à *dolia* de la fouille de l'école des Poiriers à Fréjus, Var: Une installation viticole dans la ville." In *Nouvelles recherches sur les dolia: L'exemple de la Méditerranée nord-occidentale à l'époque romaine (Ier s. av. J.-C.–IIIe s. ap. J.-C.)*, edited by C. Carrato and F. Cibecchini, 125–140. Montpellier: Editions de l'Association de la Revue archéologique de Narbonnaise.

Fant, J.C., B. Russell, and S.J. Barker. 2013. "Marble Use and Reuse at Pompeii and Herculaneum: The Evidence from the Bars." *Papers of the British School at Rome* 81: 181–209.

Fasciato, M. 1947. "Ad quadrigam Fori Vinarii: Autour du port au vin d'Ostie." *Les Mélanges de l École française de Rome—Antiquité* 59: 65–81.

Fatucci, G. 2012. "Region II: *Caelimontium*." In *The Atlas of Ancient Rome*, edited by A. Carandini, 342–358. Princeton, NJ: Princeton University Press.

Favro, D. 2011. "Construction Traffic in Imperial Rome." In *Rome, Ostia, Pompeii: Movement and Space*, edited by R. Laurence and D.J. Newsome, 332–360. Oxford: Oxford University Press.

Feige, M. 2021. "Decorative Features and Social Practices in Spaces for Agricultural Production in Roman Villas." In *Principles of Decoration in the Roman World*, edited by A. Haug and M.T. Lauritsen, 33–52. Berlin: De Gruyter.

Fentress, E., ed. 2003. *Cosa V: An Intermittent Town*. Memoirs of the American Academy in Rome. Ann Arbor: University of Michigan Press.

Fentress, E., C. Goodson, and M. Maiuro, eds. 2017. *Villa Magna: An Imperial Estate and Its Legacies: Excavations 2006–2010*. London: British School of Rome.

Fentress, E., and M. Maiuro. 2011. "Villa Magna near Anagni: The Emperor, His Winery and the Wine of Signia." *Journal of Roman Archaeology* 24: 333–369.

Fentress, E., and P. Perkins. 2016. "Cosa and the *Ager Cosanus*." In *A Companion to Roman Italy*, edited by A.E. Cooley, 378–400. Chichester: Wiley-Blackwell.

Ferdière, A. 2020. "Agriculture in Roman Gaul." In *A Companion to Ancient Agriculture*, edited by D. Hollander and T. Howe, 447–477. Chichester: Wiley-Blackwell.

Fernández-Götz, M., D. Maschek, and N. Roymans. 2020. "The Dark Side of the Empire: Roman Expansionism between Object Agency and Predatory Regime." *Antiquity* 94.378: 1630–1639.

Finley, M. 1965. "Technological Innovation and Economic Progress in the Ancient World." *Economic History Review* 18.1: 29–45.

Finley, M. 1973. *The Ancient Economy*. Berkeley: University of California Press.

Fiorelli, G. 1860. *Pompeianarum antiquitatum historia*. Vol. 1. Naples.

Fiorelli, G. 1862. *Pompeianarum antiquitatum historia*. Vol. 2. Naples.

Fiorelli, G. 1864. *Pompeianarum antiquitatum historia*. Vol. 3. Naples.

Fiorelli, G. 1873. *Gli scavi di Pompei dal 1861 al 1872*. Naples: Tipografia italiana nel liceo V. Emanuele.

Fiorelli, G. 1875. *Descrizione di Pompei*. Naples: Tipografia Italiana.

Fiorelli, G. 1880. "Aprile: XX. Roma." *Notizie degli Scavi di Antichità*: 127–128, 140–141.

Flohr, M. 2013a. "Tanning, Tanners." In *The Encyclopedia of Ancient History*, edited by R.S. Bagnall, K. Brodersen, C.B. Champion, A. Erskine, and S.R. Huebner, 6526–6527. Malden, MA: Blackwell.

Flohr, M. 2013b. *The World of the Fullo: Work, Economy, and Society in Roman Italy*. Oxford Studies on the Roman Economy. Oxford: Oxford University Press.

Flohr, M. 2016. "Constructing Occupational Identities in the Roman World." In *Work, Labour and Professions in the Roman World*, edited by K. Verboven and C. Laes, 147–172. Leiden: Brill.

Floriani Squarciapino, M. 1954. "Forme ostiensi." *Archeologia Classica* 6: 83–99.

Flory, M.B. 1984. "Sic exempla parantur: Livia's Shrine to Concordia and the Porticus Liviae." *Historia: Zeitschrift für Alte Geschichte* 33.3: 309–330.

Forbes, H., and L. Foxhall. 1995. "Ethnoarchaeology and Storage in the Ancient Mediterranean: Beyond Risk and Survival." In *Food in Antiquity*, edited by J. Wilkins, D. Harvey, and M. Dobson, 69–86. Exeter: University of Exeter Press.

Foxhall, L. 1990. "The Dependent Tenant: Land Leasing and Labour in Italy and Greece." *Journal of Roman Studies* 80: 97–114.

Foxhall, L. 1993. "Oil Extraction and Processing Equipment in Classical Greece." In *La production du vin et de l'huile en Méditerranée*, edited by M.-C. Amouretti, J.-P. Brun, and D. Eitam,183–200. Athens: École francaise d'Athènes.

Foxhall, L. 1995. "Bronze to Iron: Agricultural Systems and Political Structures in Late Bronze Age and Early Iron Age Greece." *Annual of the British School at Athens* 90: 239–250

Foxhall, L., and H. Forbes. 1982. "Sitometria: The Role of Grain as a Staple Food in Classical Antiquity." *Chiron* 12: 41–90.

France, J., and A. Hesnard. 1995. "Une statio du quarantième des Gaules et les opérations commerciales dans le port romain de Marseille (place Jules-Verne)." *Journal of Roman Archaeology* 8: 78–93.

Frank, T. 1938. "A New Advertisement at Pompeii." *American Journal of Philology* 59: 224–225.

Frazzoni, L. 2016. "Mortaria." In *Made in Roma: marchi di produzione e di possesso nella società antica*, edited by L. Ungaro, M. Milella, and S. Pastore, 25–27. Rome: Gangemi Editore spa.

Freu, C. 2016. "*Disciplina, Patrocinium, Nomen*: The Benefits of Apprenticeship in the Roman World." In *Urban Craftsmen and Traders in the Roman World*, edited by A.I. Wilson and M. Flohr, 183–199. Oxford: Oxford University Press.

Frier, B.W. 1983. "Roman Law and the Wine Trade: The Problem of 'Vinegar Sold as Wine.'" In *Zeitschrift der Savigny-Stiftung für Rechtsgeschichte: Romanistische Abteilung*, edited by T. Mayer-Maly, D. Nott, A. Laufs, W. Ogris, M. Heckel, P. Mikat, and K.W. Norr, 257–295. Weimar: Hermann Böhlaus Nacht.

Fröhlich, T. 1991. *Lararien und Fassadenbilder in den Vesuvstädten*. Mainz: von Zabern.

Gaitzsch. W. 1980. *Eiserne römische Werkzeuge: Studien zur römischen Werkzeugkunke in Italien und den nördlichen Provinzen des Imperium Romanum*. British Archaeological Reports Vol. 78. Oxford: British Archaeological Reports.

Gallimore, S. 2010. "Amphora Production in the Roman World. A View from the Papyri." *Bulletin of the American Society of Papyrologists* 47: 155–184.

Ganzert, J. 1996. *Der Mars-Ultor-Tempel auf dem Augustusforum in Rom*. Mainz am Rhine: Philipp von Zabern.

Garachon, I. 2010. "Old Repairs of China and Glass." *Rijksmuseum Bulletin* 58.1: 34–55.

Garcia, D. 1987. "Observations sur la production et le commerce des céréales en Languedoc méditerranéen durant l'Age du Fer: Les formes de stockage des grains." *Revue archéologique de Narbonnaise* 20: 43–98.

Garcia, D. 1997. "Les structures de conservation des céréales en Méditerranée nord-occidentale au premier millénaire avant J.-C.: Innovations techniques et rôle économique." In *Techniques et économie antiques et médiévales: Le temps de l'innovation: Colloque d'Aix-en-Provence (21 au 23 mai 1996)*, edited by D. Garcia and D. Meeks, 88–95. Paris: Éditions Errance.

Garcia, D. 2004. *La Celtique méditerranéenne*. Paris: Éditions Errance.

Garnsey, P. 1988. *Famine and Food Supply in the Graeco-Roman World*. Cambridge: Cambridge University Press.

Garnsey, P. 1998. *Cities, Peasants, and Food in Classical Antiquity: Essays in Social and Economic History*. Cambridge: Cambridge University Press.

Garnsey, P. 1999. *Food and Society in Classical Antiquity*. Cambridge: Cambridge University Press.

Garnsey, P., and I. Morris. 1989. "Risk and the Polis: The Evolution of Institutionalized Responses to Food Supply Problems in the Ancient Greek State." In *Bad Year Economics: Cultural Responses to Risk and Uncertainty*, edited by P. Halstead and J. O'Shea, 98–105. Cambridge: Cambridge University Press.

Gasperoni, T. 2003. *Le fornaci dei Domitii: ricerche topografiche a Mugnano in Teverina*. Vol. 5. Viterbo: Università degli studi della Tuscia.

Gatti, G. 1903. "Ostia—Rinvenimento di dolii frumentarii." *Notizie degli Scavi di Antichità* 5: 201–202.

Gatti, G. 1934. "Saepta Iulia e Porticus Aemilia nella forma severiana." *Bulletino Communale* 62: 123–144.

Gazda, E. 1987. "The Port and Fishery: Description of the Extant Remains and Sequence of Construction." In *The Roman Port of Cosa: A Center of Ancient Trade*, edited by A.M. McCann, J. Bourgeois, J.P. Oleson, and E.L. Will, 74–97. Princeton, NJ: Princeton University Press.

Gazda, E., and A.M. McCann. 1987. "Reconstruction and Function: Port, Fishery and Villa." In *The Roman Port of Cosa: A Center of Ancient Trade*, edited by A.M. McCann, J. Bourgeois, J.P. Oleson, and E.L. Will, 137–159. Princeton, NJ: Princeton University Press.

George, R. 2013. *Ninety Percent of Everything: Inside Shipping, the Invisible Industry That Puts Clothes on Your Back, Gas in Your Car, and Food on Your Plate*. New York: Metropolitan Books.

Geraci, G. 2018. "Feeding Rome: The Grain Supply." In *A Companion to the City of Rome*, edited by C. Holleran and A. Claridge, 219–245. Hoboken, NJ: John Wiley and Sons.

Ghisleni, M., E. Vaccaro, K. Bowes, A. Arnoldus, M. MacKinnon, and F. Marani. 2011. "Excavating the Roman Peasant I: Excavations at Pievina (GR)." *Papers of the British School at Rome* 79: 95–145.

Gianfrotta, P.A. 1998. "Nuovi rinvenimenti subacquei per lo studio di alcuni aspetti del commercio marittimo del vino (I sec. a.C.-I sec. d.C)." In *El vi a l'antiguitat economic, producció: i comerç al Mediterrani occidental: Actes*, 105–112. Badalona: Museu de Badalona.

Gianfrotta, P.A. 2008. "Il commercio marittimo in età tardo-repubblicana: Merci, mercanti, infrastrutture." In *Comercio, redistribución y fondeaderos: La navegación a vela en el Mediterráneo*, edited by J. Pérez Ballester and G. Pascual Berlanga, 65–78. Jornadas Internacionales de Arqueologìa Subacuàtica. Valencia: Universidad de Valencia.

Gianfrotta, P.A. 2012. "Da Baia agli *Horrea* del Lucrino: Aggiornamenti." *Archeologia Classica* 63: 277–296.

Gianfrotta, P.A., and A. Hesnard. 1987. "Due Relitti Augustei Carichi di Dolia: Quelli di Ladispoli del Grand Ribaud D." In *El vi a l'antiguitat: Economia, producció i comerç al Mediterrani occidental, I Colloqui d'Arqueologia Romana*, 285–297. Badalona: Museu de Badalona Monogradies Badalonines.

Giannopoulou, M. 2010. *Pithoi: Technology and History of Storage Vessels through the Ages*. Oxford: Archaeopress.

Giardina, A., and A. Schiavone, eds. 1981. *Società romana e produzione schiavistica*. Vol. 2, *Merci, mercati e scambi nel Mediterraneo*. Rome-Bari: Laterza.

Giardino, C. 2012. "Archaeometric Analyses of Metal, Glass, and Plaster." In *The Chora of Metaponto 4: The Late Roman Farmhouse at San Biagio*, edited by E. Lapadula, 194–207. Austin: University of Texas Press.

Gilding, B. 1971. *The Journeymen Coopers of East London: Workers' Control in an Old London Trade, with Historical Documents and Personal Reminiscences by One Who Has Worked at the Block, and an Account of Unofficial Practices Down the Wine Vaults of the London Dock*. Oxford: History Workshop.

Gliozzo, E. 2007. "The Distribution of Bricks and Tiles in the Tiber Valley: The Evidence from Piammiano (Bomarzo, Viterbo)." In *Supplying Rome and the Empire*, edited by E. Papi, 59–72. Journal of Roman Archaeology Supplementary Series 69. Portsmouth, RI: Journal of Roman Archaeology.

Gliozzo, E. 2013. "Stamped Bricks from the *Ager Cosanus* (Orbetello, Grosseto): Integrating Archaeometry, Archaeology, Epigraphy, and Prosopography." *Journal of Archaeological Science* 40: 1042–1058.

Gliozzo, E., F. Iacoviello, and L.M. Foresi. 2014. "Geosources for Ceramic Production: The Clays from the Neogene-Quaternary Albegna Basin (Southern Tuscany)." *Applied Clay Science* 91–92: 105–116.

Godyn, S. 2017, April 13. "What's the Point in Concrete Eggs?" Bibendum Wine. https://www.bibendum-wine.co.uk/news-stories/articles/wine/whats-the-point-in-concrete-eggs/.

González-Vázquez, M. 2019. "Food Storage among the Iberians of the Late Iron Age Northwest Mediterranean (ca. 225–50 B.C.)." *Journal of Mediterranean Archaeology* 32.2: 149–172.

Goodman, P. 2016. "Working Together: Clusters of Artisans in the Roman City." In *Urban Craftsmen and Traders in the Roman World*, edited by A. Wilson and M. Flohr, 301–333. Oxford: Oxford University Press.

Graham, S. 2006. *Ex Figlinis: The Network Dynamics of the Tiber River Valley Brick Industry in the Hinterland of Rome*. BAR International Series 148. Oxford: BAR.

Graham, S. 2009. "The Space Between: The Geography of Social Networks in the Tiber Valley." In *Mercator Placidissimus: The Tiber Valley in Antiquity New Research in the Upper and Middle River Valley*, edited by F. Coarelli and H. Patterson, 671–686. Rome: Quasar.

Gralfs, B. 1988. *Metallverarbeitende Produktionsstätten in Pompeji*. Oxford: BAR.

Greene, K. 2000. "Technological Innovation and Economic Progress in the Ancient World: M.I. Finley Reconsidered." *Economic History Review* 53.1: 29–59.

Gregori, G.L. 1994. "Un nuovo bollo doliare di Q. Tossius Cimber." *Publications de l'École Française de Rome* 193.1: 547–553.

Groen-Vallinga, M.J., and L.E. Tacoma. 2017. "The Value of Labour: Diocletian's Prices Edict." In *Work, Labour, and Professions in the Roman World*, edited by K. Verboven and C. Laes, 104–132. Leiden: Brill.

Güven, N. 1993. "Molecular Aspects of Clay-Water Interactions." In *Clay-Water Interface and Its Rheological Implications*, edited by N. Güven and R.M. Pollastro, 2–79. CMS Workshop Lectures 4. Boulder, CO: Clay Minerals Society.

Halstead, P. 1987. "Traditional and Ancient Rural Economy in Mediterranean Europe: Plus ça change?" *Journal of Hellenic Studies* 107: 77–87.

Halstead, P. 1989. "The Economy Has a Normal Surplus: Economic Stability and Social Change among Early Farming Communities of Thessaly, Greece." In *Bad Year Economics: Cultural Responses to Risk and Uncertainty*, edited by P. Halstead and J. O'Shea, 68–80. Cambridge: Cambridge University Press.

Halstead, P. 2011. "Redistribution in Aegean Palatial Societies. Redistribution in Aegean Palatial Societies: Terminology, Scale, and Significance." *American Journal of Archaeology* 115.2: 229–235.

Halstead, P., and C. Frederick, eds. 2000. *Landscape and Land Use in Postglacial Greece*. Sheffield: Sheffield Academic Press.

Halstead, P., and J. O'Shea. 1982. "A Friend in Need Is a Friend Indeed: Social Storage and the Origins of Social Ranking." In *Ranking, Resource, and Exchange: Aspects of the Archaeology of Early European Society*, edited by C. Renfrew and S. Shennan, 92–99. Cambridge: Cambridge University Press.

Halstead, P., and J. O'Shea. 1989. "Introduction: Cultural Responses to Risk and Uncertainty." In *Bad Year Economics: Cultural Responses to Risk and Uncertainty*, edited by P. Halstead and J. O'Shea, 1–7. Cambridge: Cambridge University Press.

Hamer, F., and J. Hamer. 2004. *The Potter's Dictionary of Materials and Techniques*. Philadelphia: University of Pennsylvania Press.

Harper, D. 1987. *Working Knowledge: Skill and Community in a Small Shop*. Chicago: University of Chicago Press.

Harris, W.V., ed. 1999. *The Transformations of Urbs Roma in Late Antiquity*. Portsmouth, RI: Journal of Roman Archaeology.

Hasaki, E. 2012. "Craft Apprenticeship in Ancient Greece: Reaching Beyond the Masters." In *Archaeology and Apprenticeship: Body Knowledge, Identity, and Communities of Practice*, edited by W. Wendrich, 171–202. Tucson: University of Arizona Press.

Hastorf, C., and L. Foxhall. 2017. "The Social and Political Aspects of Food Surplus." *World Archaeology* 49.1: 26–39.

Hawkins, C. 2016. *Roman Artisans and the Urban Economy*. Cambridge: Cambridge University Press.

Hawkins, C. 2017. "Contracts, Coercion, and the Boundaries of the Roman Artisanal Firm." In *Work, Labour, and Professions in the Roman World*, edited by K. Verboven and C. Laes, 36–61. Leiden: Brill.

Haynes, I., P. Liverani, T. Ravasi, S. Kay, and I. Peverett. 2018. "The Lateran Project: Interim Report for the 2017–18 Season (Rome)." *Papers of the British School at Rome* 86: 320–325.

Healy, J.F. 1978. *Mining and Metallurgy in the Greek and Roman World*. London: Thames and Hudson.

Healy, J.F. 1999. *Pliny the Elder on Science and Technology*. Oxford: Oxford University Press.

Heinzelmann, M. 2010. "Supplier of Rome or Mediterranean Marketplace? The Changing Economic Role of Ostia after the Construction of Portus in the Light of New Archaeological Evidence." *Bollettino di Archeologia Online* 1: 5–10.

Helbig, W. 1966. *Führer durch die öffentlichen Sammlungen klassischer Altertümer in Rom*. 5th ed. Vol. 2. Tübingen: E. Wasmuth.

Helttula, A. 2007. *Le iscrizioni sepolcrali latine nell'Isola sacra*. Rome: Institutum Romanum Finlandiae.

Hemelrijk, E.A. 2020. *Women and Society in the Roman World: A Sourcebook of Inscriptions from the Roman West*. Cambridge: Cambridge University Press.

Hermansen, G. 1978. "The Population of Imperial Rome: The Regionaries." *Historia: Zeitschrift für Alte Geschichte* 27.1: 129–168.

Hermansen, G. 1981. *Ostia: Aspects of Roman City Life*. Edmonton: University of Alberta Press.

Heslin, K. 2011. "Dolia Shipwrecks and the Wine Trade in the Roman Mediterranean." In *Maritime Archaeology and Ancient Trade in the Mediterranean*, edited by D. Robinson and A. Wilson, 157–168. Oxford: Oxford Centre for Maritime Archaeology.

Hesnard, A. 1977. "Note sur un atelier d'amphores Dr. 1 et Dr. 2–4 près de Terracine." *Mélanges de l'École française de Rome* 89.1: 157–168.

Hesnard, A. 1981. "Les Dressel 2–4, amphores à vin de la fin de la République et du début de l'Empire: Un essai de construction typologique." PhD dissertation. Université de Provence.

Hesnard, A. 1994. "Une nouvelle fouille du port de Marseille, place Jules-Verne." *Comptes rendus des séances de l'Académie des Inscriptions et Belles-Lettres* 138.1: 195–217.

Hesnard, A. 1995. "Les ports antiques de Marseille, Place Jules-Verne 1." *Journal of Roman Archaeology* 8: 65–77.

Hesnard, A. 2004. "Vitruve, *De architectura*, V, 12 et le port romain de Marseille." In *Le strutture dei porti e degli approdi antichi*, edited by A. Gallina Zevi and R. Turchetti, 175–203. Soveria Mannelli: Rubbettino.

Hesnard, A., M.B. Carre, M. Rival, B. Dangréaux, M. Thinon, M., Blaustein, A. Chéné, P. Foliot, and H. Bernard-Maugiron. 1988. "*L'épave romaine* Grand Ribaud D (Hyères, Var)." *Archaeonautica* 8.1: 5–180.

Hesnard, A., and P. Gianfrotta. 1989. "Les bouchons d'amphore en Pouzzolane." In *Amphores romaines et histoire économique: Dix ans de recherche: Actes du colloque de Sienne (22–24 mai 1986)*, 393–441. Publications de l'École française de Rome 114. Rome: Ecole française de Rome; Paris: Diffusion de Boccard.

Hickey, T.M. 2012. *Wine, Wealth, and the State in Late Antique Egypt*. Ann Arbor: University of Michigan Press.

Hilgers, W. 1969. *Lateinische Gefässnamen: Bezeichnung, Funktion und Form römischer Gefässe*. Beihefte der BJb 31. Dusseldorf: Rheinland-Verlag.

Hin, S. 2013. *The Demography of Roman Italy: Population Dynamics in an Ancient Conquest Society, 201 BCE–14 CE*. Cambridge: Cambridge University Press.

Hodder, I. 2012. *Entangled: An Archaeology of the Relationships between Humans and Things*. Malden, MA: Wiley-Blackwell.

Holleran, C. 2012. *Shopping in Ancient Rome: The Retail Trade in the Late Republic and the Principate*. Oxford: Oxford University Press.

Holleran, C. 2016. "Getting a Job: Finding Work in the City of Rome." In *Work, Labour, and Professions in the Roman World*, edited by K. Verboven and C. Laes, 87–103. Leiden: Brill.

Holleran, C. 2019. "Market Regulation and Intervention in the Urban Food Supply." In *The Routledge Handbook of Diet and Nutrition in the Roman World*, edited by P. Erdkamp and C. Holleran, 283–295. London: Routledge.

Hong, S., J.P. Candelone, C.C. Patterson, and C.F. Boutron. 1994. "Greenland Ice Evidence of Hemispheric Lead Pollution Two Millennia Ago by Greek and Roman Civilizations." *Science* 265 5180: 1841–1844.

Hopkins, K. 1978. *Conquerors and Slaves*. Cambridge: Cambridge University Press.

Hopkins, K. 1980. "Taxes and Trade in the Roman Empire." *Journal of Roman Studies* 70: 101–125.

Horden, P., and N. Purcell. 2000. *The Corrupting Sea*. Oxford: Wiley-Blackwell.

Howard, M. 1996. "Coopers and Casks in the Whaling Trade 1800–1850." *Mariner's Mirror* 82.4: 436–450.

Howe, T. 2008. *Pastoral Politics: Animals, Agriculture and Society in Ancient Greece*. Claremont, CA: Regine Books.

Hunter-Anderson, R.L. 1977. "A Theoretical Approach to the Study of House Form." In *For Theory Building in Archaeology*, edited by L.R. Binford, 287–315. New York: Academic Press.

Indelicato, M. 2020. "Columella's Wine: A Roman Enology Experiment." *EXARC Journal*. https://exarc.net/ark:/88735/10485.

Ingold, T. 2013. *Making: Anthropology, Archaeology, Art and Architecture*. London: Routledge.

Innocenti, P., and M.C. Leotta. 1996. "Horti Sallustiani." In *Lexicon Topographicum Urbis Romae III (H–O)*, edited by E.M. Steinby, 79–81. Rome: Quasar.

James, P. 2020. *Food Provisions for Rome: A Supply Chain Approach*. London: Routledge.

Jarvis, T. 2019, June 29. "The Complete Guide to Egg Fermentation." Wine Searcher. https://www.wine-searcher.com/m/2019/06/the-complete-guide-to-egg-fermentation.

Jashemski, W. 1967a. "The Caupona of Euxinus at Pompeii." *Archaeology* 20: 36–44.

Jashemski, W. 1967b. "A Pompeian Vinarius." *Classical Journal* 5.2: 193–204.

Jashemski, W. 1968. "Excavations in the 'Foro Boario' at Pompeii: A Preliminary Report." *American Journal of Archaeology* 72: 69–73.

Jashemski, W. 1973a. "The Discovery of a Large Vineyard at Pompeii: University of Maryland Excavations, 1970." *American Journal of Archaeology* 77: 27–41.

Jashemski, W. 1973b. "Large Vineyard Discovered in Ancient Pompeii." *Science* 180: 821–830.

Jashemski, W. 1974. "The Discovery of a Market-Garden Orchard at Pompeii: The Garden of the 'House of the Ship Europa.'" *American Journal of Archaeology* 78: 391–404.

Jashemski, W. 1977. "The Excavations of a Shop-House Garden at Pompeii (I.xx.5)." *American Journal of Archaeology* 81: 217–227.

Jasehmski, W. 1979a. "The Garden of Hercules at Pompeii (II.viii.6): The Discovery of a Commercial Flower Garden." *American Journal of Archaeology* 83: 403–411.

Jashemski, W. 1979b. *The Gardens of Pompeii, Herculaneum and the Villas Destroyed by Vesuvius*. Vol. 1, New Rochelle, NY: Caratzas Brothers.

Jashemski, W. 1993. *The Gardens of Pompeii, Herculaneum and the Villas Destroyed by Vesuvius*. Vol. 2, *Appendices*. New Rochelle, NY: Caratzas Brothers.

Johnson, J. 1933. *Excavations at Minturnae*. Philadelphia: University of Pennsylvania Press.

Joncheray, J.-P. 1994. "L'épave Dramont C." *Cahiers d'Archeologie Subaquatique* 12: 5–51.

Jongman, W.M. 2007. "The Early Roman Empire: Consumption." In *The Cambridge Economic History of the Graeco-Roman World*, edited by W. Scheidel, I. Morris, and R. Saller, 592–618. Cambridge: Cambridge University Press.

Joshel, S.R. 1992. *Work, Identity, and Legal Status at Rome: A Study of the Occupational Inscriptions*. Norman: University of Oklahoma Press.

Joyce, R.A., and S.D. Gillespie. 2015a. "Making Things Out of Objects That Move." In *Things in Motion*, edited by R.A. Joyce and S.D. Gillespie, 3–19. Santa Fe, NM: School for Advanced Research Press.

Joyce, R.A., and S.D. Gillespie, eds. 2015b. *Things in Motion*. Santa Fe, NM: School for Advanced Research Press.

Kaiser, A. 2011. "Cart Traffic Flow in Pompeii and Rome." In *Rome, Ostia, Pompeii: Movement and Space*, edited by R. Laurence and D.J. Newsome, 174–193. Oxford: Oxford University Press.

Kang, D.J. 2015. *Life and Learning of Korean Artists and Craftsmen*. New York: Routledge.

Kaster, R.A. 2006. *Marcus Tullius Cicero: "Speech on Behalf of Publius Sestius."* Clarendon Ancient History. Oxford: Clarendon.

Keay, S.J. 1984. *Late Roman Amphorae in the Western Mediterranean: A Typology and Economic Study, the Catalan Evidence*. Oxford: BAR.

Keay, S. 2013. "The Port System of Imperial Rome." In *Rome, Portus, and the Mediterranean*, edited by S. Keay, 33–67. London: British School at Rome.

Keenan, J.G. 2016. "Cargo Checking at Alexandria and the Late Antique *Annona*: *P.Turner* 45." In *Mélanges Jean Gascou textes et études papyrologiques (P.Gascou)*, edited by J.-L. Fournet and A. Papaconstantinou, 579–589. Paris: Association des Amis du Centre d'Histoire et Civilisation de Byzance.

Keenan, J.G. 2017. "*P.Oxy.* 24.240: A Revised Edition." *Bulletin of American Society of Papyrologists* 54: 237–248.

Kehoe, D.P. 1988. *The Economics of Agriculture on Roman Imperial Estates*. Hypomnemata 89. Göttingen: Vandenhoeck und Ruprecht.

Kehoe, D.P. 1989. "Approaches to Economic Problems in the 'Letters' of Pliny the Younger: The Question of Risk in Agriculture." In *Aufstieg und Niedergang der römischen Welt*, edited by H. Temporini, 33.1: 555–590. Berlin: De Gruyter.

Kehoe, D.P. 2007. "The Early Roman Empire: Production." In *The Cambridge Economic History of the Graeco-Roman World*, edited by W. Scheidel, I. Morris, and R. Saller, 543–569. Cambridge: Cambridge University Press.

Kehoe, D.P. 2013. "The State and Production in the Roman Agrarian Economy." In *The Roman Agricultural Economy: Organization, Investment and Production*, edited by A. Bowman and A.I. Wilson, 33–53. Oxford: Oxford University Press.

Kilby, K. 1971. *The Cooper and His Trade*. London: J. Baker.

Kleberg, T. 1957. *Hôtels, restaurants et cabarets dans l'antiquité romaine: études historiques et philologiques*. Uppsala: Almqvist & Wiksells boktr.

Kloppenborg, J.S. 2006. *The Tenants in the Vineyard: Ideology, Economics, and Agrarian Conflict in Jewish Palestine*. Tübingen: Mohr Siebeck.

Klose, A. 2015. *The Container Principle: How a Box Changes the Way We Think*. Cambridge, MA: MIT Press.

Knappett, C. 2020. *Aegean Bronze Age Art: Meaning in the Making*. Cambridge: Cambridge University Press.

Knappett, C., L. Malafouris, and P. Tomkins. 2010. "Containers as Ceramics." In *The Oxford Handbook of Material Culture Studies*, edited by D. Hicks and M.C. Beaudry, 582–606. Oxford: Oxford University Press.

Kneissl, P. 1981. "Die utricularii: Ihr Rolle im gallo-römischen Transportwesen und Weinhandel." *Bonner Jahrbücher* 181: 169–204.

Komar, P. 2020. *Eastern Wines on Western Tables: Consumption, Trade and Economy in Ancient Italy*. Leiden: Brill.

Komar, P. 2021. "Wine Imports and Economic Growth in Rome between the Late Republic and Early Empire." *Historia* 70.4: 437–462.

Kondoleon, C. 2000. *Antioch: The Lost City*. Princeton, NJ: Princeton University Press in Association with the Worcester Art Museum.

Koob, S. 1998. "Obsolete Fill Materials Found on Ceramics." *Journal of the American Institute for Conservation* 37: 49–67.

Kopytoff, I. 1986. "The Cultural Biography of Things: Commoditization as Process." In *The Social Life of Things: Commodities in Cultural Perspective*, edited by A. Appadurai, 64–91. Cambridge: Cambridge University Press.

Korn, P. 2003. *Woodworking Basics: Mastering the Essentials of Craftsmanship*. Newtown, CT: Taunton.

Korn, P. 2013. *Why We Make Things and Why It Matters: The Education of a Craftsman*. Boston: David R. Godine.

Küster, H. 1994. "The Economic Use of Abies Wood as Timber in Central Europe during Roman Times." *Vegetation History and Archaeobotany* 3: 25–32.

Laes, C. 2015. "Masters and Apprentices." In *A Companion to Ancient Education*, edited by W.M. Bloomer, 475–482. Chichester: John Wiley and Sons.

Lancaster, L.C. 2005. *Concrete Vaulted Construction in Imperial Rome: Innovations in Context*. Cambridge: Cambridge University Press.

Lancaster, L.C. 2012. "A New Vaulting Technique for Early Baths in Sussex: The Anatomy of a Romano-British Invention." *Journal of Roman Archaeology* 25: 419–440.

Lancaster, L.C. 2015. "Armchair Voussoirs and Hollow Voussoirs: An Investigation into the Diffusion of Vaulting Technology for Bath Building in the Western Roman Empire." In *Proceedings of the 5th International Congress on Construction History (Chicago, 3–7 June 2015)*. Raleigh, NC: Lulu.

Lanciani, R.A. 1868. "Ricerche topografiche sulla città di Porto." *Annali dell'Instituto di Corrispondenza Archeologica* 40: 144–195.

Lanciani, R.A. 1880. "Roma." *Notizia degli Scavi di Antichità*: 127–142.

Lanciani, R.A. 1885. "Ostia." *Notizia degli Scavi di Antichità*: 77.

Lanciani, R.A. 1890. *Ancient Rome in the Light of Recent Discoveries*. Boston: Houghton, Mifflin.

Lanciani, R.A. 1897. *The Ruins and Excavations of Ancient Rome*. Boston: Houghton, Mifflin.

Langellotti, M. 2020. *Village Life in Roman Egypt: Tebtunis in the First Century AD*. Oxford: Oxford University Press.

La Rocca, E. 1986. "Il lusso come espressione di potere." In *Le tranquille dimore degli dei: La residenza imperiale degli Horti Lamiani*, edited by M. Cima and E. La Rocca, 3–35. Venice: Catologhi Marsilio.

Laubenheimer, F. 1985. *La production des amphores en Gaule Narbonnaise*. Paris: Les Belles Lettres.

Laubenheimer, F. 2001. "Le vin gaulois de Narbonnaise exporté dans le monde romain." In *Vingt ans de recherches à Sallèles d'Aude*, edited by F. Laubenheimer, 51–65. Besançon: Presses Universitaires Franc-Comtoises.

Laubenheimer, F. 2003. "Amphorae and Vineyards from Burgundy to the Seine." *Journal of Roman Pottery Studies* 10: 32–44.

Laubenheimer, F. 2004. "Inscriptions peintes sur les amphores gauloises." *Gallia* 61: 153–192.

Laubenheimer, F., and J.A. Gisbert Santonja. 2001. "La standardisation des amphores Gauloise 4, des ateliers de Narbonnaise à la production de Denia." In *Vingt ans de recherches à Sallèles d'Aude*, edited by F. Laubenheimer, 33–50. Besançon: Presses Universitaires Franc-Comtoises.

Laurence, R. 1994. *Roman Pompeii: Space and Society*. London: Routledge.

Lawall, M.L. 2011. "Imitative Amphoras in the Greek World." In *Markburger Beiträge zur antiken Handels-, Wirtshafts- und Sozialgeschichte Band 28 2010*, edited by H.-J. Drexhage, T. Mattern, R. Rollinger, K. Ruffing, and C. Schäfer, 45–88. Rahden: Verlag Marie Leidorf.

Lawall, M.L., and J. Lund, eds. 2011. *Pottery in the Archaeological Record: Greece and Beyond: Acts of the International Colloquium Held at the Danish and Canadian Institutes in Athens, June 20–22, 2008*. Aarhus: Aarhus University Press.

Lazzeretti, A. 1998. "Un dolium di M'. Codonius e i dolia prodotti a Minturno rinvenuti a terra." *Bullettino della Commissione Archeologica Comunale di Roma* 99: 338–346.

Lazzeretti, A., and S. Pallecchi. 2005. "Le *figlinae* 'polivalenti': La produzione di *dolia* e di *mortaria* bollati." In *Interpretare I bolli laterizi di Rome r della valle del Tevere: Produzione, storia economica e topografia*, edited by C. Bruun, 213–228. Rome: Institutum Romanum Finlandiae.

Le Gall, J. 1953. *Le Tibre, fleuve de Rome dans l'antiquité*. Paris: Presses Universitaires de France.

Le Guin, U.K. 1989. "The Carrier Bag Theory of Fiction (1986)." In *Dancing at the Edge of the World: Thoughts on Words, Women, Places*, edited by U.K. Le Guin, 165–170. New York: Grove Press.

Leidwanger, J. 2020. *Roman Seas: A Maritime Archaeology of Eastern Mediterranean Economics*. New York: Oxford Univresity Press.

Lequément, R. 1975. "Étiquettes de plomb sur des amphores d'Afrique." *Mélanges de l'École française de Rome—Antiquité* 87: 669–680.

Leveau, P. 2004. "La cité romaine d'Arles et le Rhône: La romanisation d'un espace deltaïque." *American Journal of Archaeology* 108.3: 349–375.

LeVine, T., ed. 1992. *Inka Storage Systems*. Norman: University of Oklahoma Press.

Levinson, M. 2006. *The Box: How the Shipping Container Made the World Smaller and the World Economy Bigger*. Princeton, NJ: Princeton University Press.

Levinson, M. 2020. *Outside the Box: How Globalization Changed from Moving Stuff to Spreading Ideas*. Princeton, NJ: Princeton University Press.

Lewit, T. 2020. "Invention, Tinkering, or Transfer? Innovation in Oil and Wine Presses in the Roman Empire." In *Capital, Investment and Innovation in the Roman World*, edited by P. Erdkamp, K. Verboven, and A. Zuiderhoek, 307–353. Oxford Studies on the Roman Economy. Oxford: Oxford University Press.

Liebeschuetz, J.H.W.G. 2006. *Decline and Change in Late Antiquity: Religion, Barbarians and Their Historiography*. Aldershot: Ashgate/Variorum.

Liou, B. 1987. "Inscriptions peintes sur amphores: Fos (suite), Marseille, Toulon, Port-la-Nautique, Arles, Saint-Blaise, Saint-Martin-de-Crau, Mâcon, Calvi." *Archaeonautica* 7: 55–139.

Liou, B., and R. Marichal. 1979. "Les inscriptions peintes sur amphores de l'anse Saint-Gervais à Fos-sur-Mer." *Archaeonautica* 2: 109–181.

Liu, J. 2009. *Collegia Centonariorum: The Guilds of Textile Dealers in the Roman West*. Columbia Studies in the Classical Tradition. Leiden: Brill.

Liverani, P. 1996. "Horti Domitiae Lucillae." In *Lexicon Topographicum Urbis Romae III (H–O)*, edited by E.M. Steinby, 58–59. Rome: Quasar.

Liverani, P., ed. 1998. *Laterano I: Scavi sotto la Basilica di S. Giovanni in Laterano: I Materiale*. Monumenta Sanctae Sedis 1. Vatican City: Direzione Generale dei Monumenti, Musei e Gallerie Pontificie.

Liverani, P. 2004. "L'area lateranense in età tardoantica e le origini del Patriarchio." *Mélanges de l'École française de Rome—Antiquité* 116.1: 17–49.

Loughton, M.E., and L. Alberghi. 2015. "Body Piercing during the Late Iron Age: The Case of Roman Amphorae from Toulouse (France)." *HEROM* 4.1: 52–105.

Lowe, B. 2009. *Roman Iberia: Economy, Society and Culture*. London: Duckworth.

Lowe, B. 2020. "Agriculture in Roman Iberia." In *A Companion to Ancient Agriculture*, edited by D. Hollander and T. Howe, 479–497. Hoboken, NJ: John Wiley and Sons.

Lugli, G. 1957. *La tecnica edilizia romana, con particolare riguardo a Roma e Lazio*. Rome: G. Bardi.

Mac Mahon, A. 2005. "The *Taberna* Counters of Pompeii and Herculaneum." In *Roman Working Lives and Urban Living*, edited by A. Mac Mahon and J. Price, 70–87. Oxford: Oxbow Books.

Mac Mahon, A. 2006. "Fixed-Point Retail Location in the Major Towns of Roman Britain." *Oxford Journal of Archaeology* 25.3: 289–309.

Mac Mahon, A., and J. Price, eds. 2005. *Roman Working Lives and Urban Living*. Oxford: Oxbow Books.

Magi, F. 1972. *Il calendario dipinto sotto Santa Maria Maggiore*. Rome: Tipografia poliglotta vaticana.

Maiuri, A. 1927. "Pompei: Relazione sui lavori di scavo dal marzo 1924 al marzo 1926." *Notizie degli Scavi di Antichità*, ser. 6, vol. 3: 3–83.

Maiuri, A. 1928. *Pompeii*. Novara: Instituto Geografico de Agostini.

Maiuri, A. 1956. *Pompei ed Ercolano: Fra case ed abitanti*. Milan: Aldo Martello Editore.

Maiuri, A. 1958. *Ercolano: I nuovi scavi (1927–1958)*. Rome: Libreria dello Stato.

Maiuri, A. 1959. "Navalia Pompeiana." *Rendiconti della Accademia di Archeologia Lettere e Belle Arti* 33: 18–22.

Manacorda, D. 1978. "The Ager Cosanus and the Production of the Amphorae of Sestius: New Evidence and a Reassessment." *Journal of Roman Studies* 68: 122–131.

Manacorda, D. 1980. "'L'ager cosanus' tra tarda Repubblica e Impero: forme di produzione e assetto della proprietà." *Memoirs of the American Academy in Rome* 36: 173–184.

Manacorda, D. 1981. "Produzione agricola, produzione ceramica e proprietari nell'-*ager cosanus* nel I a.C." In *Società romana e produzione schiavistica*, vol. 2, *Merci, mercati e scambi nel Mediterraneo*, edited by A. Giardina and A. Schiavone, 3–54. Bari: Laterza.

Manacorda, D. 1993. "Appunti sulla bollatura in età romana." In *The Inscribed Economy: Production and Distribution in the Roman Empire in the Light of Instrumentum Domesticum: The Proceedings of a Conference Held at the American Academy in Rome on 10–11 January 1992*, edited by W.V. Harris, 37–54. Ann Arbor: University of Michigan Press.

Manca, R., L. Pagliantini, E. Pecchioni, A.P. Santo, F. Cambi, L. Chiarantini, A. Corretti, P. Costagliola, A. Orlando, and M. Benvenuti. 2016. "The Island of Elba (Tuscany, Italy) at the Crossroads of Ancient Trade Routes: An Archaeometric Investigation of Dolia Defossa from the Archaeological Site of San Giovanni." *Mineralogy and Petrology* 110.6: 693–711.

Mandowsky, E., and C. Mitchell. 1963. *Pirro Ligorio's Roman Antiquities: The Drawings in MS XIII. B. 7 in the National Library in Naples*. Studies of the Warburg Institute, Vol. 28. London: Warburg Institute.

Manning, W.H., and R.P. Wright. 1961. "Some New Graffiti on a Barrel from Silchester." *Antiquaries Journal* 41: 238.

Marabini Moevs, M.T. 2006. *Cosa: The Italian Sigillata*. Memoirs of the American Academy in Rome. Ann Arbor: University of Michigan Press.

Marlier, S. 2008. "Architecture et espace de navigation des navires à dolia." *Archaeonautica* 15: 153–173.

Marlier, S. 2020. "L'architecture et l'espace de navigation des navires à *dolia*: Une synthèse sur la question." In *Nouvelles recherches sur les dolia: L'exemple de la Méditerranée nord-occidentale à l'époque romaine (Ier s. av. J.-C.–IIIe s. ap. J.-C.)*, edited by C. Carrato and F. Cibecchini, 207–220. Montpellier: Editions de l'Association de la Revue archéologique de Narbonnaise.

Marlier, S., and P. Sibella. 2008. "La Giraglia, a *Dolia* Wreck of the 1st Century BC from Corsica, France: Study of Its Hull Remains." *International Journal of Nautical Archaeology* 31.2: 161–171.

Marlière, E. 2001. "Le tonneau en Gaule romaine." *Gallia* 58: 181–201.

Marlière, E. 2002. *L'outre et le tonneau dans l'Occident romain*. Montagnac: Éditions Monique Mergoil.

Marlière, E., and J. Torres Costa. 2007. "Transport et stockage des denrées dans l'Afrique romaine: Le rôle de l'outre et du tonneau." In *In Africa et in Hispania: Études sur l'huile africaine*, edited by A. Mrabet and J. Remesal Rodríguez, 85–106. Barcelona: Universitat de Barcelona.

Martelli, E. 2013. *Sulle spalle dei* saccarii*: Le rappresentazioni di facchini e il trasporto di derrate nel porto di Ostia in epoca imperiale*. BAR International Series 2467. Oxford: BAR.

Martin, A. 1999. "Amphorae." In *A Roman Villa and a Late Roman Infant Cemetery: Excavation at Poggio Gramignano Lugnano in Teverina*, edited by D. Soren and N. Soren, 329–362. Rome: L'Erma di Bretschneider.

Martin, S., ed. 2019. *Rural Granaries in Northern Gaul (6th Century BCE–4th Century CE): From Archaeology to Economic History*. Radboud Studies in Humanities, Volume 8. Leiden: Brill.

Martínez Ferreras, V., C. Capelli, R. Cabella, and X. Nieto Prieto. 2013. "From Hispania Tarraconensis (NE Spain) to Gallia Narbonensis (S France): New Data on Pascual 1 Amphora Trade in the Augustan Period." *Applied Clay Science* 82: 70–78.

Martin-Kilcher, S. 1994. *Die römischen amphoren aus Augst und Kaiseraugst: Ein Beitrag zur römischen Handels- und Kulturgeschichte II: Die Amphoren für Wein, fischsauce, Südfrüchte (Gruppen 2-24) und Gesamtauswertung*. Forschungen in Augst. Augst: Römermuseum Augst.

Martín Ocaña, S., and I. Cañas Guerrero. 2006. "Comparison of Analytical and On Site Temperature Results on Spanish Traditional Wine Cellars." *Applied Thermal Engineering* 26.7: 700–708.

Marzano, A. 2007. *Roman Villas in Central Italy: A Social and Economic History*. Leiden: Brill.

Marzano, A. 2013a. "Agricultural Production in the Hinterland of Rome: Wine and Olive Oil." In *The Roman Agricultural Economy: Organization, Investment, and Production*, edited by A. Bowman and A. Wilson, 85–106. Oxford: Oxford University Press.

Marzano, A. 2013b. "Capital Investment and Agriculture: Multi-press Facilities from Gaul, Iberian Peninsula and the Black Sea Region." In *The Roman Agricultural Economy: Organization, Investment, and Production*, edited by A. Bowman and A. Wilson, 107–142. Oxford: Oxford University Press.

Marzano, A., and G.P.R. Métraux, eds. 2018. *The Roman Villa in the Mediterranean Basin: Late Republic to Late Antiquity*. Cambridge: Cambridge University Press.

Mattingly, D. 1988. "Oil for Export? A Comparison of Libyan, Spanish and Tunisian Olive Oil Production in the Roman Empire." *Journal of Roman Archaeology* 1: 33–56.

Mattingly, D., and G. Aldrete. 2000. "The Feeding of Imperial Rome: The Mechanics of the Food Supply System." In *Ancient Rome: The Archaeology of the Ancient City*, edited by J. Coulston and H. Dodge, 142–165. Oxford: Oxford University School of Archaeology.

Mau, A. 1889. "Scavi di Pompei 1886–88: Insula IX, 7." *Mitteilungen des Deutschen Archäologischen Instituts: Römische Abteiluing* 4: 3–31.

Mau, A. 1890. "Scavi di Pompei. Insula IX, 7." *Mitteilungen des Deutschen Archäologischen Instituts: Römische Abteiluing* 5: 228–284.

Mauné, S. 2003. "La villa gallo-romaine de Vareilles à Paulhan (Hérault; fouille de l'Autoroute A75) un centre domanial du Haut-Empire spécialisé dans la viticulture?" In *Cultivateurs, éleveurs et artisans dans les campagnes de Gaule romaine, matières premières et produits transformés: Actes du Ve colloque AGER, Compiègne 2002*, edited by S. Lepetz and V. Matterne, 309–337. Amiens: Revue Archéologique de Picardie.

Mauné, S., R. Bourgaut, J. Lescure, C. Carrato, and C. Santran. 2006. "Nouvelles données sur les productions céramiques de l'atelier de Dourbie à Aspiran (Hérault)." In *"Actes du Congrès de Pézenas" organisé par la Société française d'étude de la céramique antique en Gaule du 25–28 mai 2006*, 157–188. Marseille: Société française d'étude de la céramique antique en Gaule.

Mauné, S., and C. Carrato. 2012. "Le complexe domanial et artisanal de Saint-Bézard (Aspiran, Hérault) au début du Ier s. ap. J.-C. Fondation et genèse." *Revue archéologique de Narbonnaise* 45.1: 21–38.

Mauné, S., B. Durand, C. Carrato, and R. Bourgaut. 2010. "La villa de Quintus Iulius Pri(. . .) à Aspiran (Hérault): Un centre domanial de Gaule Narbonnaise (Ier–Ve s. apr. J.-C.)." *Pallas: Revue d'études antiques* 84: 111–143.

Mazarron, F.R., and I. Canas. 2008. "Exponential Sinusoidal Model for Predicting Temperature inside Underground Wine Cellars from a Spanish Region." *Energy and Buildings* 40.10: 1931–1940.

McCallum, M. 2010. "The Supply of Stone to the City of Rome: A Case Study of the Transport of Anician Building Stone and Millstone from the Santa Trinità Quarry (Orvieto)." In *Trade and Exchange: Archaeological Studies from History and Prehistory*, edited by C. White and C. Dillian, 75–94. New York: Springer.

McCann, A.M. 1987. *The Roman Port and Fishery of Cosa: A Center of Ancient Trade*. Princeton, NJ: Princeton University Press.

McCann, A.M. 2002. *The Roman Port and Fishery of Cosa: A Short Guide*. Rome: The American Academy in Rome.

McConnell, J.R., A.I. Wilson, A. Stohl, M.M. Arienzo, N.J. Chellman, S. Eckhardt, E.M. Thompson, A.M. Pollard, and J.P. Steffensen. 2018. "Lead Pollution Recorded in Greenland Ice Indicates European Emissions Tracked Plagues, Wars, and Imperial Expansion during Antiquity." *Proceedings of the National Academy of Sciences* 115.22: 5726–5731.

McCormick, M. 2012. "Movements and Markets in the First Millennium: Information, Containers and Shipwrecks." In *Trade and Markets in Byzantium*, edited by C. Morrisson, 51–98. Washington, DC: Dumbarton Oaks Research Library and Collection.

McGovern, P.E. 2003. *Ancient Wine: The Search for the Origins of Viniculture*. Princeton, NJ: Princeton University Press.

McGovern, P.E., B.P. Luley, N. Rovira, A. Mirzoian, M.P. Callahan, K.E. Smith, G.R. Hall, T. Davidson, and J.M. Henkin. 2013. "Beginning of Viniculture in France." *Proceedings of the National Academy of Sciences* 110.25: 10147–10152.

Meiggs, R. 1973. *Roman Ostia*. Oxford: Clarendon Press.

Métraux, G.P.R. 1998. "Villa Rustica Alimentaria et Annonaria." In *The Roman Villa: Villa Urbana*, edited by A. Frazer, 21–28. Williams Symposium on Classical Architecture. Philadelphia: University Museum, University of Pennsylvania.

Metzler, J. 1991. "Sanctuaires gaulois en territoire trévire." In *Les Sanctuaires celtiques et leurs rapports avec le monde méditerranéen*, edited by J.-L. Brunaux, 28–39. Paris: Errance.

Meyer, F.G. 1980. "Carbonized Food Plants of Pompeii, Herculaneum, and the Villa at Torre Annunziata." *Economic Botany* 34.4: 401–437.

Millar, F. 1984. "Condemnation to Hard Labour in the Roman Empire, from the Julio-Claudians to Constantine." *Papers of the British School at Rome* 52: 124–147.

Miller, H.M.-L. 2009. *Archaeological Approaches to Technology*. Amsterdam: Elsevier/Academic Press.

Miller, H.M.-L. 2012. "Types of Learning in Apprenticeship." In *Archaeology and Apprenticeship: Body Knowledge, Identity, and Communities of Practice*, edited by W. Wendrich, 224–239. Tucson: University of Arizona Press.

Minutoli, D. 2014. "Stampigliature su coperture d'anfora in argilla provenienti da Antinoupolis." *Analecta Papyrologica* 26: 323–358.

Miró, J. 1988. *La producción de ánforas Romanas en Catalunya: Un estudio sobre el comercio del vino de la Tarraconense (Siglos I a.C.–1 d.C.)*. BAR International Series 473. Oxford: BAR.

Mogetta, M. 2021. *The Origins of Concrete Construction in Roman Architecture*. Cambridge: Cambridge University Press.

Montana, G., L. Randazzo, D. Barca, and M. Carroll. 2021. "Archaeometric Analysis of Building Ceramics and '*Dolia Defossa*' from the Roman Imperial Estate of Vagnari (Gravina in Puglia, Italy)." *Journal of Archaeological Science: Reports* 38 (2–3). https://doi.org/10.1016/j.jasrep.2021.103057.

Monteix, N. 2008. "La conservation des denrées dans l'espace domestique à Pompéi et Herculanum." *Mélanges de l'École française de rome* 120.1: 121–138.

Monteix, N. 2010. *Les lieux de métier: Boutiques et ateliers d'Herculanum*. Collection du Centre Jean Bérard 34. Rome: École française de Rome.

Monteix, N. 2016. "Perceptions of Technical Culture among Pompeian Élites, Considering the Cupids Frieze of the Casa dei Vettii." In *Antike Wirtschaft und ihre kulturelle Prägung (2000 v.Chr.–500 n.Chr.)—Antike Wirtschaft und ihre kulturelle Prägung*, edited by K. Droß-Krüpe, S. Föllinger and K. Ruffing, 199–221. Philippika 98. Wiesbaden: Harrassowitz Verlag.

Monteix, N. 2017. "Urban Production and the Pompeian Economy." In *The Economy of Pompeii*, edited by A. Wilson and A. Bowman, 209–242. Oxford: Oxford University Press.

Monteix, N., and E. Rosso. 2008. "L'artisanat du plomb à Pompéi." *Mélanges de l'École française de Rome—Antiquité* 120: 241–247.

Monticone, A. 2017/2018. "Tre Relitti a Dolia nel Golfo di Olbia." PhD dissertation, Università degli Studi di Sassari.

Moore, J. 1995. "A Survey of the Italian Dressel 2–4 Wine Amphorae." MA thesis, McMaster University.

Moore, J. 2011. "When Not Just Any Wine Will Do . . . ? The Proliferation of Coan-Type Wine and Amphoras in the Greco-Roman World." In *Markburger Beiträge zur antiken Handels-, Wirtshafts- und Sozialgeschichte Band 28 2010*, edited by H.-J. Drexhage, T. Mattern, R. Rollinger, K. Ruffing, and C. Schäfer, 89–122. Rahden: Verlag Marie Leidorf.

Morel, J.-P. 1998. "Que buvaient les Carthaginois?" In *El Vi a l'antiguitat. Economia, producció I comerç al Mediterrani occidental: II Col· loqui Internacional d'Arqueologia Romana, actes (Barcelona 6-9 de maig de 1998)*, 29–38. Badalona: Museu de Badalona.

Morel, J.-P. 2004. "Les Amphores importées à Carthage punique." In *La circulatió d'amfores al Mediterrani durant la Protohistòria*, edited by J. Sanmartí-Grego, D. Ugolini, J. Ramón, and D. Asensio, 11–24. Barcelona: Universitat de Barcelona.

Morley, N. 1996. *Metropolis and Hinterland: The City of Rome and the Italian Economy, 200 B.C.–A.D. 200*. Cambridge: Cambridge University Press.

Morley, N. 2007. "The Early Roman Empire: Distribution." In *The Cambridge Economic History of the Graeco-Roman World*, edited by W. Scheidel, I. Morris, and R. Saller, 570–591. Cambridge: Cambridge University Press.

Morley, N. 2013. "Population Size and Social Structure." In *The Cambridge Companion to Ancient Rome*, edited by P. Erdkamp, 29–44. Cambridge: Cambridge University Press.

Mulder, S.F. 2007. "Jar Stoppers, Seals, and Lids, 2000 Season." In *Berenike 1999/2000: Report of the Excavations at Berenike, Including Excavations in Wadi Kalalat and Siket, and the Survey of the Mons Smaragdus Region*, edited by S.E. Sidebotham and W.Z. Wendrich, 270–284. Los Angeles: Cotsen Institute of Archaeology.

Munteanu, C. 2013. "Roman Military Pontoons Sustained on Inflated Animal Skins." *Archäologisches Korrespondenzblatt* 43.4: 545–552.

Murgatroyd, P. 2005. "Erotic Play in Apuleius' Tale of the Tub." *Latomus* 64.1: 121–124.

Murphy, E.A. 2016. "Roman Workers and Their Workplaces: Some Archaeological Thoughts on the Organization of Workshop Labour in Ceramic Production." In *Work, Labour, and Professions in the Roman World*, edited by K. Verboven and C. Laes, 133–146. Leiden: Brill.

Murphy, E.A. 2017. "Rethinking Standardisation through Late Antique Sagalassos Ceramic Production: Tradition, Improvisation and Fluidity." In *Materialising Roman Histories*, edited by A. Van Oyen and M. Pitts, 101–122. Oxford: Oxbow Books.

Muslin, J. 2019. "Between Farm and Table: Oplontis B and the Dynamics of Amphora Packaging, Design, and Reuse on the Bay of Naples." PhD dissertation, University of Texas, Austin.

Nachtergael, G. 2000. "Sceaux et timbres de bois d'Égypte: I. En marge des archives d'Hèroninos: Cachets et bouchons d'amphores de Théadelphie." *Chronique d'Egypte* 75: 150–173.

Nachtergael, G. 2001. "Sceaux et timbres de bois d'Égypte: II. Les sceaux de grand format." *Chronique d'Égypte* 76: 231–257.

Nachtergael, G. 2003. "Sceaux et timbres de bois d'Égypte: III. La Collection Froehner (suite et fin)." *Chronique d'Égypte* 78: 277–293.

Nakagawa, T., J.-L. De Beaulieu, and H. Kitagawa. 2000. "Pollen-Derived History of Timber Exploitation from the Roman Period Onwards in the Romanche Valley, Central French Alps." *Vegetation History and Archaeobotany* 9: 85–89.

Nakassis, D., W.A. Parkinson, and M.L. Galaty. 2011. "Redistribution in Aegean Palatial Societies: Redistributive Economies from a Theoretical and Cross-Cultural Perspective." *American Journal of Archaeology* 115.2: 177–184.

Nappo, S.C. 1988. "Regio I, insula 20." *Rivista dei Studi Pompeiani* 2: 186–192.

Nappo, S.C. 1997. "Urban Transformation at Pompeii in the Late 3rd and Early 2nd c. BC." In *Domestic Space in the Roman World: Pompeii and Beyond*, edited by R. Laurence and A. Wallace-Hadrill, 91–120. Journal of Roman Archaeology Supplementary Series 22. Portsmouth, RI: Journal of Roman Archaeology.

Niccolini, F. and F. Niccolini. 1854. *Le case ed i monumenti di Pompei designati e descritti*. Vol. 1. Naples.

Niccolini, F. and F. Niccolini. 1862. *Le case ed i monumenti di Pompei designati e descritti*. Vol. 2. Naples.

Niccolini, F. and F. Niccolini. 1890. *Le case ed i monumenti di Pompei designati e descritti*. Vol. 3. Naples.

Niccolini, F. and F. Niccolini. 1896. *Le case ed i monumenti di Pompei designati e descritti*. Vol. 4. Naples.

Nicoletta, N. 2007. "Dolia e mortaria: Studio morfologico e ipotesi funzionali." In *Scoppieto 1: Il territorio e i materiali*, edited by M. Bergamini, 153–186. Florence: All'Insegna del Giglio.

Norton, T. 1999. *Stars beneath the Sea: The Pioneers of Diving*. London: Century.

O'Connell, E.R., ed. 2014. "Catalogue of British Museum Objects from the Egypt Exploration Fund's 1913/14 Excavation at Antinoupolis (Antinoë)." In *Scavi e materiali*, vol. 3, *Antinoupolis II*, edited by R. Pintaudi, 467–504. Edizioni dell'Istituto papirologico "G. Vitelli." Florence: Firenze University Press.

Odiot, T. 1996. "Donzère: Le Molard." In *Formes de l'habitat rural en Gaule Narbonnaise*, edited by C. Pellecuer, 3:1–25. Juan-les-Pins: Editions APDCA.

Olcese, G. 2012. *Atlante dei siti di produzione ceramica (Toscana, Lazio, Campania e Sicilia): Con le tabelle dei principali relitti del Mediterraneo occidentale con carichi dall'Italia centro meridionale, IV secolo a.C.–I secolo d.C.* Rome: Quasar.

Olcese, G., and C. Coletti. 2016. *Ceramiche da contesti repubblicani del territorio di Ostia*. Rome: Quasar.

Oleson, J., ed. 2014. *The Oxford Handbook of Engineering and Technology in the Classical World*. Oxford: Oxford University Press.

Olesti Vilà, O. 1995. *El Territori del Maresme en Època Republicana (s. III–I a.C.)*. Mataró: Caixa d'Estalvis Laietana.

Olesti Vilà, O. 1997. "El origen de las villae romanas en Cataluña." *Archivo Español de Arqueología* 70: 71–90.

Olesti Vilà, O. 1998. "Els inicis de la producció vinícola a Catalunya: El paper del món indígena." In *El vi a l'Antiguitat: Economia, producció i comerç al Mediterrani Occidental*, 246–257. Monografies Badalonines 14. Badalona: Museu de Badalona.

Orr, D.G. 1972. "Roman Domestic Religion: A Study of the Roman Household Deities and Their Shrines at Pompeii and Herculaneum." PhD dissertation, University of Maryland.

Orton, C., and M. Hughes. 1993. *Pottery in Archaeology*. Cambridge: Cambridge University Press.

Osborne, R. 2017. "Discussion, Material Standards." In *Materialising Roman Histories*, edited by A. Van Oyen and M. Pitts, 123–130. Oxford: Oxbow Books.

Packer, J.E. 1967. "Housing and Population in Imperial Ostia and Rome." *Journal of Roman Studies* 57.1–2: 80–95.

Packer, J.E. 1971. *The Insulae of Imperial Ostia*. Memoirs of the American Academy 31. Rome: American Academy in Rome.

Packer, J.E. 1978. "Inns at Pompeii: A Short Survey." *Cronache pompeiane* 4: 5–53.

Pagliaro, F., E. Bukowiecki, F. Gugliermetti, and F. Bisegna. 2014. "The Architecture of Warehouses: A Multidisciplinary Study on Thermal Performances of Portus' Roman Store Buildings." *Journal of Cultural Heritage* 16.4: 560–566.

Pagliaro, F., F. Nardecchia, F. Gugliermetti, and F. Bisegna. 2016. "CFD Analysis for the Validation of Archaeological Hypotheses—The Indoor Microclimate of Ancient Storage-Rooms." *Journal of Archaeological Science* 73: 107–119.

Pallarés, F. 1985. "VII Campagna di Scavo sul Relitto del Golfo Dianese (IM)." *Rivista di Studi Liguri* 51: 612–622.

Pallecchi, S. 2005. *I mortaria di produzione centro-italica: Corpus dei bolli*. Rome: Quasar.

Palma, B. 1983. *Museo Nazionale Romano: Le Sculture I.4: I Marmi Ludovisi: Storia della Collezione*. Rome: De Luca Editore.

Palmer, R. 2001. "Bridging the Gap: The Continuity of Greek Agriculture from the Mycenaean to the Historical Period." In *Prehistory and History: Ethnicity, Class, and Political Economy*, edited by. D. W. Tandy, 41–84. Montreal: Black Rose Books.

Palmer, R.E.A. 1974. *Roman Religion and Roman Empire: Five Essays*. Philadelphia: University of Pennsylvania Press.

Palmer, R.E.A. 1978. "Silvanus, Sylvester, and the Chair of St. Peter." *Proceedings of the American Philosophical Society* 122.4: 222–247.

Palmer, R.E.A. 1980. "Customs on Market Goods Imported into the City of Rome." *Memoirs of the American Academy in Rome* 36: 217–233.

Palmer, R.E.A. 1990. *Studies of the Northern Campus Martius in Ancient Rome*. Philadelphia: American Philosophical Society.

Panella, C. 1981. "Merci destinate al commercio transmarino: Il vino: La distribuzione e i mercati." In *Società romana e produzione schiavistica*, vol. 2, *Merci, mercati e scambi nel Mediterraneo*, edited by A. Giardina and A. Schiavone, 55–80. Bari: Laterza.

Panella, C. 1989. "Le anfore italiche del II secolo D.C." In *Amphores romaines et histoire économique: Dix ans de recherche: Actes du colloque de Sienne (22–24 mai 1986)*, edited by University of Siena, 139–178. Collection de l'École Française de Rome 114. Rome: École française de Rome; Paris: Diffusion de Boccard.

Paparazzo, E. 2003. "Pliny the Elder on the Melting and Corrosion of Silver with Tin Solders: *Prius liquescat argentum ... ab eo erodi argentum (HN* 34.161)." *Classical Quarterly* 53.2: 523–529.

Paparazzo, E. 2008. "Pliny the Elder on Metals: Philosophical and Scientific Issues." *Classical Philology* 103: 40–54.

Parca, M. 2001. "Local Languages and Native Cultures." In *Epigraphic Evidence: Ancient History from Inscriptions*, edited by J. Bodel, 57–72. London: Routledge.

Parker, A. 1992. *Ancient Shipwrecks of the Mediterranean and the Roman Provinces*. BAR-IS 580. Oxford: Tempus Reparatum.

Parkin, T.G. 1992. *Demography and Roman Society*. Baltimore: Johns Hopkins University Press.

Paroli, L. 1996. "Ostia alla fine del mondo antico: Nuovi dati dallo scavo di un magazzino doliare." In *Roman Ostia Revisited: Archaeological and Historical Papers in Honor of Russell Meiggs*, edited by A. Gallina Zevi, 249–264. London: British School at Rome, in collaboration with Soprintendenza Archeologica di Ostia.

Pasqui, A. 1897. "La Villa Pompeiana della Pisanella presso Boscoreale." *Monumenti Antichi* 7: 397–554.

Pasqui, A. 1906. "Ostia—Nuove scoperte presso il Casone." *Notizie degli Scavi di Antichità* 10: 357–373.

Patterson, H., A. Bosquet, S. Fontana, R. Witcher, and S. Zampini. 2005. "Late Roman Common Wares and Amphorae in the Middle Tiber Valley: The Preliminary Results of the Tiber Valley Project." In *LRCW I: Late Roman Coarse Wares, Cooking Wares and Amphorae in the Mediterranean: Archaeology and Archaeometry*, edited by J. Ma. Gurt i Esparraguera, J. Buxeda i Garrigós, and M.A. Cau Ontiveros, 369–384. British Archaeological Reports International Series 1340. Oxford: Archaeopress.

Patterson, O. 1982. *Slavery and Social Death*. Cambridge, MA: Harvard University Press.

Pavolini, C. 1983. *La vita quotidiana a Ostia*. Bari: Editori Laterza.

Peacock, D.P.S. 1980. "The Roman Millstone Trade: A Petrological Sketch." *World Archaeology* 12.1: 43–53.

Peacock, D.P.S. 1984. "Seawater, Salt and Ceramics." In *Excavations at Carthage: The British Mission*, vol. I.2, *The Avenue du Président Habib Bourguiba, Salammbo: The Pottery and Other Ceramic Objects from the Site*, edited by M.G. Fulford and D.P.S. Peacock, 263–264. Sheffield: British Academy from the Department of Prehistory and Archaeology, University of Sheffield.

Peacock, D.P.S., and D.F. Williams. 1986. *Amphorae and the Roman Economy: An Introductory Guide*. London: Longman.

Pecci, A. 2020. "Produzione e consumo di alimenti in area vesuviana: Dati dalle analisi dei residui organici nelle ceramiche." In *Fecisti cretaria: Produzione e circolazione ceramica a Pompei: Stato degli studi e prospettive di ricercar*, edited by M. Osanna and L. Toniolo, 57–62. Studi e ricerche del Parco archeologico di Pompei 40. Rome: L'Erma di Bretschneider.

Pecci, A., J. Clarke, M. Thomas, J. Muslin, I. van der Graaff, L. Toniolo, D. Miriello, G.M. Crisci, M. Buonincontri, and G. Di Pasquale. 2017. "Use and Reuse of Amphorae: Wine Residues in Dressel 2–4 Amphorae from Oplontis Villa B (Torre Annunziata, Italy)." *Journal of Archaeological Science: Reports* 12: 515–521.

Pellecuer, C., ed. 1996. *Formes de l'habitat rural en Gaule Narbonnaise*. Vol. 3. Juan-les-Pins: Editions APDCA.

Pellecuer, C. 2000. "La villa des Près-Bas (Loupian, Hérault) dans son environnement: Contribution à l'étude des villae et de l'économie domaniale en Narbonnaise." 2 vols. PhD thesis, Université of Provence Aix-Marseille I.

Pellegrino, A., and A. Licordari. 2018. "Il forum vinarium ad Ostia." In *Ostia Antica: Nouvelles études et recherches sur les quartiers occidentaux de la cité*, edited by C. De Ruyt, Th. Morard, and F. Van Haeperen, 261–272. Brussels: Belgisch Historisch Instituut te Rome; Rome: Instituto Storico Belga di Roma.

Peña, J.T. 1998. "The Mobilization of State Olive Oil in Roman Africa: The Evidence of Late Fourth Century Ostraca from Carthage." In *Carthage Papers: The Early Colony's Economy, Water Supply, a Public Bath, and the Mobilization of State Olive Oil*, edited by J.T. Peña, J.J. Rossiter, A.I. Wilson, and C. Wells, 116–238. Journal of Roman Archaeology Supplementary Series 28. Portsmouth, RI: Journal of Roman Archaeology.

Peña, J.T. 1999. *The Urban Economy during the Early Dominate: Pottery Evidence from the Palatine Hill*. BAR International Series 784. Oxford: British Archaeological Reports.

Peña, J.T. 2007a. "A Reinterpretation of Two Groups of *Tituli Picti* from Pompeii and Environs: Sicilian Wine, Not Flour and Hand-Picked Olives." *Journal of Roman Archaeology* 20: 233–254.

Peña, J.T. 2007b. *Roman Pottery in the Archaeological Record*. Cambridge: Cambridge University Press.

Peña, J.T. 2014. "The Pompeii Artifact Life History Project: Conceptual Background and First Season's Results." *Rei Cretariae Romanae Fautorum Acta* 43: 297–304.

Peña, J.T., and C. Cheung. 2015. "The Pompeii Artifact Life History Project: Conceptual Basis and Results of First Three Seasons." In *Proceedings of XIII International Forum Le Vie dei Mercanti, Aversa, Capri, June 11–13, 2015*, 2115–2123. Naples: La Scuola di Pitagora.

Peña, J.T., and M. McCallum. 2009a. "The Production and Distribution of Pottery at Pompeii: A Review of the Evidence; Part 1, Production." *American Journal of Archaeology* 113.1: 57–79.

Peña, J.T., and M. McCallum. 2009b. "The Production and Distribution of Pottery at Pompeii: A Review of the Evidence; Part 2, The Material Basis for Production and Distribution." *American Journal of Archaeology* 113.2: 165–201.

Peña Cervantes, Y. 2005/2006. "La producción de vino y aceite en los asentamientos rurales de Hispania durante la Antigüedad Tardia (s. IV–VII d.C.)." *Cuadernos de Prehistoria y Arqueología de la Universidad Autónoma de Madrid* 31–32: 103–116.

Peña Cervantes, Y. 2010. *Torcularia: La producción de vino y aceite en Hispania.* Tarragona: Institut Català d'Arqueologia Clàssica.

Peña Cervantes, Y. 2020. "Wine Making in the Iberian Peninsula during the Roman Period: Archaeology, Archaeobotany and Biochemical Analysis." In *Archaeology, Archaeobotany and Biochemical Analysis: A. Making Wine in Western-Mediterranean, B. Production and the Trade of Amphorae: Some New Data from Italy,* edited by J.P. Brun, N. Garnier, and G. Olcese, 73–87. Heidelberg: Propylaeum.

Perkins, P. 2012. "Production and Commercialization of Etruscan Wine in the Albegna Valley." In *Archeologia della vite e del vino in Toscana e nel Lazio: Dalle tecniche dell'indagine archeologica alle prospettive della biologia molecolare,* edited by A. Ciacci, P. Rendini, and A. Zifferero, 413–426. Florence: All'Insegna del Giglio.

Perkins, P. 2021. "The Etruscan *Pithos* Revolution." In *Making Cities: Economies of Production and Urbanization in Mediterranean Europe, 1000–500 BCE,* edited by M. Gleba, B. Marín-Aguilera, and B. Dimova, 231–258. McDonald Institute Conversations. Cambridge, UK: McDonald Institute.

Pernice, E.A. 1932. *Hellenistische Tische, Zisternenmündungen, Beckenuntersätze, Altäre und Truhen.* Berlin; Leipzig: W. de Gruyter & Company.

Philippon, A., and L. Védrine. 2009. "Paysage urbain, patrimoine et musées: Le port antique de Marseille." *Les nouvelles de l'archéologie* 117: 40–46.

Piccarreta, F. 1977. *Forma Italiae: Astura.* Vol. I.13. Florence: Casa Editrice Leo S. Olschki.

Pirandello, L. 1927. *La giara: Novelle per un anno.* Florence: Bemporad.

Platner, S. 2015. Reprint. *A Topographical Dictionary of Ancient Rome.* Edited by Thomas Ashby. Cambridge: Cambridge University Press. Original edition, Oxford: Oxford University Press, 1929.

Poehler, E. 2011. "Where to Park? Carts, Stables and the Economics of Transport in Pompeii." In *Rome, Ostia and Pompeii: Movement and Space,* edited by R. Laurence and D. Newsome, 194–214. Oxford: Oxford University Press.

Poehler, E. 2017. *The Traffic Systems of Pompeii.* New York: Oxford University Press.

Poggesi, G. 2001. *Mura di cinta di Cosa: L'intervento di restauro 1999/2000.* Padua: Tecnostudio.

Ponsich, M. 1974. *Implantation rurale antique sur le Bas-Guadalquivir.* Vol. 1. Publications de la Casa de Velázquez 2. Paris: Casa de Velázquez.

Ponsich, M. 1991. *Implantation rurale antique sur le Bas-Guadalquivir.* Vol. 4. Publications de la Casa de Velázquez 33. Paris: Casa de Velázquez.

Porter, M.E. 1998. "Clusters and the New Economics of Competition." *Harvard Business Review* 76.6: 77–90.

Poux, M. 2004a. "De Midas à Luern, le vin des banquets." In *Le vin, nectar des dieux, génie des hommes,* edited by J.-P. Brun, M. Poux, and A. Tchernia, 68–95. Gollion: Infolio éditions.

Poux, M. 2004b. *L'âge du vin: Rites de boisson, festins et libations en Gaule indépendante.* Montagnac: Éditions Monique Mergoil.

Poux, M., J.-P. Brun, and M. Hervé-Monteil. 2011. *Gallia 68.1: La vigne et le vin dans les Trois Gaules.* Lattes: CNRS.

Privitera, S. 2014. "Long-Term Grain Storage and Political Economy in Bronze Age Crete: Contextualizing Ayia Triada's Silo-Complexes." *American Journal of Archaeology* 118.3: 429–449.

Pullen, D.J., ed. 2010. *Political Economy of the Aegean Bronze Age.* Oxford: Oxbow Books.

Pullen, D.J. 2011. "Redistribution in Aegean Palatial Societies: Before the Palaces: Redistribution and Chiefdoms in Mainland Greece." *American Journal of Archaeology* 115.2: 185–195.

Purcell, N. 1985. "Wine and Wealth in Ancient Italy." *Journal of Roman Studies* 75: 1–19.

Purcell, N. 1988. "Town in Country and Country in Town." In *The Ancient Roman Villa-Garden*, edited by E. MacDougall, 185–203. Washington, DC: Dumbarton Oaks.

Purcell, N. 1994. "The City of Rome and the *Plebs Urbana* in the Late Republic." In *Cambridge Ancient History*, vol. 9, *The Last Age of the Roman Republic, 146–43* B.C., 2nd ed., edited by J.A. Crook, A. Lintott, E. Rawson, 644–688. Cambridge: Cambridge University Press.

Purcell, N. 1995. "The Roman Villa and the Landscape of Production." In *Urban Society in Roman Italy*, edited by T. Cornell and K. Lomas, 151–179. London: Routledge.

Purcell, N. 1996. "The Ports of Rome: Evolution of a *Façade Maritime*." In *"Roman Ostia" Revisited: Archaeological and Historical Papers in Memory of Russell Meiggs*, edited by A.G. Zevi and A. Claridge, 267–279. Rome: British School at Rome and Soprintendenza di Ostia.

Purcell, N. 2007. "The Horti of Rome and the Landscape of Property." In *Res bene gestae: Ricerche di storia urbana su Roma antica in onore di Eva Margaret Steinby (Festschrift M. Steinby)*, edited by A. Leone, D. Palombi, and S. Walker, 362–378. Rome: Quasar.

Py, M. 1993. *Les Gaulois du Midi: De la fin de l'Age du Bronze à la conquête romaine*. Paris: Hachette.

Radić Rossi, I. 2020. "A *Dolia* Shipwreck in the Adriatic and Other Underwater *Dolia* Finds in Dalmatia." In *Nouvelles recherches sur les dolia: L'exemple de la Méditerranée nord-occidentale à l'époque romaine (Ier s. av. J.-C.–IIIe s. ap. J.-C.)*, edited by C. Carrato and F. Cibecchini, 251–259. Montpellier: Editions de l'Association de la Revue archéologique de Narbonnaise.

Rando, G. 1996. "Le antiche riparazioni in piombo sui *dolia* provenienti dal relitto della nave romana del Golfo di Diano Marina." In *Atti del convegno internazionale della ceramic, Albisola superiore, maggio 1996*. http://www.pinorando.com/Atti/atti-convegno-albisola.pdf.

Rathbone, D.W. 1981. "The Development of Agriculture in the *Ager Cosanus* during the Roman Republic: Problems of Evidence and Interpretation." *Journal of Roman Studies* 71: 10–23.

Ravasi, T., P. Liverani, I. Haynes, and S. Kay. 2020. "San Giovanni in Laterano 2 Project (sgl2)." *Papers of the British School at Rome* 88: 350–354.

Rebay-Salisbury, K., A. Brysbaert, and L. Foxhall, eds. 2015. *Knowledge Networks and Craft Traditions in the Ancient World: Material Crossovers*. New York: Routledge.

Rehder, J.E. 2000. *The Mastery and Uses of Fire: A Sourcebook on Ancient Pyrotechnology*. Montreal: McGill-Queen's University Press.

Remesal Rodríguez, J. 1998. "Baetican Olive Oil and the Roman Economy." In *The Archaeology of Early Roman Baetica*, edited by S. Keay, 183–199. Portsmouth, RI: Journal of Roman Archaeology.

Remesal Rodríguez, J., V. Revilla Calvo, C. Carreras Monfort, and P. Berni Millet. 1997. "Arva: Prospecciones en un centro productor de ánforas Dressel 20 (Alcolea del Río, Sevilla)." *Pyrenae* 28: 151–178.

Rice, C. 2016. "Shipwreck Cargoes in the Western Mediterranean and the Organization of Roman Maritime Trade." *Journal of Roman Archaeology* 29: 165–192.

Rice, C. 2019. "Rivers, Roads, and Ports." In *A Companion to the City of Rome*, edited by C. Holleran and A. Claridge, 197–217. Malden, MA: Wiley Blackwell.

Richardson, L. 1992. *A New Topographical Dictionary of Ancient Rome*. Baltimore: Johns Hopkins University Press.

Rickman, G. 1971. *Roman Granaries and Store Buildings*. Cambridge: Cambridge University Press.

Rickman, G. 1980. *The Corn Supply of Ancient Rome*. Cambridge: Cambridge University Press.

Rickman, G. 1996. "Portus in Perspective." In *"Roman Ostia" Revisited: Archaeological and Historical Papers in Memory of Russell Meiggs*, edited by A.G. Zevi and A. Claridge, 281–291. Rome: British School at Rome.

Rickman, G. 2002. "Rome, Ostia, Portus: The Problem of Storage." *Mélanges de l'École française de Rome* 114.1: 353–362.

Ridgway, F. 2010. *Pithoi stampigliati ceretani: Una classe originale di ceramica etrusca*. Edited by Lisa C. Pieraccini. Rome: L'Erma di Bretschneider.

Riggsby, A.M. 2019. *Mosaics of Knowledge: Representing Information in the Roman World*. Oxford: Oxford University Press.

Riley, F.R. 1999. *The Role of the Traditional Mediterranean Diet in the Development of Minoan Crete: Archaeological, Nutritional, and Biochemical Evidence*. BAR International Series 810. Oxford: British Archaeological Reports.

Rizos, E. 2013. "Centres of the Late Roman Military Supply Network in the Balkans: A Survey of Horrea." *Jahrbuch des Römisch-Germanischen Zentralmuseums* 60: 659–696.

Rizzo, G. 2003. *Instrumenta Urbis I: Ceramiche fini da mensa lucerne ed anfore a Roma nei primi due secoli dell'impero*. Rome: École française de Rome.

Robb, J. 2018. "Contained within History." *History and Anthropology* 29.1: 32–36.

Robinson, D., C. Rice, and K. Schörle. 2020. "Ship Losses and the Growth of Harbour Infrastructure." *Journal of Mediterranean Archaeology* 33.1: 102–125.

Robinson, D., and A.I. Wilson, eds. 2011. *Maritime Archaeology and Ancient Trade in the Mediterranean*. Oxford: Oxford Centre for Maritime Archaeology.

Rodríguez Almeida, E. 1984. *Il Monte Testaccio: Ambiente, storia, materiali*. Rome: Quasar.

Rodríguez Almeida, E. 1989. *Los tituli picti de las ánforas olearias de la Bética: I, Tituli picti de los Severos y la Ratio fisci*. Madrid: Universidad Complutense de Madrid.

Rodríguez Almeida, E. 1993. "Cellae Vinariae Nova et Arruntiana." In *Lexicon Topographicum Urbis Romae I (A–C)*, edited by E.M. Steinby, 259. Rome: Quasar.

Rogers, E. 2003. *Diffusion of Innovations*. 5th ed. New York: Free Press.

Roldán Gómez, L., and M. Bustamante Álvarez. 2015. "The Production, Dispersion and Use of Bricks in Hispania." In *Il laterizio nei cantieri imperiali: Roma e il Mediterraneo: Archeologia dell'Architectura XX: Atti del I workshop "Laterizio" (Roma, 27–28 Novembre 2014)*, edited by E. Bukowiecki, R. Volpe, and U. Wulf-rheidt, 135–144. Florence: All'Insegna del Giglio.

Roldán Gómez, L., and M. Bustamante Álvarez. 2017. "El material latericio en Hispania." In *Manual de cerámica romana III: Cerámicas romanas de época altoimperial III: Cerámica común de mesa, cocina y almacenaje, imitaciones hispanas de series romanas, otras producciones*, edited by C. Fernández Ochoa, Á. Morillo Cerdán, and M. del Mar Zarzalejos Prieto, 435–475. Madrid: Museo Arqueológico de la comunidad de Madrid, Colegio de Doctores y Licenciados en Filosofía y Letras y en Ciencias de la Comunidad de Madrid.

Romanazzi, L., and A.M. Volonté. 1986. "Gli scarichi tra Porta Nola e la Torre." In *Nuovi contributi sulle fortificazioni pompeiane*, edited by C. Chiaramonte Treré, 55–113. Quaderni di Annali della Facoltà di Lettere e Filosofia dell'Università degli Studi di Milano 6. Milan: Cisalpino-Goliardica.

Romero, A., and S. Cabasa. 1999. *La tinajería tradicional en la cerámica Española*. Barcelona: Ediciones CEAC.

Roselaar, S.T. 2019. *Italy's Economic Revolution: Integration and Economy in Republican Italy*. Oxford: Oxford University Press.

Roselaar, S.T. 2020. "Agriculture in Republican Italy." In *A Companion to Ancient Agriculture*, edited by D. Hollander and T. Howe, 417–430. Chichester: John Wiley and Sons.

Rosenfeld, A. 1965. *The Inorganic Raw Materials of Antiquity*. London: Weidenfeld and Nicolson.

Ross, L.A. 1985. "16th Century Spanish Basque Coopering." *Historical Archaeology* 19.1: 1–31.

Rossiter, J.J. 1981. "Wine and Oil Processing at Roman Farms in Italy." *Phoenix* 35.4: 345–361.

Rossiter, J.J. 2007. "Wine-Making after Pliny: Viticulture and Farming Technology in Late Antique Italy." In *Technology in Transition: A.D. 300–650*, edited by L. Lavan, E. Zanini, and A. Sarantis, 93–118. Leiden: Brill.

Rossiter, J.J., and E. Haldenby. 1989. "A Wine-Making Plant in Pompeii *Insula* II.5." *Echos du Monde Classique/Classical Views* 33.8: 229–239.

Rostovtzeff, M. 1911. "Die hellenistisch-römische Architekturlandschaft." *Römische Mitteilungen* 26: 1–186.

Rostovtzeff, M. 1941. *The Social and Economic History of the Hellenistic World*. 3 vols. Oxford: Clarendon Press.

Rotroff, S.I. 2011. "Mended in Antiquity: Repairs to Ceramics at the Athenian Agora." In *Pottery in the Archaeological Record: Greece and Beyond*, edited by M.L. Lawall and J. Lund, 117–134. Aarhus: Aarhus University Press.

Rougé, J. 1957. "Ad Ciconias Nixas." *Révue des études anciennes* 59.3: 320–328.
Rougé, J. 1966. *Recherches sur l'organisation du commerce maritime en Méditerranée sous l'empire romain.* École pratique des hautes études. VIe section. Ports, routes, traffics. Centre de recherches historiques 21. Paris: S.E.V.P.E.N.
Rowan, E. 2019a. "Olives and Olive Oil." In *The Routledge Handbook of Diet and Nutrition in the Roman World,* edited by P. Erdkamp and C. Holleran, 129–139. London: Routledge.
Rowan, E. 2019b. "Same Taste, Different Place: Looking at the Consciousness of Food Origins in the Roman World." *Theoretical Roman Archaeology* 2.1: 1–18.
Rowlandson, J. 1996. *Landowners and Tenants in Roman Egypt: The Social Relations of Agriculture in the Oxyrhynchite Nome.* Oxford: Oxford University Press.
Rowlandson, J. 1999. "Agricultural Tenancy and Village Society in Roman Egypt." In *Agriculture in Egypt,* edited by A.K. Bowman and E. Rogan, 139–518. Oxford: Oxford University Press.
Ruegg, S.D. 1995. *Underwater Investigations at Roman Minturnae: Liris-Garigliano River.* Jonsered: Paul Åströms Förlag.
Salido Domínguez, J. 2011. *Horrea militaria: El aprovisionamiento de grano al ejército en el occidente del Imperio Romano.* Madrid: Consejo Superior de Investigaciones Científicas.
Salido Domínguez, J. 2017. "Los *dolia* en *Hispana*: Caracterización, funcionalidad y tipología." In *Manual de cerámica romana III: Cerámicas romanas de época altoimperial III: Cerámica común de mesa, cocina e almacenaje, imitaciones hispanas de series romanas, otras producciones,* edited by C. Fernández Ochoa, Á. Morillo Cerdán, and M. del Mar Zarzalejos Prieto, 239–309. Madrid: Museo Arqueológico de la comunidad de Madrid, Colegio de Doctores y Licenciados en Filosofía y Letras y en Ciencias de la Comunidad de Madrid.
Salomonson, J.W. 1972. "Röemische Tonformen mit Inschriften: Ein Beitrag zum Problem der sogenannten 'Kuchenformen' aus Ostia." *BABESCH* 47: 88–113.
Salzman, M.R. 2021. "A New 'Topography of Devotion': Aurelian and Solar Worship in Rome." In *Urban Religion in Late Antiquity,* edited by A. Lätzer-Lasar and E. Urciuoli, 149–168. Berlin: De Gruyter.
Sands, R., and E. Marlière. 2020. "Produce, Repair, Reuse, Adapt, and Recycle: The Multiple Biographies of a Roman Barrel." *European Journal of Archaeology* 23.3: 356–380.
Santa Maria Scrinari, V. 1995. *Il Laterano imperiale.* Vol. 2, *Dagli "horti Domitiae" alla Cappella cristiana.* Vatican City: Pontificio Istituto di Archeologia Cristiana.
Saul, S., and D. Hakim. 2018, June 7. "The Most Powerful Conservative Couple You've Never Heard Of." *New York Times.* https://www.nytimes.com/2018/06/07/us/politics/liz-dick-uihlein-republican-donors.html.
Schatzberg, E. 2018. *Technology: Critical History of a Concept.* Chicago: University of Chicago Press.
Scheidel, W. 1997. "Quantifying the Sources of Slaves in the Early Roman Empire." *Journal of Roman Studies* 87: 156–169.
Scheidel, W. 1999. "The Slave Population of Roman Italy: Speculation and Constraints." *Topoi* 9.1: 129–144.
Scheidel, W. 2001. "Roman Age Structure: Evidence and Models." *Journal of Roman Studies* 91: 1–26.
Scheidel, W. 2004. "Human Mobility in Roman Italy, I: The Free Population." *Journal of Roman Studies* 94: 1–26.
Scheidel, W. 2005. "Human Mobility in Roman Italy, II: The Slave Population." *Journal of Roman Studies* 95: 64–79.
Scheidel, W. 2007. "Demography." In *The Cambridge Economic History of the Graeco-Roman World,* edited by W. Scheidel, I. Morris, and R. Saller, 38–85. Cambridge: Cambridge University Press.
Scheidel, W., I. Morris, and R. Saller, eds. 2007. *The Cambridge Economic History of the Graeco-Roman World.* Cambridge: Cambridge University Press.
Schiffer, M. 1972. "Archaeological Context and Systemic Context." *American Anthropologist* 37: 156–165.
Schiffer, M. 1996. *Formation Processes of the Archaeological Record.* Albuquerque: University of New Mexico Press.

Schmitt, A. 1993. "Apports et limites de la petrographie quantitative: Application au cas des amphores de Lyon." *Revue d'Archéometrie* 17: 51–63.

Schreiber, T. 1896. "Die hellenistischen Reliefbilder und die augusteische Kunst." *Jahrbuch des Deutschen Archäologischen Instituts* 11: 79–101.

Schwinden, L. 2019. "Die Weinschiffe der römischen Grabmäler von Neumagen." *Funde und Ausgrabungen im Bezirk Trier* 51: 27–45.

Sciallano, M., and S. Marlier. 2008. "L'épave à dolia de l'île de la Giraglia (Haute-Corse)." *Archaeonautica* 15: 113–151.

Scott, A.R. 2008. *Cosa: The Black-Glaze Pottery* 2. Memoirs of the American Academy in Rome. Ann Arbor: University of Michigan Press.

Scott, J.C. 2017. *Against the Grain: Deep History of the Earliest States*. New Haven, CT: Yale University Press.

Scott, R.T., A.U. De Giorgi, S. Crawford-Brown, A. Glennie, and A. Smith. 2015. "Cosa Excavations: The 2013 Report." *Orizzonti* 16: 11–22.

Sebastiani, R., and M. Serlorenzi. 2011. "Nuove scoperte dall'area di Testaccio (Roma): Tecniche costruttive, riuso e smaltimento dei contenitori anforici pertinenti ad horrea e strutture utilitarie di età imperiale." In *Horrea d'Hispanie et de la méditerranée romaine*, edited by J. Arce Martínez and B. Goffaux, 67–96. Madrid: Casa de Velázquez.

Senatore, F. 1998. "*Ager Pompeianus*: Viticoltura e territorio nella Piana del Sarno nel I sec. d.C." In *Pompei il Sarno e la Penisola Sorrentina: Atti del primo ciclo di conferenze di geologica, storia e archeologica*, edited by F. Senatore, 135–166. Pompei: Rufus.

Sennett, R. 2008. *The Craftsman*. New Haven, CT: Yale University Press.

Serlorenzi, M. 2010. "La costruzione di un complesso horreario a Testaccio: Primi indizi per delineare l'organizzazione del cantiere edilizio." In *Arqueología de la construcción*, vol. 2, *Los procesos constructivos en el mundo romano: Italia y provincias orientales (Certosa di Pontignano, Siena, 13–15 de noviembre de 2008)*, edited by S. Camporeale, H. Dessales, and A. Pizzo, 105–126. Mérida: Consejo Superior de Investigaciones Científicas, Instituto de Arqueología.

Setälä, P. 1977. *Private Domini in Roman Brick Stamps of the Empire: A Historical and Prosopographical Study of Landowners in the District of Rome*. Vol. 10. Institutum Romanum Finlandiae. Helsinki: Suomen Tiedeakatemia.

Shaw, B.D. 2013. *Bringing in the Sheaves: Economy and Metaphor in the Roman World*. Toronto: University of Toronto Press.

Shaw, B.D. 2019. "Grape Expectations." In *Uomini, istituzioni, mercati. Studi di storia per Elio Lo Cascio*, edited by M. Maiuro, with G.D. Merola, M. De Nardis, and G. Soricelli, 535–551. Bari: Edipuglia.

Shryock, A., and D.L. Smail. 2018b. "On Containers: A Forum: Introduction." *History and Anthropology* 29.1: 1–6.

Silver, M. 2009. "Glimpses of Vertical Integration/Disintegration in Ancient Rome." *Ancient Society* 39: 171–184.

Silver, M. 2013. "Vertical Integration/Disintegration in the Roman Economy: A Reply to Wim Broekaert." *Ancient Society* 43: 309–315.

Simon Reig, J., J. Tremoleda Trilla, P. Castanyer Masoliver, and V. Martínez Farreras. 2020. "La producción de *dolia* del alfar de Ermedàs (Cornellà del Terri)." In *Nouvelles recherches sur les dolia: L'exemple de la Méditerranée nord-occidentale à l'époque romaine (Ier s. av. J.-C.–IIIe s. ap. J.-C.)*, edited by C. Carrato and F. Cibecchini, 63–82. Montpellier: Editions de l'Association de la Revue archéologique de Narbonnaise.

Sirks, A.J.B. 1991. *Food for Rome: The Legal Structure of the Transportation and Processing of Supplies for the Imperial Distributions in Rome and Constantinople*. Amsterdam: J.C. Gieben.

Sissa, G. 1990. *Greek Virginity*. Trans. A. Goldhammer. Cambridge, MA: Harvard University Press.

Skibo, J.M. 1992. *Pottery Function: Use-Alteration Perspective*. New York: Plenum.

Slane, K.W. 2011. "Repair and Recycling in Corinth and the Archaeological Record." In *Pottery in the Archaeological Record: Greece and Beyond*, edited by M.L. Lawall and J. Lund, 95–106. Aarhus: Aarhus University Press.

Slatcher, S. 2017, July 4. "Qvevri—Manufacture, Use, and Rise to Fame." Winenous. http://www.winenous.co.uk/wp/archives/10018.

Smith, M.E. 2004. "The Archaeology of Ancient State Economies." *Annual Review of Anthropology* 3: 73–102.

Smith, W. 1890. *A Dictionary of Greek and Roman Antiquities*. 3rd ed. London: John Murray.

Sogliano, A. 1888. "Pompei: Degli edificii recentemente scoperti, e degli oggetti raccolti negli scavi dal dicembre 1887 al giugno 1888: Relazione dell'ispettore prof. A. Sogliano." *Notizie degli Scavi di Antichità* 13: 509–530.

Sogliano, A. 1889. "Pompei: Degli edificii recentemente coperti e degli oggetti raccolti negli scavi dal settembre 1888 al marzo 1889." *Notizie degli Scavi di Antichità* 14: 114–136.

Solin, H. 1987. "Analecta epigraphica CXIII–CXX." *Arctos–Acta Philologica Fennica* 21: 119–138.

Solin, H. 1998. *Analecta epigraphica, 1970–1997*. Helsinki: Acta Instituti Romani Finlandiae.

Spanu, M., ed. 2015. *Opus doliare tiberninum: atti dell Girnate di Studio (Viterbo 25–26 ottobre 2012)*. Tuscia: Università della Tuscia.

Spinazzola, V. 1953. *Pompei alla luce degli nuovi scavi dell'Abbondanza (Anni 1910–1923)*. Rome: Libreria dello Stato.

Steinby, E.M. 1974/1975. "La cronologia delle figlinae doliari urbane dalla fine dell'età repubblicana fino all'inizio dellII sec." *Bullettino della Commissione Archeologica Comunale di Roma* 84: 7–132.

Steinby, E.M. 1981. "La diffusione dell'opus doliare urbano." In *Società romana e produzione schiavistica*, vol. 2, *Merci, mercati e scambi nel Mediterraneo*, edited by A. Giardina and A. Schiavone, 237–245. Bari: Laterza.

Steinby, E.M. 1982. "I senatori e l'industria laterizia urbana." *Tituli* 4: 227–237.

Steinby, E.M. 1987. *Indici complementari ai bolli doliari urbani (CIL. XV,1)*. Acta Instituti Romani Finlandiae 11. Rome: Institutum Romanum Finlandiae,

Steinby, E.M. 1993. "Ricerche sull'industria doliare nelle aree di Roma and e Pompei: Un possibile modello interpretativo." In *I Laterizi di Età Romana nell'Area Nordadriatica*, edited by C. Zaccaria, 9–14. Rome: L'Erma di Bretschneider.

Steinby, E.M., ed. 1993–2000. *Lexicon Topographicum Urbis Romae*. 6 vols. Rome: Edizioni Quasar.

Steiner, D.T. 2013. "The Priority of Pots: Pandora's Pithos Re-viewed." *Mètis—Anthropologie des mondes grecs anciens* 11: 211–238.

Stevens, S. 2005. "Reconstructing the Garden Houses at Ostia: Exploring Water Supply and Building Height." *Bulletin van de Antieke Beschaving* 80: 113–123.

Stöger, H. 2008. "Roman Ostia: Space Syntax and the Domestication of Space." In *Layers of Perception: Proceedings of the 35th International Conference on Computer Applications and Quantitative Methods in Archaeology (CAA), Berlin, 2–6 April 2007*, edited by A. Posluschny, K. Lambers, and I. Herzog, 322–327. Kolloquien zur Vor- und Frühgeschichte 10. Bonn: Dr. Rudolf Habelt.

Stöger, H. 2014. "The Spatial Signature of an Insula Neighborhood of Roman Ostia." In *Spatial Analysis and Social Spaces: Interdisciplinary Approaches to the Interpretation of Prehistoric and Historic Built Environments*, edited by E. Paliou, U. Lieberwirth, and S. Polla, 297–316. Berlin: De Gruyter.

Storey, G. 1997. "The Population of Ancient Rome." *Antiquity* 71: 966–978.

Storey, G. 2001. "Regionaries-Type Insulae 1: Architectural/Residential Units at Ostia." *American Journal of Archaeology* 105.3: 389–401.

Storey, G. 2002. Regionaries-Type Insulae 2: Architectural/Residential Units at Rome." *American Journal of Archaeology* 106.3: 411–434.

Storey, G. 2004. "The Meaning of 'Insula' in Roman Residential Terminology." *Memoirs of the American Academy in Rome* 49: 47–84.

Strickland, M. 2016, May 4. "Why Refrigerators Were So Slow to Catch On in China." *Atlantic*, Object Lessons series. https://www.theatlantic.com/technology/archive/2016/05/why-refrigerators-were-so-slow-to-catch-on-in-china/481029/.

Sundelin, L.K.R. 1996. "Plaster Jar Stoppers." In *Berenike 1995: Preliminary Report of the 1995 Excavations at Berenike (Egyptian Red Sea Coast) and the Survey of the Eastern Desert*, edited by S.E. Sidebotham and W.Z. Wendrich, 297–308. Leiden: Brill.

Taglietti, F. 2015. "*Dolia* e coperchi di *dolia*: Problematici assortimenti." In *Opus Doliare Tiberinum: atti delle giornate di studio (Vitero 25–26 ottobre 2012)*, edited by M. Spanu, 267–291. Viterbo: Dipartimento di Scienze dei Beni Culturali, Università degli Studi della Tuscia.

Taglietti, F., and C. Zaccaria. 1994. "Bolli laterizi." In *Enciclopedia dell'Arte Antica*, Treccani 2: 705–713. Rome: Istituto della Enciclopedia italiana.

Taylor, R.M. 2010. "Bread and Water: Septimius Severus and the Rise of the *Curator aquarum et Miniciae*." *Memoirs of the American Academy in Rome* 55: 199–220.

Taylor, R.M., K. Rinne, and S. Kostof. 2016. *Rome: An Urban History from Antiquity to the Present*. New York: Cambridge University Press.

Tchernia, A. 1986. *Le vin de l'Italie romaine: Essai d'histoire économique d'après les amphores*. Rome: École française de Rome.

Tchernia, A. 1997. "Le tonneau, de la bière au vin." In *Techniques et économie antiques et médiévales: Le temp de l'innovation*, edited by D. Garcia and D. Meeks, 121–129. Paris: Errance.

Tchernia, A. 2016. *The Romans and Trade*. Oxford: Oxford University Press.

Tchernia, A. 2022a. "Les avatars des vraquiers du Ier au IIIe siècle, hypothèses et énigmes." Paper read at "Bateaux et entrepôts à dolia à l'époque romaine: Problèmes de chronologie et d'évolution du transport maritime en vrac," March 28, Paris.

Tchernia, A. 2022b. "Les bas-reliefs du *Cassegiato di Annio* à Ostie." Paper read at "Bateaux et entrepôts à dolia à l'époque romaine: Problèmes de chronologie et d'évolution du transport maritime en vrac," March 28, Paris.

Teichner, F. 2011/2012. "La producción de aceite y vino en la villa romana de Milreu (Estói): El éxito del modelo catoniano en la Lusitania." *Anales de Prehistoria y Arqueología* 27–28: 471–484.

Teichner, F. 2018. "Roman Villas in the Iberian Peninsula (Second Century BCE–Third Century CE)." In *The Roman Villa in the Mediterranean Basin: Late Republic to Late Antiquity*, edited by A. Marzano and G.P.R. Métraux, 235–254. Cambridge: Cambridge University Press.

Terrenato, N. 2012. "The Enigma of 'Catonian' Villas: The *De Agricultura* in the Context of 2nd c. BC Rural Italian Architecture." In *Roman Republican Villas: Architecture, Context, and Ideology*, edited by J.A. Becker and N. Terrenato, 69–93. Ann Arbor: University of Michigan Press.

Teyssonneyre, Y., J. Planchon, and C. Ronco. 2020. "Les *dolia* de Pontaix et de la vallée de la Drôme: Des marques de capacité au rapport contenant contenu." In *Nouvelles recherches sur les dolia: L'exemple de la Méditerranée nord-occidentale à l'époque romaine (Ier s. av. J.-C.–IIIe s. ap. J.-C.)*, edited by C. Carrato and F. Cibecchini, 43–62. Montpellier: Editions de l'Association de la Revue archéologique de Narbonnaise.

Thomas, M.L. 2015. "Oplontis B: A Center for the Distribution and Export of Vesuvian Wine." *Journal of Roman Archaeology* 28: 403–412.

Thomas, M.L. 2016. "Oplontis B and the Wine Industry in the Vesuvian Area." In *Leisure and Luxury in the Age of Nero: The Villas of Oplontis near Pompeii*, edited by E.K. Gazda and J.R. Clarke, 160–165. Kelsey Museum publication 14. Ann Arbor, MI: Kelsey Museum of Archaeology.

Thomas, R.I. 2011. "Roman Vessel Stoppers." In *Myos Hormos—Quseir al-Qadim: Roman and Islamic Ports on the Red Sea*, vol. 2, *Finds from the Excavations 1999–2003*, edited by D. Peacock and L. Blue, 11–34. British Archaeological Reports International Series 2286. University of Southampton Series in Archaeology 6. Oxford: Oxbow Books; Archaeopress.

Thomas, R.I. 2014. "Nos 1–10, 42, 46–58." In "Catalogue of British Museum Objects from the Egypt Exploration Fund's 1913/14 Excavation at Antinoupolis (Antinoë)," edited by E.R. O'Connell, in *Scavi e materiali*, vol. 3, *Antinoupolis II*, edited by R. Pintaudi, 467–504. Florence: Firenze University Press.

Thomas, R.I., and R. Tomber. 2006. "Vessel Stoppers." In *Mons Claudianus: 1987–1993: Survey and Excavation*, vol. 3, *Ceramic Vessels and Related Objects*, edited by V. Maxfield and D. Peacock, 239–258. Fouilles de l'IFAO 54. Cairo: Institut français d'archéologie orientale.

Thornton, J. 1998. "A Brief History and Review of the Early Practice and Materials of Gap-Filling in the West." *Journal of the American Institute for Conservation* 37: 3–22.

Thurmond, D.L. 2006. *A Handbook of Food Processing in Classical Rome: For Her Bounty No Winter*. Leiden: Brill.

Thurmond, D.L. 2017. *From Vines to Wines in Classical Rome: A Handbook of Viticulture and Oenology in Rome and the Roman West*. Leiden: Brill.

Thylander, H. 1952. *Inscriptions du Port D'Ostie*. Lund: CWK Gleerup.

Tol, G., and B. Borgers. 2016. "The Study of Local Production and Exchange in the Lower Pontine Plain." *Journal of Roman Archaeology* 29.1: 349–370.

Tomlin, R. 2016. *Roman London's First Voices: Writing Tablets from the Bloomberg Excavations, 2010–14*. London: Museum of London Archaeology.

Tran, N. 2016. "*Ars* and *Doctrina*: The Socioeconomic Identity of Roman Skilled Workers (First Century BC–Third Century AD)." In *Work, Labour, and Professions in the Roman World*, edited by K. Verboven and C. Laes, 246–261. Leiden: Brill.

Treggiari, S. 1975. "Jobs in the Household of Livia." *Papers of the British School at Rome* 43: 48–77.

Treggiari, S. 1976. "Jobs for Women." *American Journal of Ancient History* 1: 76–104.

Treggiari, S. 1980. "Urban Labour in Rome: *Mercennarii* and *Tabernarii*." In *Non-slave Labour in Graco-Roman Antiquity*, edited by P. Garnsey, 48–64. Cambridge Philological Society Suppl. Vol. 6. Cambridge: Cambridge Philological Society.

Tremoleda Trilla, J. 2020. "Los *dolia* de Catalunya: Producción y prosopografía." In *Nouvelles recherches sur les dolia: L'exemple de la Méditerranée nord-occidentale à l'époque romaine (Ier s. av. J.-C.–IIIe s. ap. J.-C.)*, edited by C. Carrato and F. Cibecchini, 83–124. Montpellier: Editions de l'Association de la Revue archéologique de Narbonnaise.

Tsakirgis, B. 1984. "The Domestic Architecture of Morgantina in the Hellenistic and Roman Periods." PhD dissertation, Princeton University.

Tucci, P.L. 2004. "Eight Fragments of the Marble Plan of Rome Shedding New Light on the Transtiberim." *Papers of the British School at Rome* 72: 185–202.

Twede, D. 2002a. "Commercial Amphoras: The Earliest Commercial Packaging?" *Journal of Macromarketing* 22.1: 98–108.

Twede, D. 2002b. "The Packaging Technology and Science of Ancient Transport Amphoras." *Packaging Technology and Science* 15.4: 181–195.

Twede, D. 2005a. "Baskets, Barrels and Boxes: The History of Wooden Shipping Containers." In *22nd IAPRI Symposium Proceedings*. Campinas: Centro de Tecnologia de Embalagem.

Twede, D. 2005b. "The Cask Age: The Technology and History of Wooden Barrels." *Packaging Technology and Science* 18.5: 253–264.

Twede, D. 2009. "Packaging Economics." In *Wiley Encyclopedia of Packaging Technology*, 3rd ed., edited by K. Yam, 383–389. New York: Wiley Interscience.

Twede, D., R. Clarke, and J. Tait. 2000a. "Packaging Postponement: A Global Packaging Strategy." *Packaging Technology and Science* 13.3: 105–115.

Twede, D., R. Clarke, and J. Tait. 2000b. "Perspective on a Global Packaging Strategy—Packaging Postponement." *Journal for Packaging Professionals* 1.1: 15–18.

Twede, D., and B. Harte. 2011. "Logistical Packaging for Food Marketing Systems." In *Food and Beverage Packaging Technology*, 2nd ed., edited by R. Coles and M. Kirwan, 85–105. Chichester: Wiley-Blackwell.

Uboldi, M. 2005. "Laterizi e opus doliare." In *La ceramica e i materiali di età romana: Classi, produzioni, commerci e consumi*, edited by D. Gandolfi, 479–490. Bordighera: Scuola Interdisciplinare delle Metodologie Archeologiche.

Ulrich, R.B. 2007. *Roman Woodworking*. New Haven, CT: Yale University Press.

Ulrich, R.B. 2013. "Horrea and Insulae." In *A Companion to Roman Architecture*, edited by R.B. Ulrich and C. Quenmoen, 324–341. Blackwell Companions to the Ancient World. London: Wiley Blackwell.

Ural, A., and S. Uslu. 2014. "Shear Tests on Stone Masonry Walls with Metal Connectors." *European Journal of Environmental and Civil Engineering* 18.1: 66–86.

Vaccaro, E., M. Ghisleni, A. Arnoldus-Huyzendveld, C. Grey, K. Bowes, M. MacKinnon, A.M. Mercuri, A. Pecci, M.A. Cau Ontiveros, E. Rattigheri, and R. Rinaldi. 2013. "Excavating the Roman Peasant II: Excavations at Case Nuove, Cinigiano (GR)." *Papers of the British School in Rome* 81: 129–179, 408–413.

van Andel, T.H., and C. Runnels. 1987. *Beyond the Acropolis: A Rural Greek Past.* Stanford, CA: Stanford University Press.

Van der Werff, J. 1989. "Sekundare graffiti auf römischen Amphoren." *Archäologische Korrespondenzblatt* 19: 361–376.

Van der Werff, J. 2003. "The Second and Third Lives of Amphoras in Alfen Aan Den Rijn, the Netherlands." *Journal of Roman Pottery Studies* 10: 109–116.

Vandorpe, K. 2005. "Sealing Containers in Greco-Roman Egypt: The Inscriptional and Papyrological Evidence." In *Oggetti in argilla dall'area templare di Bakchias (El-Fayyum, Egitto): Catalogo dei rinvenimenti delle campagne di scavo 1996–2002*, edited by P. Davoli, 163–175. Pisa: Giardini.

Vandorpe, K., and W. Clarysse. 1997. "Viticulture and Wine Consumption in the Arsinoite Nome (P.Köln. V 221)." *Ancient Society* 28: 67–73.

van Driel-Murray, C. 2008. "Tanning and Leather." In *The Oxford Handbook of Engineering and Technology in the Classical World*, edited by J.P. Oleson, 483–495. Oxford: Oxford University Press.

Van Limbergen, D. 2020. "Wine, Greek and Roman." *Oxford Research Encyclopedia of Classics*. https://doi.org/10.1093/acrefore/9780199381135.013.6888.

Van Oyen, A. 2015a. "The Moral Architecture of Villa Storage in Italy in the 1st c. BC." *Journal of Roman Archaeology* 28: 97–123.

Van Oyen, A. 2015b. "The Roman City as Articulated through Terra Sigillata." *Oxford Journal of Archaeology* 34.3: 279–299.

Van Oyen, A. 2019. "Rural Time." *World Archaeology* 51.2: 191–207.

Van Oyen, A. 2020a. "Innovation and Investment in the Roman Rural Economy through the Lens of Marzuolo (Tuscany, Italy)." *Past and Present* 248: 3–40.

Van Oyen, A. 2020b. *The Socio-economics of Roman Storage: Agriculture, Trade, and Family*. Cambridge: Cambridge University Press.

Van Oyen, A. 2023. "Roman Failure: Privilege and Precarity at Early Imperial Podere Marzuolo, Tuscany." *Journal of Roman Studies*. Online ahead of print. https://doi.org/10.1017/S0075435822000958.

Van Oyen, A., and M. Pitts, eds. 2017a. *Materialising Roman Histories*. Oxford: Oxbow Books.

Van Oyen, A., and M. Pitts. 2017b. "What Did Things Do in the Roman World? Beyond Representation." In *Materialising Roman Histories*, edited by A. Van Oyen and M. Pitts, 3–20. Oxford: Oxbow Books.

Van Oyen, A., R.G. Vennarucci, A.L. Fischetti, and G. Tol. 2019. "Un centro artigianale di epoca romana: Terzo anno di scavo a Podere Marzuolo (Cinigiano, GR)." *Bollettino di Archeologia Online* 10.3–4: 71–84.

Varga, R. 2020. *Carving a Professional Identity: The Occupational Epigraphy of the Roman Latin West.* Oxford: Archaeopress Archaeology.

Vennarucci, R.G., A. Pecci, S. Mileto, G.W. Tol, and A. Van Oyen. Forthcoming. "Two Cylindrical Wine Tanks at Podere Marzuolo (Cinigiano, GR, Italy)." In *Vine-Growing and Winemaking in the Roman World*, edited by D. Van Limbergen, E. Dodd, and M. S. Busana. Leuven: Peeters Publishers.

Vera, D. 2006. "Un'iscrizione sulle distribuzioni pubbliche di vino a Roma (*CIL* VI 1785 = 31931)." In *Studi in onore di Francesco Grelle*, edited by F. Grelle, M. Silvestrini, T. Spagnuolo Vigorita, and G. Volpe, 303–317. Insulae Diomedeae 5, Scavi e ricerche 16. Bari: Edipuglia.

Vilucchi, S. 1993. "Cella Nigriniana." In *Lexicon Topographicum Urbis Romae I (A–C)*, edited by E.M. Steinby, 257. Rome: Quasar.

Virlouvet, C. 1995. *Tessera frumentaria: Les procédures de la distribution de blé public à Rome*. Rome: École française de Rome.

Virlouvet, C. 2011. "Les entrepôts dans le monde romain antique, formes et fonctions." In *Horrea d'Hispanie et de la Méditerranée romaine*, edited by J. Arce and B. Goffaux, 7–21. Collection de la Casa de Velázquez 125. Madrid: Casa de Velázquez.

Vitali, D., ed. 2007. *Le fornaci e le anfore di Albinia: Primi dati su produzioni e scambi dalla costa tirrenica al mondo gallico: Atti del seminario internazionale, Ravenna, 6–7 maggio 2006*. Bologna: Università di Bologne.

Vitali, D., F. Laubenheimer, and L. Benquet. 2012. "La produzione e il commercio del vino nell'Etruria romana: Le fornaci ci Albinia, Orbetello (Grosseto)." In *Archeologia della vite e del vino in Toscana e nel Lazio: Dalle tecniche dell'indagine archeologica alle prospettive della biologia molecolare*, edited by A. Ciacci, P. Rendini, and A. Zifferero, 427–436. Florence: All'Insegna del Giglio.

Vitelli, G. 1980. "Grain Storage and Urban Growth in Imperial Ostia: A Quantitative Study." *World Archaeology* 12.1: 54–68.

Waliszewski, T. 2014. *ELAION: Olive Oil Production in Roman and Byzantine Syria–Palestine*. Polish Archaeology in the Mediterranean Monograph Series Volume 6. Warsaw: Polish Centre of Mediterranean Archaeology, University of Warsaw Press.

Wallace-Hadrill, A. 1995. "Public Honor and Private Shame." In *Urban Society in Roman Italy*, edited by T.J. Cornell and K. Lomas, 39–62. London: Routledge.

Wallace-Hadrill, A. 1998. "Horti and Hellenization." In *Horti romani: Atti del Convegno Internazionale, Roma, 4–6 maggio 1995*, edited by M. Cima and E. La Rocca, 1–12. Rome: L'Erma di Bretschneider.

Wallace-Hadrill, A. 2008. *Rome's Cultural Revolution*. Cambridge: Cambridge University Press.

Ward-Perkins, B. 1984. *From Classical Antiquity to the Middle Ages: Urban Public Building in Northern and Central Italy AD 300–850*. Oxford: Oxford University Press.

Warscher, T. 1935–1960. "Codex Topographicus Pompeianus." Rome.

Weaver, P. 1998. "Imperial Slaves and Freedmen in the Brick Industry." *Zeitschrift für Papyrologie und Epigraphik* 122: 238–246.

Weller, J. 1999. "Roman Traction Systems." http://www.humanist.de/rome/rts/index.html.

Wells, B., ed. 1992. *Agricultural in Ancient Greece: Proceedings of the Seventh International Symposium at the Swedish Institute in Athens, 16–17 May 1990*. Stockholm: Swedish Institute in Athens.

Wendrich, W., ed. 2012a. *Archaeology and Apprenticeship: Body Knowledge, Identity, and Communities of Practice*. Tucson: University of Arizona Press.

Wendrich, W. 2012b. "Recognizing Knowledge Transfer in the Archaeological Record." In *Archaeology and Apprenticeship: Body Knowledge, Identity, and Communities of Practice*, edited by W. Wendrich, 255–262. Tucson: University of Arizona Press.

Westermann, W.L. 1914. "Apprentice Contracts and the Apprentice System in Roman Egypt." *Classical Philology* 9.3: 295–315.

White, K.D. 1975. *Farm Equipment of the Roman World*. Cambridge: Cambridge University Press.

White, P. 2016. "Qvevri Culture." *World of Fine Wine* 53: 117–123.

Widemann, F., F. Laubenheimer, and J. Leblanc. 1979. "Amphorae Workshops in Western Narbonnensis: The Non-resolution Space Problem." In *XIXth Symposium on Archaeometry and Archaeological Prospection*, edited by P.T. Craddock, 57–71. London: London University.

Will, E.L. 1979. "The Sestius Amphoras: A Reappraisal." *Journal of Field Archaeology* 6.3: 339–350.

Will, E.L. 1982a. "Ambiguity in Horace *Odes* 1.4." *Classical Philology* 77.3: 240–245.

Will, E.L. 1982b. "Greco-Italic Amphoras." *Hesperia* 51.3: 338–356.

Will, E.L. 1987. "The Roman Amphoras." In *The Roman Port and Fishery of Cosa*, edited by A.M. McCann, 171–220. Princeton, NJ: Princeton University Press.

Will, E.L. 2001. "Defining the 'Regna Vini' of the Sestii." In *New Light from Ancient Cosa: Studies in Honor of Cleo Rickman Fitch*, edited by N. Goldman, 35–47. New York: Peter Lang Publishers.

Will, E.L., and K.W. Slane. 2019. *Cosa: The Roman and Greek Amphoras*. Ann Arbor: University of Michigan Press.

Williams, J. 2018. *Stand Out of Our Light: Freedom and Resistance in the Attention Economy*. Cambridge: Cambridge University Press.

Wilson, A.I. 2002. "Machines, Power and the Ancient Economy." *Journal of Roman Studies* 92: 1–32.

Wilson, A.I. 2009. "Indicators for Roman Economic Growth: A Response to Walter Scheidel." *Journal of Roman Archaeology* 22: 71–82.

Wilson, A.I. 2011. "Developments in Mediterranean Shipping and Maritime Trade from the Hellenistic Period to AD 1000." In *Maritime Archaeology and Ancient Trade in the Mediterranean*, edited by D. Robinson and A.I. Wilson, 33–60. Oxford Centre for Maritime Archaeology Monograph 6. Oxford: Oxford Centre for Maritime Archaeology.

Wilson, A.I., K. Schörle, and C. Rice. 2012. "Roman Ports and Mediterranean Connectivity." In *Rome, Portus and the Mediterranean*, edited by S. Keay, 367–391. London: The British School at Rome.

Wilson, H.L. 1910. "Latin Inscriptions at the Johns Hopkins University IV." *American Journal of Philology* 31: 25–42.

Witcher, R. 2005. "The Extended Metropolis: Urbs, Suburbium and Population." *Journal of Roman Archaeology* 18: 120–138.

Woolf, G. 1992. "Imperialism, Empire, and the Integration of the Roman Economy." *World Archaeology* 23.3: 283–293.

Wootton, W., B. Russell, and P. Rockwell. 2013. "Stoneworking Techniques and Processes (Version 1.0)." The Art of Making in Antiquity: Stoneworking in the Roman World. http://www.artofmaking.ac.uk/content/essays/3-stoneworking-techniques-and-processes-w-wootton-b-russell-p-rockwell/.

Work, H.H. 2014. *Wood, Whiskey, and Wine: A History of Barrels*. Chicago: Reaktion Books.

Zeitlin, F.I. 1997. *Playing the Other: Gender and Society in Classical Greek Literature*. Chicago: University of Chicago Press.

Ziccardi, A. 2000. "Il ruolo dei circuiti di mercati periodici nell'ambito del sistema di scambio dell'Italia romana." In *Mercati permanenti e mercati periodici nel mondo romano: Atti degli Incontri capresi di storia dell'economia antica (Capri 13–15 ottobre 1997)*, edited by E. Lo Cascio, 131–148. Bari: Edipuglia.

Zimmer, G. 1982. *Römische Berufsdarstellungen*. Berlin: Gebrüder Mann Verlag.

INDEX

Photos, drawings, and tables have page locators in **bold**.

abandonment of dolia: discarding dolia, 179, 181; fictional story of, 173–74; issues relating to, 174; repurposing dolia, 174–79, 181; sale on the second-hand market, 143, 174–76; specialized container system, moving away from, 179–93, 203–4 (*see also* barrel(s)); in tanker ships, 179–80, 187; at villas and farms, 187–89

Aelius Aristides (Publius Aelius Aristides Theodorus), 117

agriculture: advantages of using dolia in, 67–68; depiction of workers treading grapes, *208*; dolia and, 52–53; dolium included in lists of equipment for, 5; food scarcity, responses to, 12n43; olive oil production (*see* olive oil); storage in Iberia and Gaul and, 103–5; storage of surplus production, 11–12, 69–73 (*see also* storage); technological advances and surplus production, 69–73; villas and farms with dolia in central Italy, *66*; viticulture and wine production (*see* viticulture; wine). *See also* villas, estates, houses, and farms

Alexander the Great, 144, *145*

Alma-Tadema, Lawrence, *210*, 211

amphorae: barrels, compared to, 184; buyers' expectation of keeping, 205; compared to barrels and cullei, *185*; compared to dolia, 2; continued use of, 206; in Cosa, 119, 121–22; cost of, 97; development of flat-bottomed, 182; diversity of, 203; Dressel 1, 84, 86, 88, 197; Dressel 1B, *80*; Dressel 2/4, *80*, 87–88, 95–96, 100, 102, 104–6, 115–16, 182, 197; Dressel 20, 106; Empoli, 182; found on the Madrague shipwreck, 81n31; Gauloise 4, 105, 182, *183*; labeling of, 79n25, 84; Lamboglia 2, 87n62; logistics of compared to barrels, dolia, and cullei, *195*; for long-distance trade, 79–83; manufacture of, long-distance wine trade and, 83, 87–88; olive oil transported in, 113; in Pompeii, 124, 126; pouring wine from, *72*, 72–73, 126; produced by the Sestius family enterprise, 83–87, 97–98; reuse of, 138–39; shape, capacity, material, and usage of, 79–81, 86–87; as single-use vessels, 82–83, 97; Spello, 182, *183*; study of, 7; taxes associated with importing, 203; unlined for olive oil, 80n28; waste associated with single-use, 202–3; wine distribution, role in, 14, 46, 75, 136; wine storage in, 64, 81–82; workers filling with wine, *78*, 117, *118*

Anatolios, 27, 30–32, 35, 53, 59

animal-hide containers, 76–79, 126, 136, 203

Annius Serapiodorus, 134

Apuleius (Lucius Apuleius Madaurensis), 173

Augustine of Hippo, Saint, 30, 186

Aurelian (emperor of Rome), 189, 194

Aurelius Sabinus, Lucius, 24, 25, 51

Bain, Read, 9n27

Barker, Simon, 176

barrel(s): adoption as official containers for the Roman Empire, 193–94, 203; capacity of, 174n3, 184; coexistence with dolia and amphorae, 206; compared to dolia, amphorae, and cullei, *185*; as Diogenes' dwelling, 209–10; illustration of and the process of cooperage, *192*; leakiness of and suitability as a storage container, 207; logistics of compared to dolia, amphorae, and cullei, *195*; maintenance and repair of, 193, 204; moving wine in, *206*; as a new container technology, 19, 174, 183–84, 186–87, 196, 203–4; production of, 191–92; status/ownership of, issue regarding, 205–6; viticulture and, 188–90

Basilica of Santa Sabina, representation of the Miracle of Cana, *208*, 208–9

Beckham, Andrew, 26n9

Beresford, James, 82

Brown, Frank, 179n27
Brun, Jean-Pierre, 91, 189

Caillaud, Christophe, 27n12
Calateus, 48
Calpurnius Piso Caesoninus, Lucius, 139
capacity of dolia: average, 5, 144n3; climate and, 106; in Cosa, *215–16*; in Gaul, 107–8, 114; in Iberia, 107; incisions indicating (*see* incisions); in Ostia and Rome, 43, 131, *224–28*; in Pompeii, 38, 39n34, *217–20*; range of, 1; in tanker ships, 94–95; at urban retail shops, 126
Carrato, Charlotte, 103, 106, 108–10, 113, 190
Carroll, Maureen, 109
Carthage: leather worker present in olive oil storehouse, 79n23
Cato the Elder (Marcus Porcius Cato): animal-hide containers, use of, 77–78; dolia included as essential equipment for the farm, 5, 52; dolium, cost of, 54; dolium repair, advice on, 153–54, 168; fermentation of wine, 14n55; ladles to be provided in wine and oil rooms, 76; major production centers for dolia, recommendation of, 30–31; materials used in dolium repairs, 148; moving heavy farm equipment, discussion of, 54; recipes for wines, 67; sale of agricultural surplus, advice on, 75; sealing dolia after fermentation, advice on, 146; treatment of a new oil dolium, 63
Caupona of Salvius, 250
Caupona of Spatulus, *127*, 128
Caupona of the Gladiators, *127*, 129, 178
Ceci, Monica, 134n61
cellae oleariae (olive oil cellar): construction of in central Italy between the second century BCE and the first century CE, 65; in Iberia, 113; as storeroom containing dolia, 12
cellae vinariae (wine cellar): barrels used in, 188–90, 206–7; clustered along the Tiber, 138; construction of in central Italy between the second century BCE and the first century CE, 65; construction of in Iberia, 104; for dolium tanker ships, 97; in Gaul, *114*, 114–15; in Iberia, 112–13, *114*; options for vintners and size of, 69; plans for, *110*; repair of dolia, significance of, 171; in Rome and Ostia, 22, 118, 132–43; size and organization of, 66–67; specialized, abandonment of, 204; as storeroom containing dolia, 12, 206; transformations of, 181; as unroofed courtyards, 61; in urban areas, 142–43; wealth, power, and prestige associated with, 18, 117–18, 136–37, 140–42, 199
Cicero (Marcus Tullius Cicero), 84, 98, 139–40
Cimber, 48
Cluentius Ampliatus, C., 42
Columella (Lucius Junius Moderatus Columella): animal-hide containers, use of, 77–78; maintenance of dolia, recommendations on, 63; recipes for wines, 67; sealing dolia after fermentation, advice on, 146; smoke used in the aging of wine, 82
Comisius Successus, Gaius, 138
Constantine (emperor of Rome), 208
containers. *See* storage
Corinthus, 42, 47
Cornelianum dolium, 209, 210
Cornelius Felix, C., 48
Cosa: as a case study, 19–21; descriptions of select dolia from, 245–46; dolia repurposed at, 175–78, 181–82; dolium-based storage, trial and abandonment of, 119, 121–22, 142; dolium dimensions, *215–16*; dolium fragments, photos of, *38*; dolium lids, photos of, *34*; dolium production at, 33–38; dolium repairs at, 155–60, 171, *239*; microphotographs of dolium ceramic fabrics from, *35*; photos of dolium repairs, *156–59*; plan of, *120*; port modifications to facilitate agricultural exports from, 83; profile drawings of dolia from, *34*; proportion of repaired dolia at, *155*; shift from exporting to importing at the port of, 87, 122; stamps on dolia from, 36, *37–38*, *216*; tavern with dolium, *120*
cost of dolia: most expensive pottery in antiquity, 26; transportation costs, 54–56, 119 (*see also* transporting dolia)
Cousteau, Jacques, 81
culleus, 14, 115, 182–84, *185*, *195*
cupa. See barrel(s)
cylindrical jars, 125–27, *128*, 129–30. See also *orcae/(h)orcae*

data and case studies, 19–23, 213–14; capacity of Cosan dolium, **215–16**; capacity of Ostian dolium, **224–27**; capacity of Pompeiian dolium, **217–20**; capacity of Roman dolium, **228**; descriptions of select dolia from Cosa, 245–46; descriptions of select dolia from Ostia, 251–53; descriptions of select dolia from Pompeii, 246–50; descriptions of select dolia from Rome, 253–54; dolium production sites in west-central Italy, **232–33**; dolium repairs according to stage of execution,

238; dolium repairs at Cosa, **239**; dolium repairs at Ostia, **241–43**; dolium repairs at Pompeii, **239–40**; dolium repairs at Rome, **243**; map of Italy including case study sites, **20**; sources for understanding the use of dolia and storage in Rome, 130; stamps from Cosan dolium, **218**; stamps from Ostian dolium, **229–31**; stamps from Pompeiian dolium, **221–23**; villas and farms with dolia in west-central Italy, **234–36**; volume incisions on dolia from Ostia, **237**

de Angelis, Francesco, 73n78

De Sena, Eric C., 126n35

Diocletian (emperor of Rome), 26, 54n7, 55, 77, 144, 207

Diodorus Siculus, 85–86

Diogenes the Cynic, 144, *145*, *209–10*, 209–11

Djaoui, David, 187

Dodd, Emlyn, 71, 109

dolia defossa: abandonment of, 182, 187, 190, 203 (*see also* abandonment of dolia); in central Italian villas, 65–66, 72, 207; coexistence with barrels, 206; decoration in the presentation of, 70–73; definition/placement of, 1, 5; in Gaul, 114–15, 182; in Iberia, 112–13; installation of, 112; in Ostia, 22, 132, 143; photo of, 3; protection of during hotter days, 61; at Saint-Bézard à Aspiran, 109; small farms, not found on, 64; in urban warehouses, 116, 131n53, 134, 181; at the Villa of the Mysteries, 250

doliarus (dolium maker), 24–25, 51; funerary altar for, 25

dolium, production of: brick and tile production, combined with, 46–48, 191, 198–99; challenges and rewards of, 17, 25; challenges and risks of, 49–50; coils, photos of, *29*; construction techniques, 27–30; in Cosa, 33–38 (*see also* Cosa); cracks formed during, *28*, 146–47, 149–50; *doliarii* as potters specializing in, 198–99; drop in, 190–91; economies of scale in, 48–49, 199; hypothetical example of, 7; improvements in, 198–99; investors and entrepreneurs involved in, 200–201; logistics of, 32; in the northwest, 106–11; in Ostia, 43–49; phases of, *8*; by the Piranus family, 91–92, 94–95, 98–99, 172, 180; in Pompeii, 38–43 (*see also* Pompeii); raw materials, 27, 33, 35, 39–40, 44; repair during, 152–53, 162–68; repair innovations, success of the industry and, 168–72; in Rome, 43–49; seams and paddling marks, photos of, *31*; seasonal limitations for, 191–93, 199; sites in Tuscany, Latium, and Campania, *6*; as a specialist craft, 25–33; stamps, 32–33, 200 (*see also* stamps); standardization in (*see* standardization); for tanker ships, 92, 94; in the Tiber River Valley, 44, 48–51; time required for, 30, 191; vertical integration in, 50, 201; in west-central Italy, *232–33*; on wine-producing estates, 200–201; workshops for, 30–32, 46–47 (*see also* opus doliare workshops)

dolium/dolia, xiv–xv, 197–98; abandonment of/discarding (*see* abandonment of dolia); agriculture, use in association with (*see* agriculture; villas, estates, houses, and farms); barrels, compared to, 184, *185*; capacity incisions, 43, *44*; capacity of (*see* capacity of dolia); coexistence with barrels, 206; consumption of wine from, 121; continued use of, 206–9; cost of (*see* cost of dolia); descriptions of select, 245–54; development of, 5–7, 201–2; differences between/separation of wine and olive oil, 61, 65; as Diogenes' dwelling, 144, *145*, *209–10*, 209–11; distinguished from other pottery/storage vessels, *2*, 2–3, 5; economic development/expansion of productive power associated with, 1, 7, 52–53, 65, 75, 198, 201; Gallic, *109*; hypothetical life of a, 7–9; Iberian, *107*; installation of, 56–57; legacy of, 207–11; lids (*see* lids); "life expectancy" of, 64; listed in Diocletian's *Price Edict*, 207; logistics of compared to barrels, amphorae, and cullei, *195*; maintenance of (*see* maintenance and repair of dolia); olive oil storage (*see* olive oil); photos of, *2–3*; in Pompeii (*see* Pompeii); preservation/archaeological record of, 1–2; production of (*see* dolium, production of); purchasing, 31–32, 53; quality of, 36, 42–43; repair of (*see* maintenance and repair); representations of in artworks, 208–11; for retail activities, 122, 124–30; shape of (*see* shape of dolia); shipwrecked, 95; stamps (*see* stamps); as a storage container technology, 9, 12–13, 16, 23, 56, 202; transferring liquids from, 76, 96–97, 117, *118*; transportation of wine in, 75, 88–89 (*see also* dolium tanker ships; trade and transportation); transporting (*see* transporting dolia); as understudied, 2, 7; urban diet and lifestyle, contribution to, 198 (*see also* urban areas); villas and (*see* villas, estates, houses, and farms); waste associated with the use of, 202–3; wine, as storage container for, xiv, 1, 5–6, 23; wine production, role in, 14, 16; wine storage (*see* wine)

dolium tanker ships: advantages of, 95–97; as a container technology, 97–100; decline in use of, 179–80, 187; drawings of, *89*, *93*; found in shipwrecks, 88–89, 91, *93*; limitations of, 180; map of dolium shipwrecks, *90*; production and design of, 91–92, 94–95; repair and reinforcement of dolia on, 171–72; saving time and resources through use of, 203; Spanish wine carried in, 113; trade networks extended by use of, 198; trade patterns including Iberia and Gaul using, 102–3; volume of wine transported on, *97*. *See also* shipwrecks
Domitia Lucilla (Minor), 49–50, 141, 201

Egyptian merchants, 81n29
Ellis, Steven: architectural fixtures as products of urban investment, 124–25; argument against bars as places of deviant behavior, 132n56; bars, decoration of, 129; counters, description of, 98; disappearance of specialized bar counters, 203; marble-clad bars, dating of, 177n16; retail in urban areas, emergence of, 88; subelites as operators of bars, 176
Euphrastus, 46

Fant, J. Clayton, 176
Feige, Michael, 71
Fernández-Götz, Manuel, 199n2
food supply: dolium storage technology and the expansion of production of, 100; grain imports, annual amount of, xiii; Ostia as a pivotal point for imports, 135–36; significance of containers for, 197; storage of, 11–13 (*see also* cylindrical jars; pithos/pithoi; storage); system for, 1; transporting (*see* trade and transportation); urban growth and, 10–12. *See also* agriculture
Frontoni, Riccardo, 109
Fulvius, Marcus, 42, 48

Galli, Giuliana, 109
Gandolfi, Mauro, *209*
Gatti, Giuseppe, 135
Gaul: agriculture and storage in, 104–5; amphorae originating from Cosa found in, 20; dolia defossa abandoned at southern ports of, 182; dolia from the Tossius workshop found in, 46; dolium from, *109*; dolium production in, 106–11; limited viticulture in, 85–86; maintenance of dolia in, 63; as a market for wine, 85–87, 100; moving wine using barrels and amphorae, *206*; use of dolia and expansion of agricultural production at villas in, 113–16; *utriclarius* found in inscriptions in, 77n10; the wine trade, position in, 102–3
Gérôme, Jean-Léon, *210*, 211
Gliozzo, Elisabetta, 33n25
Gordian (emperor of Rome), 207
Graham, Shawn, 49

Heslin, Karen, 99
Holleran, Claire, 130n47
Horace (Quintus Horatius Flaccus), 82
Horden, Peregrine, 86n60

Iberia: agriculture and storage in, 103–5; amphorae used for food transportation from, 81; dolium from, **107**; dolium production in, 106–7, 111; as a market for wine, 100; use of dolia and expansion of agricultural production at villas in, 111–13, 116; the wine trade, position in, 102–3. *See also* Spain
Ikäheimo, Janne P., 126n35
incisions: on dolia at Cosa, *246*; on dolia at Gaul, 114; on dolia at Iberia, 107; on dolia at Ostia and Rome, 43, 135, *228*, *237*; on dolia in shipwrecks, 94, 147; on pithoi, 4
Iulius Primus/Priscus, Quintus, 109, 200

James, Paul, 202n14
Jashemski, Wilhemina, 247
Junius (cooper in Londinium), 204
Junius Brutus, Marcus, 84
Juvenal (Decimus Junius Juvenalis), 144, 148

Kang, Dae Joong, 27n11
Kelly, John B., Sr., 201
Kilby, Kenneth, 192n65, 210n34

lagonae, 138
Lazzeretti, Alessandra, 48
Le Guin, Ursula, 197, 209
Leidwanger, Justin, 82
Lewit, Tamara, 206
lids: broken in the fermentation of wine, 57; Cosan, photos of, *34*; found on shipwrecks, 94; inner (*operculum*) and outer (*tectorium*), 39, 56; Ostian and Roman, 43; outer (*tectorium*), photos of, *40*; Pompeiian, photos of, *39*; production of as a separate industry, 39; standardization of, ease of replacement due to, 57n25

Liebeschuetz, Wolf, 190
Lucan (Marcus Annaeus Lucanus), 186

Macrobius (Ambrosius Theodosius Macrobius), 121, 207
Madrague shipwreck, 81n31
maintenance and repair of barrels, 193, 204
maintenance and repair of dolia, 144–45; appearance of repairs, aesthetic preferences and, 154; breakage during production and use, 146–47; by *cellarii,* 140; comparison of types of repairs, *243;* cost of a broken dolium, 144; development of innovative repairs, 169–72; of Diogenes' dolium, 144, *145;* dolium production industry and, 168–72, 193; hypothetical example of repair, 8–9; illustration of various repairs, *150;* lead used in repairs, 146, 148–49, 160–61, 163–64; neglect, consequences of, 173–74; photos showing repairs, *148, 151, 153, 156–59, 161–63, 165–68;* proportion of repaired dolia at selected sites, *155;* repairs made during production, 152–53, 162–63, 193; repairs made during use and during production, *148, 238;* repairs made in Cosa, 155–60, 171, *239;* repairs made in Ostia, 164–68, 171, *241–43;* repairs made in Pompeii, 160–63, 171, *239–40;* repairs made in Rome, 164–68, 171, *243;* routine maintenance and installation, 52–53, 63–65, *64,* 130; successful winemaking and, 72; techniques and materials used by repairers, 146–55; timing of, 68
Marcus Aurelius (emperor of Rome), 59
Marlière, Elise, 183
Marsigli, Giuseppe, *78*
Martial (Marcus Valerius Martialis), 101
Mausoleum of Santa Costanza, 208, *208*
McCallum, Myles, 21n72
Memmius Auctus, M., 249
Métraux, Guy, 136
Minturnae, 53, 91, 99, 109, 172
Monteix, Nicolas, 73n78, 126n35
Mucius Scaevola, Gaius, 187–88, 205
Mussolini, Benito, 181

olive oil: consumption of, 65, 202; dolia not embedded in the ground for storage of, 62–63; investment by large estates in dolia for, 65; maintenance of dolia for storage of, 63; process of making, 14, 16, 62; from Spain, 81, 105, 107; storage of, 1, 12, 53, 79n23 (see also *cellae oleariae* (olive oil cellar)); unlined amphorae for, 80n28; waste from single-use oil amphorae, 202–3
onggi (Korean dolium-like vessel), 26, *32,* 170n50
opus doliare workshops: building industry of Rome supplied by, 7; combining production of dolium, brick, and tiles in, 46–48, 198–99, 201; dark side of, 199–200; decline of, 190–91; *doliarii* (specialist potters) working in, 198–99; of Domitia Lucilla, 108, 201; hypothesized producing multiple products, *49;* innovative repairs developed by, 169–72; investors in, 17, 99, 200–201; location of, 169; maintenance and repair by, 18; mortarium/mortaria produced in, 156; number of, *47;* of the Piranus family, 92, 99; as places conducive to learning and development of skills and techniques, 199; preemptive production repairs, 164–68; specializing in heavy terracotta materials, 41; stamps of/stamping by, 41–42, 44–46, 168; in the Tiber River Valley, 44, 48–51; Tossius workshop, 45–46, 108, 199; vertical integration into wine production of, 201; workers and labor organization in, 42, 45–46, 48, 198–200
orcae/(h)orcae, 59, 107, 113
Ostia: capacity incisions on dolia in, 135; as a case study, 22; concentration of dolia in cellae vinariae, 132–34, 142–43; descriptions of select dolia from, 251–53; dolia, decision not to salvage, 182; dolium dimensions, *224–27;* dolium production in, 43–49; dolium repairs, photos of, *165–67;* dolium repairs at, *241–43;* dolium repairs in, 149n18, 164–68, 171; dolium use in, 130–36; food supply, pivotal point for imports, 135–36; incisions on dolia from, *237;* microphotographs of dolium ceramic fabrics from, *35;* plan of with cellae marked, *131;* population of, 132; profile drawings of dolia from, *34;* proportion of repaired dolia at, *155,* 164; renovations at, 181; stamps on dolia from, *229–31;* storage capacity of cellae vinaria in, 136. *See also* Rome
Ovid (Publius Ovidius Naso), 188

Palladius (Rutilius Taurus Aemilianus Palladius), 67, 206–7
Pallecchi, Silvia, 48
Peña, J. Theodore, 21n72, 136n74, 138, 148n17, 149, 152n29
Perkins, Phil, 4
Phileros, 42
Pirandello, Luigi, 154–55

Piranus family workshops, 91–92, 94–95, 98–99, 102, 172, 180, 201
pithos/pithoi: Greek, 26; Iberia, use in, 112; photos of, *4*; production of Greek, 30, 32; purchasing, 53; repair of, 149n18; shape of compared to dolia, 5; as storage vessel for food, 3–5; transportation of, 55; wealth and economic growth associated with, 4–5
Plato, 26–27
Plautus (Titus Maccius Plautus), 5
Pliny the Elder (Gaius Plinius Secundus): demand for heavy terracotta products, 41; dolia buried in the ground for storage of wine, 61; dolia invented to store wine, 5; lead alloys were common according to, 165; maintenance of dolia, recommendations and warnings about, 63; recipes for wines, 67; storage of wine, profits and, 68; storage of wine in wooden vessels, 188; story of great profit from viticulture speculation, 11; two kinds of lead noted by, 149; on varieties of wine, 74; wine production well established in Italy by the mid-second century BCE, 124
Pliny the Younger (Gaius Plinius Caecilius Secundus), 66, 70
Poehler, Eric, 54–55
Pompeii: bars of, 124–27, 129–30; broken dolium lids found in, 57; as a case study, 21–22; cylindrical jars at the Thermopolium of Vetutius Placidus, *128*; damaged dolia installed in bars at, 176, *177*; descriptions of select dolia from, 246–50; dolia and retail activities at, 122, 124–30, 132, 142–43; dolia repurposed at, 175–78; dolium dimensions, *217–20*; dolium lids, photos of, *39–40*; dolium production at, 38–43; dolium repairs, photos of, *148*, *153*, *161–63*; dolium repairs at, 160–63, 171, *239–40*; drawings of a lararium and a sign found in bars, *127*; House of Stabianus, dolia preserved at, 27; microphotographs of dolium ceramic fabrics from, *35*; plan of including properties with dolia and storage containers, *123*; plan of winery in, *58*; profile drawings of dolia from, *34*; properties with dolia (selection of), *127*; proportion of repaired dolia at, *155*; stamps on dolia from, 40–42, *41*, *221–23*; at the time of the eruption, 122, 124; types of storage jars used in, 125–27; wine production and consumption in, 124, 128–30
Poux, Matthieu, 85
Publicius, Lucius, 85n59
Purcell, Nicholas, 86n60

qvevri (Georgian dolium-like vessel): cleaning of, 63n47; cracks in, 150n23; for fermentation and aging of wine, 59: installation of, 62n36; life expectancy of, 64; logistics of building, **32**; made and used today, 26; production of, 30

Rando, Pino, 147
Randolph, Thomas, 210
Redemptus, 46
Rice, Candace, 89n74
Rickman, Geoffrey, xiii
Roman Empire: economic expansion of, dolia and, 16; food storage technology and infrastructure, 1, 12; Gaul and Iberia under Roman rule, 103–4; urban growth and the food supply in, 10–12
Rome: artificial harbor of (Portus), abandonment of, 204; capacity incisions on dolia in, 135; Capitoline Museum, 253–54; as a case study, 22; concentration of dolia in cellae vinariae, 134, 142–43; descriptions of select dolia from, 253–54; dolium dimensions, *228*; dolium production in, 43–49; dolium repairs, photos of, *165*, *168*; dolium repairs at, *243*; dolium repairs in, 164–68, 171; dolium use in, 130–36; as a market for wine, 73, 198; National Museum of Rome, *186*, 214, 254; olive oil, consumption of, 202; population of, 10, 117, 132, 202; profile drawings of dolia from, *34*; proportion of repaired dolia at, *155*, 164; recommendation as major dolia production center, 31; relocation of the Sestii to, 87; retail in, transformation of, 204; transportation of wine and oil from the hinterlands to, 77–78; as warehouse of the world, 130–43; wine, consumption and storage of, 117–18, 202. *See also* Ostia; Tiber River/Tiber River Valley
Rotroff, Susan I., 152n29
Russell, Ben, 176

Salido Domínguez, Javier, 103
Sands, Rob, 183
Schwinden, Lothar, 187
Sempronius Gracchus, Gaius, 130
Sentia Amarantis, 187, *188*
Sentius Victor, 187
Sestius family amphora enterprise, 83–87, 97–98, 119, 122, 201

Sextus Pompey (Sextus Pompeius Magnus Pius), 179
shape of dolia: cracks appearing during production due to, 147; drawings of, *34, 93*; exposure of wine to air due to, 121; of Ostian and Roman dolia, 43; of Pompeian dolia, 38; strawberry shape, 5, 27, 56, 106, 125, 132; of tanker ship dolia, 92, *93*, 94; wine fermentation and, 57, 59–61, *60*
shipwrecks: Cap Bénat B, 94; Chrétienne H, 96, *97*, 187; cracks/repairs of dolia and, 171–72; Diano Marina, *93*, 94, 96, *97*, 147, 171, 180n30; dolium from, *95*; of dolium tanker ships, 88–89, 91–92, 100, 102, 180; Grand Ribaud D, 94, 180; La Giraglia, 94, *97*; Madrague, 81n31; map of, *90*; packed with wine amphorae, 86; Petite Congloué, *93*, 96, *97*, 102n3, 171; Sud-Lavezzi 3, 96, *97*
Shop of the *Vinarius*, 128
Solin, Heikki, 140n94
Spain: amphora used to carry specific food products from, 81; olive oil produced in the Guadalquivir Valley of, 81, 105; products exported by the Sulpicinae family to, 48; *tinajas* and *talhas* (Spanish and Portuguese dolium-like vessels) from, 26, *32*, 55, 64. *See also* Iberia
Spinazzola, Vittorio, 42
stamps, 32–33; on amphorae, 79n25, 84; on Cosan amphorae, 122; on Cosan dolia, 36, *37–38, 218*; on dolia from shipwrecks, 91n81; on dolia imported to Gaul, 108; on dolia with a family connection to Minturnae, 91n84; evidence of manumission on dolium, 200; of opus doliare workshops, 44–45, 50; on Ostian and Roman dolia, 44–46, *46*, 48, *229–31*; on Piranus family dolia, 91–92; on Pompeiian dolia, 40–42, *41, 221–23*; on repaired dolia, 168
standardization: in Cosa, 33, 36; industry standards as indication of quality, 32; of lids, 39; in Ostia and Rome, 43–44; in Pompeii, 38–40
Stanford University, ORBIS project, 79n22
Stevens, Saskia, 137n79
storage: advances in food, 69–70; agriculture in Iberia and Gaul and, 103–5; buildings, xiii, 12; choosing container technologies for, 201–7; coexistence of competing technologies for, 206–7; consumption and, 121; containers, broader significance of, 197, 209–11; containers/vessels for, 12–13 (*see also* amphorae; barrel(s); cylindrical jars; dolium/dolia; pithos/pithoi); of dry goods in older dolia, 56; investing in, 53–56; as key to Rome's food supply, 11–12; long-term, problems associated with, 68–69; of olive oil, 1, 12; packaging preferences, change in, 174, 205–7 (*see also* abandonment of dolia); silos, 103; specialized container system, moving away from, 179–93, 203–4 (*see also* barrel(s)); technology and infrastructure for, 1, 9, 11–12, 118; in urban areas (*see* urban areas); waste from ceramic-based, 202–3; of wine (*see* wine)
Strabo, 186
Suetonius (Gaius Suetonius Tranquillus), xiii
Symmachus (Quintus Aurelius Symmachus), 207

Tchernia, André, 75, 84, 91
technology(ies): coexistence of competing, 206–7; definition of, 9n27; dolia as a, 9–10
Terence (Publius Terentius Afer), 121
Thomas, Michael L., 21n72
Tibbott, Gina, 147n13
Tiber River/Tiber River Valley: dolium production along the, 17, 44, 48, 50–51, 53, 198–99; navigation/travel along the, 97, 134n68, 138, 182, 194; oversight of, 10; warehouses and cellae vinariae along the, 117, 130, 134, 138
tinajas and *talhas* (Spanish and Portuguese dolium-like vessels), 26, *32*, 55, 64, 149n22
Tossius Cimber, 46, 199
Tossius family opus doliare workshop, 45–46, 108, 199
Tossius Ingenuus, 46
trade and transportation: amphorae design/manufacture and, 83–88; on boats and ships, 74, 81n35, 86, 88–89, 91–92, 94–100, *97*, 102–3; choosing container technologies for, 201–7; containers: amphorae, 79–87, 96, 97–98 (*see also* amphorae); containers: animal-hide, 76–79; containers: barrels, 184, 186–87; containers: dolium tanker ships, 88–89, *89–90*, 91–92, *93*, 94–100, 171–72, 198; containers: *lagonae*, 138; dolia design/manufacture and, 92, 94; length in time of common journeys, 79n22; long-distance commercial shipping, dolium-based storage technology and, 115; long-distance commercial shipping, frequency of, 10n37; merchants moving wine, *206*; overland, 74–76; pattern of including Iberia and Gaul, 102–3; price speculation, wine transportation and, 78–79; profitable packaging

trade and transportation (*continued*)
for, 97–100; requirements/challenges of, 76, 79; the Roman retail district/cellae vinariae and, 135–42; sailing season for, 82–83; scale of long-distance from the second century BCE onward, 74–75

transporting dolia: on boats and ships, 53–54; cost of, 54–56, 119; household vs. commercial mode for, 54–55; hypothetical example of, 7–8; legal maximum vehicle load weights, 55; only when absolutely necessary, 6; overland, 54–56, *64*, 119; special care required for, 52–53; in urban areas, 139–40

Tremoleda Trilla, Joaquim, 103

Uihlein family/ULINE, 13

Ulpian, 77, 205

urban areas: cellae vinariae in, 135–41; commercial districts in, 137–38; concentration of storage in, 132; concentration of wealth and *horti* (pleasure gardens) in, 140–42; Cosa (*see* Cosa); dolia used for commercial or communal storage in, 142; dolium use in urban retail activities, 122, 124–30, 198; investment in warehouses, 136–42; opus doliare workshops in, 44 (*see also* Tiber River/Tiber River Valley); Ostia (*see* Ostia); Pompeii (*see* Pompeii); Rome (*see* Rome); storage containers in, 117–19; transporting dolia in, 13–140; vineyards in, 129

Van Oyen, Astrid, xiii, 12, 69–70, 112, 135–36

Varro (Marcus Terentius Varro): *calpar* as vessel for holding wine in existence prior to dolium, 5; containers used for wine fermentation, 57, 59, 146; estate owners encouraged to exploit their clay pits for ceramics production, 200; importance of roads and infrastructure for the household mode of transportation, 55n11; materials used in dolium repairs, 148; wine cellars built in Rome in pursuit of profit, 136

Vedius Pollio, Publius, 142

Vesuvian towns, masonry shop counters in, 98

Vibius Crescens, C., 200

Vibius Donatus, C., 200

Vibius Fortunatus, C., 200

Villa della Pisanella (Boscoreale): arrangement of dolia at, 67; grains, nuts, and legumes stored in dolia at, 176; installation of dolia and level of the courtyard at, 62n36; mixed use of dolia at, 57n19; number of dolia for wine and olive oil at, *66*, 76, 114, 124; plan of winery and olive oil cellar, *58*; productive capabilities of, 52; wealth and status of the owner of, 66–67

Villa Magna (Anagni): arrangement of dolia at, 67; dolia at, 235; dolia reinstalled at, 207; dolia removed at, 175, 181; dolia shifted from, 109; elaborate decoration celebrating viticulture at, 70–71; opus doliare workshops supplying materials and containers for, 201; plan of winery, *58*

Villa Regina (Boscoreale): annual production of wine at, 96; dolia at, 67, 124, 125n30; dolia repurposed at, 178; lid with a production defect used at, 146n7; modest quarters and wine production at, 70; plan of, *58*; storage capabilities at, 114, 136

villas, estates, houses, and farms: Arellano (Navarra), *114*; Can Bonvilar (Catalonia), 112; Caseggiato dei Doli (Ostia), 181–82, 251–52; conspicuous production celebrated at, 70–73; Cortijo de la Marina (Sevilla), 113; decline in dolia use at, 187–89; dolia and development of, 52–53, 65–70; with dolia in central Italy, *66*; with dolia in west-central Italy, *234–36*; Els Tolegassos (Girona), 112; Els Tolegassos (Tarraconensis), 113; Garden of Hercules (Pompeii), *127*, 129, 175, 178, 249; Garden of the Fugitives (Pompeii), *127*, 161, 246; House of Annius (Ostia), 134, 252; House of D. Caprasius Primus (Pompeii), *127*; House of Diana (Cosa), 119n10; House of Ganymede (Morgantina), 50n76; House of Medusa (Pompeii), 178; House of Meleager (Pompeii), 175–76; House of Memmius Auctus (Pompeii), *127*, 249–50; House of Stabianus (Pompeii), 27, *127*, 176, 178, 247; House of the Bicentenary (Herculaneum), 178; House of the Lararium of Hercules (Pompeii), *127*; House of the senator Rosa (Ostia), *186*; House of the Skeleton (Cosa), 119; House of the Summer Triclinium (Pompeii), *127*; House of the Vettii (Pompeii), *72*, 72–73; investing in multiple, 66; Las Musas (Navarra), 113; Le Molard in Donzére (La Drôme), 114; L'Estagnol (Clermont-l'Herault), 114; location and installation of dolia, 57, *58*, 61–63; Magazzino Annonario (Ostia), *44*, *131*, 132–33, *133*, 137, *137*, 252–53; Magazzino dei Doli (Ostia), *131*, *133*, 134, *137*, 252; Olivet d'en Pujol (Girona), 112; Place Vivaux (Marseille), *114*; Prés-Bas, 114; Rasero de Luján (Cuenca), 113; Rumansil (Murça do Douro), 113; Saint-Bézard à Aspiran (Languedoc), 109, *110*,

114–15, 200–201; Saint-Martin (southern Gaul), 114; scene of cupids sampling wine at the House of the Vettii, *72*, 72–73; Sentromà (Tiana), 112; size and production capabilities in Iberian and Gallic compared to Italian, 116; Torrebonica (Catalonia), 112; Torre de Palma (Lusitania), 113; transportation of dolia from and to, 53; Vicus at Vagnari (Puglia), 53, 108–9, 175; villa at La Maladrerie (Saillans), 190, *190*; villa at Valle Lungha, *58*; Villa Augustea di Somma Vesuviana (central Italy), 207; Villa B (Gragnano), *221*; Villa B (Oplontis), 14n56, 82, 86, 124; Villa della Muracciola (Suburbium, Rome), 70, 235; Villa della Pisanella (*see* Villa della Pisanella); Villa Farnesina (Tiber River), 134; Villa Giuliana (Boscoreale), 66, *236*; Villa i Medici (Stabiae), 67, 76, *236*; Villa Magna (*see* Villa Magna); Villa of Ambrosan (San Pietro in Cariano), 189; Villa of N. Popidius Maior (Scafati), 175; Villa of N. Popidius Narcissus (Scafati), 124; Villa of Russi (Emilia Romagna), 189–90; Villa of Russi (Ravenna), 71–72; Villa of the Mysteries (Pompeii), *58*, 70, 124, 125n30, 250; Villa of the Quintilii (Rome), 71, 109, 207; villa of Vareilles, 115; Villa Regina (*see* Villa Regina (Boscoreale)); Villa Settefinestre (*see* Villa Settefinestre (Ansedonia)); Villa Stazione Ferrovia (Boscoreale), 66, *236*; Villa Venezia Nuova of Villa Bartolomea (Verona), 189; the western Mediterranean, installation of dolia and expanded agricultural production in, 111–16

Villa Settefinestre (Ansedonia): dolia at, 114, 235; excavation of, 119; layout and narrow corridor of, 69; plan of winery, *58*; Sestius family, possible operation by, 98; wine presses at, 116; wine production capability at, 83, 114

Vitellius (emperor of Rome), xiii

viticulture: barrels and, 188–90; in central Italy, 65, 124; components of, 64; dolium production and, 99; dolium use and, 69–70, 207–9; expansion of in the western Mediterranean, 111–16; limited in Gaul, 85–86; potential profit and risk associated with, 11; practiced beyond household consumption, 72–73; urban vineyards, 129. *See also* wine

Vitruvius (Marcus Vitruvius Pollio), 57, 61

Wallace-Hadrill, Andrew, 21n71

Will, Elizabeth Lyding, 84

Wilson, Andrew I., 10n37, 86n60

wine: additives and recipes for, 67, 72, 198; advantages of using dolia in the production of, 67–68, 75, 198; aging of, profits and, 68–69; barrels as a container for, 184, 203 (*see also* barrel(s)); burial of dolia for storing, 61–62; cellars (see *cellae vinariae* (wine cellar)); cleanliness of equipment for the making of, 71–72; consumption of, xiv, 1, 11, 65, 117, 121, 202 (*see also* Pompeii, dolia and retail activities at); cost of a single serving of, 180n30; dolium tanker ships for the transportation of, 92 (*see* dolium tanker ships); fermentation in dolia, 57, 59–61, *60*, 94; large-scale production in Iberia and Gaul of, 105; maintenance of dolia for storage of, 63–64, 68 (*see also* maintenance and repair of dolia); merchants moving, *206*; options for vintners, 67–69; plans of wineries, *58*; process of making, 13–14, 57, 68–69, 206; quality of, Roman elitism and, 139; retail district for, 138–40; sale of, 75–76, 117, *118*, 128–29 (*see also* trade and transportation); sampling of before sales, 73; smoke used to enhance the flavors of, 82; storage of, xiv, 1, 5–6, 12, 60–62, 68, 81–82 (see also *cellae vinariae* (wine cellar); dolium/dolia); storage of, problem of long-term, 68; storage of, quality and, 126; supply chain and containers used in, 15; taste of, aged and stored in barrels vs. dolia, 207; transporting (*see* trade and transportation); urban areas, storage and consumption in (*see* urban areas); wine merchants with barrels, *188*

wine, markets for: commercial districts associated with, 137–38; creation of sellers' markets in Rome, 137; development of from the late first century BCE onward, 87–88; Gaul, 85–87; urban markets in Italy, 105

wines, varieties of: Aminaean, 74; available to Romans of the first century CE, 101–2; Falernian, 91, 124, 126; Laietanian, 105; Massic, 124; Mentana, 74; Pompeian, 74; Setinum, 74; Surrentine, 124; Tarraconian, 105

Woolf, Greg, 84

Woolf, Virginia, 197

A NOTE ON THE TYPE

This book has been composed in Arno, an Old-style serif typeface in the classic Venetian tradition, designed by Robert Slimbach at Adobe.